Springer Series in
Surface Sciences

18

Editor: Gerhard Ertl

Springer Series in **Surface Sciences**

Editors: G. Ertl and R. Gomer Managing Editor: H. K. V. Lotsch

Volume 1: **Physisorption Kinetics** By H. J. Kreuzer, Z. W. Gortel

Volume 2: **The Structure of Surfaces** Editors: M. A. Van Hove, S. Y. Tong

Volume 3: **Dynamical Phenomena at Surfaces, Interfaces and Superlattices**
Editors: F. Nizzoli, K.-H. Rieder, R. F. Willis

Volume 4: **Desorption Induced by Electronic Transitions, DIET II**
Editors: W. Brenig, D. Menzel

Volume 5: **Chemistry and Physics of Solid Surfaces VI**
Editors: R. Vanselow, R. Howe

Volume 6: **Low-Energy Electron Diffraction**
Experiment, Theory and Surface Structure Determination
By M. A. Van Hove, W. H. Weinberg, C.-M. Chan

Volume 7: **Electronic Phenomena in Adsorption and Catalysis**
By V. F. Kiselev, O. V. Krylov

Volume 8: **Kinetics of Interface Reactions** Editors: M. Grunze, H. J. Kreuzer

Volume 9: **Adsorption and Catalysis on Transition Metals and Their Oxides**
By V. F. Kiselev, O. V. Krylov

Volume 10: **Chemistry and Physics of Solid Surfaces VII**
Editors: R. Vanselow, R. F. Howe

Volume 11: **The Structure of Surfaces II**
Editors: J. F. van der Veen, M. A. Van Hove

Volume 12: **Diffusion at Interfaces: Microsopic Concepts**
Editors: M. Grunze, H. J. Kreuzer, J. J. Weimer

Volume 13: **Desorption Induced by Electronic Transitions, DIET III**
Editors: R. H. Stulen, M. L. Knotek

Volume 14: **Solvay Conference on Surface Science** Editor: F. W. de Wette

Volume 15: **Surfaces and Interfaces of Solids** By H. Lüth

Volume 16: **Theory of the Atomic and Electronic Structure of Surfaces**
By M. Lannoo and P. Friedel

Volume 17: **Adhesion and Friction** Editors: M. Grunze and H. J. Kreuzer

Volume 18: **Auger Spectroscopy and Electronic Structure**
Editors: G. Cubiotti, G. Mondio, K. Wandelt

G. Cubiotti G. Mondio
K. Wandelt (Eds.)

Auger Spectroscopy and Electronic Structure

Proceedings of the
First International Workshop,
Giardini Naxos-Taormina, Messina, Italy,
September 10–14, 1988

With 162 Figures

Springer-Verlag Berlin Heidelberg New York
London Paris Tokyo Hong Kong

Professor Dr. Gaetano Cubiotti
Institute of Theoretical Physics, University of Messina,
Salita Sperone, N.31, I-98166 Sant'Agata-Messina, Italy

Professor Dr. Guglielmo Mondio
Institute of the Structure of Matter, University of Messina, Salita
Sperone, N.31, I-98166 Sant'Agata-Messina, Italy

Professor Dr. Klaus Wandelt
Institut für Physikalische und Theoretische Chemie, Universität Bonn,
Wegelerstr. 12, D-5300 Bonn 1, Fed. Rep. of Germany

Series Editors

Professor Dr. Gerhard Ertl

Fritz-Haber-Institut der Max-Planck-Gesellschaft, Faradayweg 4–6
D-1000 Berlin 33

Professor Robert Gomer, Ph. D.

The James Franck Institute, The University of Chicago, 5640 Ellis Avenue
Chicago, IL 60637, USA

Managing Editor: Dr. Helmut K. V. Lotsch

Springer-Verlag, Tiergartenstrasse 17
D-6900 Heidelberg, Fed. Rep. of Germany

ISBN-13:978-3-642-75068-7 e-ISBN-13:978-3-642-75066-3
DOI: 10.1007/978-3-642-75066-3

Softcover reprint of the hardcover 1st edition 1989

2154/3150-543210 – Printed on acid-free paper

Preface

This first International Workshop on Auger Spectroscopy and Electronic Structure – IWASES 1, held in Giardini-Naxos, Sicily, Italy, grew out of a number of longstanding collaborations between the various Institutes of Physics of the University of Messina, namely the Institute of the Structure of Matter, the Institute of Theoretical Physics and the Institute of General Physics, and groups in other European countries at the University of Liverpool, England, the Institute of Physical Chemistry, University of Munich, and the Fritz Haber Institute of the Max Planck Society, Berlin, FRG. This workshop was the first to be devoted solely to Auger electron spectroscopy. This initiative was motivated by the enormous evolution of the field within the last decade to a point where it now extends far beyond the mere application of this spectroscopy as an analytical tool to determine surface cleanliness and surface composition.

In fact, the Auger process, which is a multi-electron process, and which leaves the sample in a doubly (or higher) ionized state, is an invaluable probe for investigating excited states and, in particular, electron (or hole) correlation effects. These correlation effects play an important role for many physical properties of matter such as magnetism, screening processes, and electron stimulated desorption (ESD), to name but a few. Although at first glance photoelectron spectroscopy appears to be a more straightforward technique for the elucidation of the electronic structure of matter, a full understanding of photoemission spectra and, in particular, of final state relaxation processes, requires the inclusion of multi-electron and correlation phenomena, which in turn may be studied by Auger electron spectroscopy. Finally, the Auger emission current may be subject to diffraction phenomena. Accordingly, satellite structure in Auger spectra have been interpreted in terms of EXAFS.

All of the above-mentioned aspects of Auger electron spectroscopy were covered during the IWASES 1 meeting by the 11 invited and 21 contributed papers, which are presented in this volume. The majority of the papers are presented in the sections: Theory of the Auger Process and Electronic Structure Determination; Auger Spectra from Solid and Gaseous Samples; Analytical Aspects of Auger Electron Emission. A further section deals with techniques that provide complementary information to the pure Auger experiments, namely Scanning Tunneling Microscopy (STM) and Inverse Photoemission (IPE, BIS).

Needless to say, the magnificent scenery of the most ancient greek colony in Sicily and the charming surrounding of the Giardini-Naxos Beach Hotel created a relaxed atmosphere for most fruitful discussions and many new scientific friendships between the participants from nine different countries. The evident success of this workshop, witnessed by this volume, has stimulated the organizers to plan a repeat event every 3 years.

Finally, the organizers would like to acknowledge the financial and organizational help of the ECC Stimulation Action Project, the National Group Structure of Matter in the National Council of Researchers of Italy (GNSM-CNR, Italy) and the three host institutes of the University of Messina: the Institute of the Structure of Matter, the Institute of Theoretical Physics and the Institute of General Physics.

Sant'Agata-Messina
Bonn
July 1989

G. Cubiotti
G. Mondio
K. Wandelt

Contents

Part III Auger Spectra from Solid and Gaseous Samples

Part IV Analytical Aspects of Auger Electron Emission

Part V Complementary Techniques

Part I

Introduction

Experimental Probes of Electron Correlation Effects and the Influence on the Electronic Structure

G.A. Sawatzky

Department of Applied and Solid State Physics, Materials Science Center, University of Groningen, Nijenborgh 18, NL-9747 AG Groningen, The Netherlands

In this paper I discuss some interesting materials and phenomena in which electron correlation effects play an important, if not a dominating, role. I attempt to describe the effects of electron correlation in terms of simple physical pictures. Also I discuss some techniques which can be used to determine the magnitude of the electron-electron interaction and apply these to some problems of modern interest in solid state physics and chemistry.

1. Introduction

It has long been recognized that many of the most interesting and intensively studied phenomena in solid state physics and chemistry are a result of local electron-electron interactions (U) which are comparable to or considerably larger than other (one electron) energy scales like band widths (W) and band gaps. As early as 1937 [1] it was pointed out by Peierls in a comment to a paper by de Boer and Verwey that the insulating behaviour of NiO could be explained if the repulsive Ni d-d Coulomb interaction (U) was larger than the kinetic energy gained by delocalization and therefore each Ni ion would have 8 3d electrons, the O2p band would be full and only spin degrees of freedom would remain. This led to the Mott Hubbard [2-4] description of the insulating behaviour of a large class of materials and to Herrings [5] "stay at home" principle for electrons in states with a Coulomb repulsion energy larger than the corresponding band width. The stay at home principle quite naturally explained the occurrence of local magnetic moments due to the well known Hund's rule stabilization of the atomic configuration with a maximum spin and orbital magnetic moment respectively. These spins on neighboring ions would be exchange coupled because of hybridization with excited states leading to long range magnetic order. In compounds like NiO the exchange interaction between the Ni spins was explained by the Anderson theory of superexchange [6] which again involves the Coulomb interaction U. Based on symmetry arguments and basic knowledge of atomic physics Goodenough [7] and Kanamori [8] set up general rules relating the superexchange interaction to bond angles and orbital occupation.

Another area of intense research in which U plays a dominating role is that of magnetic impurities in metals and the extension to a lattice of such impurities leading to Kondo [9] and heavy fermion [10] behaviour. Friedel [11] suggested that 3d transition metal impurities would retain their local magnetic moment even when dissolved in a metal although the d states would broaden somewhat and shift due to hybridization with conduction band states. Anderson [12] introduced a much used model Hamiltonian which forms the basis for a discussion of magnetic impurities in metals. Here again the Coulomb interaction U between d electrons in transition metal impurities or f electrons in rare earth impurities plays an important role and determines whether or not a local moment exists. In

Springer Series in Surface Sciences, Vol. 18 **Auger Spectroscopy and Electronic Structure**
Editors: G. Cubiotti · G. Mondio · K. Wandelt

the limit that U is large the Anderson Hamiltonian can in certain cases be transformed [13] to a Kondo Hamiltonian in which only the spin degrees of freedom of the impurity are retained and U enters in the size of the exchange interaction with the conduction electrons.

More recently yet the attention has once again been focussed on transition metal oxides because of the discovery of the high T_c superconductors [14]. A large group of scientists is of the opinion that a large Cu-d-d Coulomb interaction plays an important role in understanding these materials.

In the above, I have tried to indicate very briefly some of the broader areas of research in which knowledge of the local Coulomb interaction is of great importance in understanding the physical properties. In general one could say that the 3d and 4d transition metals, rare earth and actinide metals, their compounds and alloys exhibit properties like localized magnetic moments, Mott-Hubbard transitions, mixed valency and valence fluctuations, heavy fermion behaviour, high T_c superconductivity, Kondo behaviour etc. all of which are most likely due to the importance of the atomic d-d and f-f electron-electron interactions.

For all of these areas of research it is of utmost importance to have methods with which the Coulomb interactions can be quite directly measured. By quite directly I mean without too much interpretation. In this paper I will discuss examples for which various forms of high energy spectroscopy can provide such information although when one gets to the details concerning the exact size of the interactions one eventually does require a detailed often complicated theoretical analysis.

2. One Electron Theory and its Breakdown

In the independent particle picture often used to describe the electronic structure of solids it is assumed that each electron moves in an average/static periodic potential set up by the nuclear charge and the other electrons of the solid. The total N electron wave function (Ψ^N) is then assumed to be a single Slater determinant of one electron Bloch states (Ψ_k). In this way we are assured that Ψ^N has the periodic symmetry of the solid but in addition we have assumed that the one electron Ψ_k also must be Bloch states. A standard example used to illustrate the problem here is a linear infinite chain of hydrogen atoms with lattice spacing **a**, and one electron per atom. In a tight binding approximation we have

$$\psi_k \simeq \frac{1}{\sqrt{N}} \Sigma\, e^{ik.R_i}\, \psi_{1s}(r-R_i)$$ and the energy

$$\varepsilon_k = \varepsilon_{1s} + 2T \cos k\mathbf{a}$$ where

R_i labels a H atom position, ε_{1s} the atomic 1s binding energy and T the "transfer" integral $T = \langle\psi_{1s}(r-R_i)|\mathbf{H}|\psi_{1s}(r-R_{i\pm1})\rangle$. For the nonmagnetic system with a band of width W=4T the Fermi level would be at the centre of the band and the system would be metallic. As we increase the interatomic distance (even to infinity) the band narrows but the system remains metallic.

Obviously there is something wrong here since we would all agree that for a large enough such a system would be an insulator. Of course we can do much better than a tight binding model using very sophisticated computational techniques [15] based on density functional (DF) theory [16].

DF theory gives us the exact ground state energy and charge density including all the effects of correlation. To carry out the calculation however the effect of exchange and correlation are replaced by an effective potential so we are back to a one particle description as far as the wave function is concerned. Ψ^N(DF) as used in a band structure calculation still is assumed to be a single Slater determinant of one electron Bloch states. If we applied DF band theory to the system just described in tight binding we would get a much better ground state energy and charge density but the system would still be metallic. The problem is that the wave function used in DF band theory does not explicitly include any correlation and therefore is not an eigenfunction of the Hamiltonian including the electron-electron interaction but is solely a mathematical tool to be able to describe the electron density and energy of the ground state. This wave function therefore has no physical significance and should formally not be used to describe the density of states below and above the Fermi energy E_F which obviously are not ground state properties.

In discussing electron-correlation effects, we will refer to those physical properties and effects which can only be described by explicitly including correlation in the many electron wave function itself. Going back to our example of the chain of hydrogen atoms it could be so that the ground state wave function could be better described by a Slater determinant of site localized electrons (one on each site) rather than one electron Bloch states. This would avoid double site occupancies. It is interesting to note that the total electron density is the same for these two extreme limits of the many electron wave function for **a** large but the predictions of the excited states and therefore the physical properties are entirely different. For the one electron Bloch state description the material would be metallic whilst for the site localized description the system would be a Mott-Hubbard antiferromagnetic insulator for **a** large enough.

Having said this one might come to the conclusion that band theory is of no use to describe the physical properties of solids. Fortunately there are scores of materials for which band theory does a remarkably good job and it is our goal to try to understand for which class of systems this is true and also to try to find better descriptions for those materials for which band theory apparently fails.

We can quite directly check the predictions of band theory with angular resolved photoelectron spectroscopy (ARUPS) [17]. With ARUPS both the energy of the electrons and the corresponding momentum (k) can be determined providing us with a powerful technique for band mapping. In Fig. 1 the result of a study on Cu is shown [18]. We see nicely the strongly dispersing 4s band crossing the Fermi level and a manifold of weakly dispersing 3d bands between 2 and 5 eV below E_F. The band theory (solid lines) does an excellent job of describing these data bringing us to the conclusion that one particle theory works well for the first electron removal states which are the states below E_F. Similarly the inverse photoelectron spectra IPS yield the states above E_F [19] which again are in reasonable agreement with band theory for Cu [20] for not too high energies [21].

We however get quite a shock upon studying the $L_{23}VV$ Auger spectrum of Cu. In Auger spectroscopy the final state has two electrons missing in the valence band and since the core L hole (2p) in the intermediate state is strongly localized and the Auger matrix elements are atomic like we expected to see the local two hole density of states [22]. In a one particle theory the local two hole density of states is just the self-convolution of the one hole density of states which in turn is closely approximated by the self-convolution of the X-ray photoelectron spectrum (XPS). The resulting broad (5 eV) structure together with the Auger

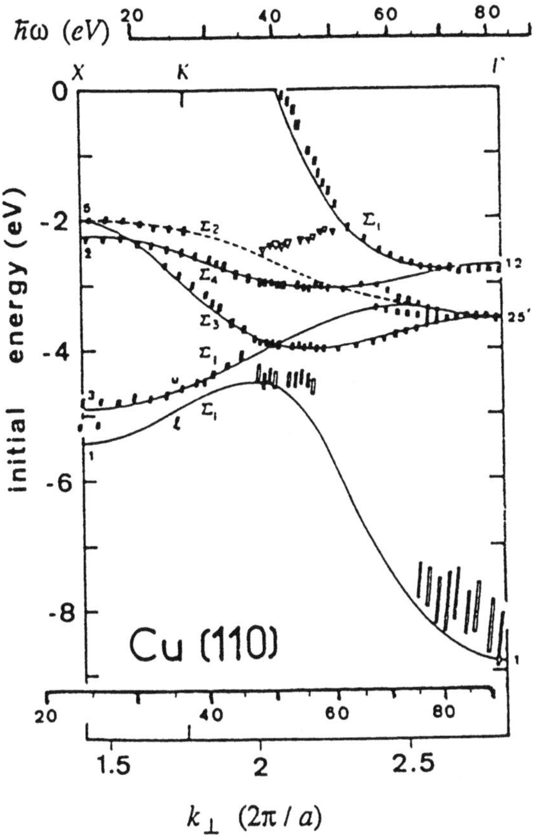

Figure 1. Experimentally determined valence band for copper along the ΓKX line. The full curves correspond to a band-structure calculation. The dashed curve (level Σ_2) is forbidden in both polarizations (s and p) for normal emission. The few points indicated along this line are obtained by moving the analyzer by a few degrees. The height of the rectangles gives the experimental uncertainty. Taken from ref. 17

spectrum [23] is shown in Fig. 2a. In Fig. 2b is shown the Auger spectrum [23] and the Auger photoelectron coincidence spectrum (APECS) [24]. (In APECS we can eliminate the Coster-Kronig preceded Auger structures and therefore get a more direct picture of the local two hole density of states.) We see two regions one is marked "band" and the other consists of the high intensity sharp structures. The low intensity band region corresponds closely to the self-convolution of the d density of states shown in Fig. 2a. The atomic part which contains most of the intensity is sharply structured and shows the multiplet structure of an atomic d^8 state as indicated by the vertical bars. This behaviour is quite nicely explained by the Cini-Sawatzky [25,26] theory in which proper account is taken of the intra-atomic d-d Coulomb interaction U_{dd}. In fact the separation of the "d^8" part and the band like part where the 2 holes are on different sites is a direct measure of U_{dd} although there are small shifts due to hybridization [27]. Since there are a number of other contributions in this proceedings describing this in more detail, I will suffice by stating that if U is larger than the one hole bandwidth the d^8 states are discrete bound states which look very atomic like [28].

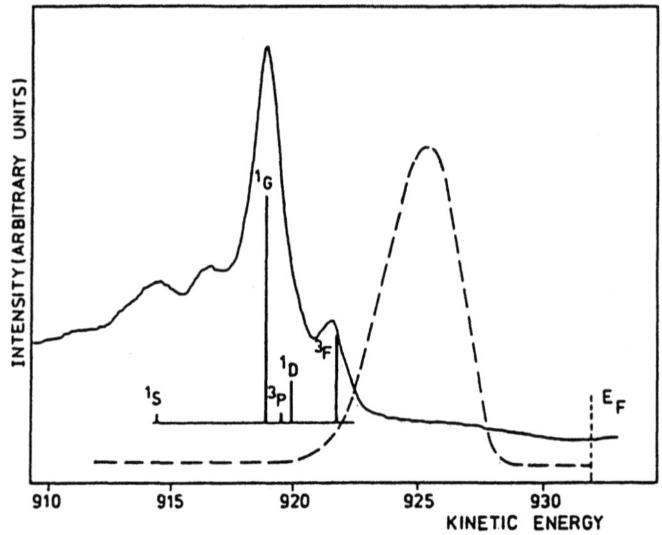

Figure 2a. The $L_3M_{45}M_{45}$ Auger spectrum of Cu (solid line). The dashed line represents the self convolution of the 3d band, E_F is the Fermi energy. The energy scale is in electron volts

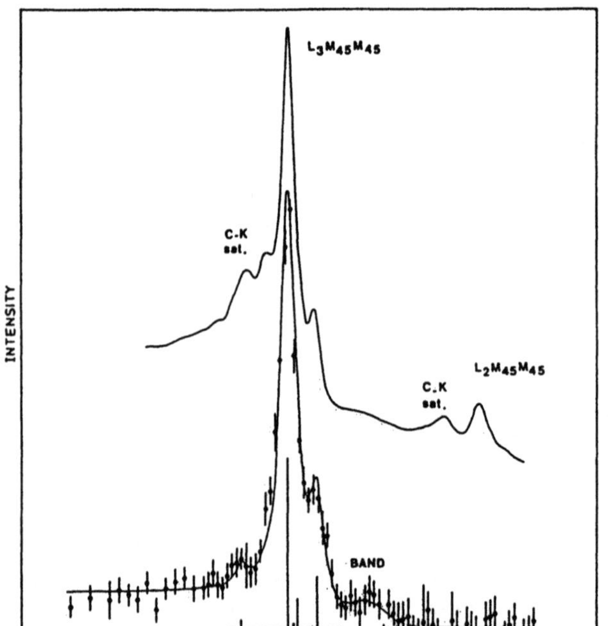

Figure 2b. The upper solid curve is the Auger spectrum as in Fig. 2a. The $L_{23}M_{45}M_{45}$ Auger spectrum in coincidence (APECS) with the L_3 ($2p_{3/2}$) line is shown as dots. The solid line fit to the experimental APECS spectrum shows both the atomic like multiplet split part and the band like part separated in energy by U_{dd}

Obviously the one particle band theory breaks down in describing the Auger spectra of Cu. We argued that this is a result of the strong repulsion of the two final state holes if they are on the same atom. In a single particle approach we can not distinguish between these two sets of states since the particles are assumed to be uncorrelated. The reason we do not notice a problem in the photoelectron spectrum is because the Cu d band is initially full so that removing one d electron results in only one d hole and therefore is a one particle problem if we can neglect the correlation with other 4s,4p etc. holes. The conclusion we reach is that for the electrons in atomic like wave functions with radial extents small compared to the interatomic distance we must explicitly take electron correlation into account in the wave function if the states we study contain more than one hole or electron per atom.

It is interesting then to look at the photoelectron spectrum of Ni metal for which the d band is not full in the ground state. In perfect analogy to the Auger spectrum of Cu metal we expect to see a breakdown of the one electron picture for describing the photoelectron spectrum of Ni metal. Upon removing one d electron we can reach states with two holes (d^8) or one hole (d^9) on the same atom if we consider the ground state to have on the average 9.4 3d electrons per atom corresponding to a magnetic moment of 0.6 Bohr magnetons. This results in the much discussed "6 eV satellite" in the XPS spectrum of Ni metal [29] shown in Fig. 3. Also shown in Fig. 3 is the Ni L_3 VV Auger spectrum which corresponds to primarily d^8 like final states since it originates from mainly the $2p^5 3d^{10}$ (fully screened) [30] core line. The correspondence of the Auger spectrum and the XPS satellites suggests that our interpretation of the origin of the satellite is correct. Even more conclusive are the resonant photoemission results [31] which clearly show the strong resonance of the satellite at the 3p absorption threshold photon energy. For a detailed description I refer to the review of Davis [29].

By comparing the energy of the satellite (d^8) to that of the main line (d^9) we can get a rough estimate of $U \approx 3$ eV which is considerably smaller than that for Cu metal (≈ 7 eV) and in fact comparable to the Ni 3d band width of also about 3 - 3.5 eV. Nonetheless U has a large influence on the electronic structure. This can be seen from Fig. 4 showing again the XPS and resonant photoemission spectra together with a band structure

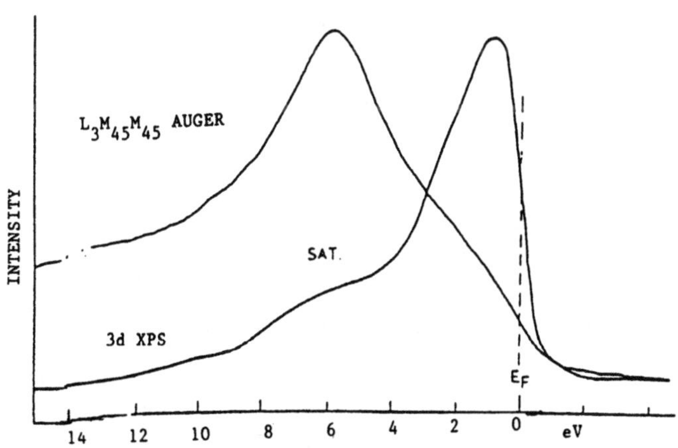

Figure 3. The $L_3M_{45}M_{45}$ Auger spectrum and the 3d valence band XPS spectrum of Ni, the Auger spectrum is shifted by the $2p_{3/2}$ core level binding energy so that E_F coincides with that of the XPS spectrum

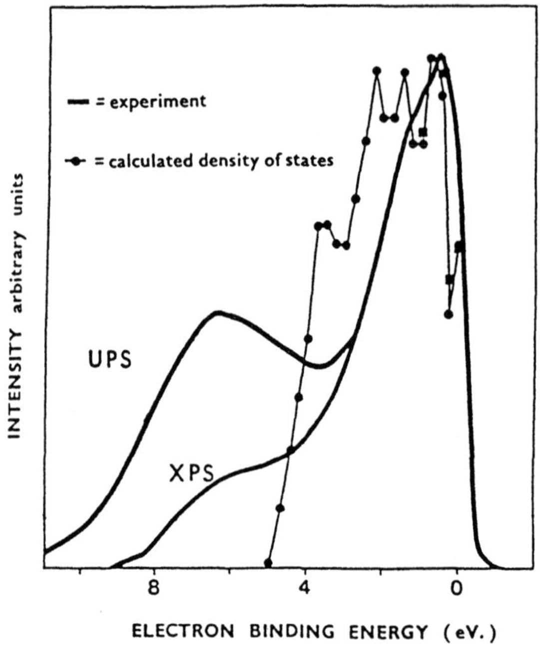

Figure 4. The heavy solid lines are the XPS and on resonance UPS spectra of Ni metal showing the "6 eV" satellite and its resonant behaviour. Also shown with dots is the calculated density of states. This figure is taken from Falicov [32]

calculation. As expected the band structure calculation does not predict the satellite but in addition the band width is quite a bit larger than that of the d^9 band like part of the spectrum. This band narrowing is a direct result of U and can be qualitatively understood as follows. A d hole corresponding to a d^9state can hop to a neighboring site provided that site is in a d^{10} configuration, i.e. $d^9{}_i d^{10}{}_{i+a} \rightarrow d^{10}_i d^9{}_{i+a}$ without an on site energy change. But if the neighboring site is also d^9 a hop would result in a d^8 state i.e. $d^9{}_i d^9{}_{i+a} \rightarrow d^{10}{}_i d^8{}_{i+a}$ which costs an energy U. For U large these hopping processes are not allowed so that the number of degrees of freedom for hopping is reduced resulting in a band narrowing. For $U \rightarrow \infty$ the expected band width would be 0.4 of that obtained from band theory.

3. Intermezzo - Ni Auger Spectrum

Before we go on, a brief intermezzo concerning the Ni L VV Auger spectrum. Although the Cu Auger spectrum could be very well described by the Cini-Sawatzky theory [25,26] I want to warn against using this for partially filled d bands. Neglecting for the moment the multiplet structure the Auger spectrum is determined by the local two hole Greens function

$$G = \frac{g_{oo}}{1-Ug_{oo}} \tag{1}$$

where g_{oo} is the self-convolution of the one-hole Greens function

$$g_{oo} = \frac{1}{2\pi i} g_o \boxtimes g_o \tag{2}$$

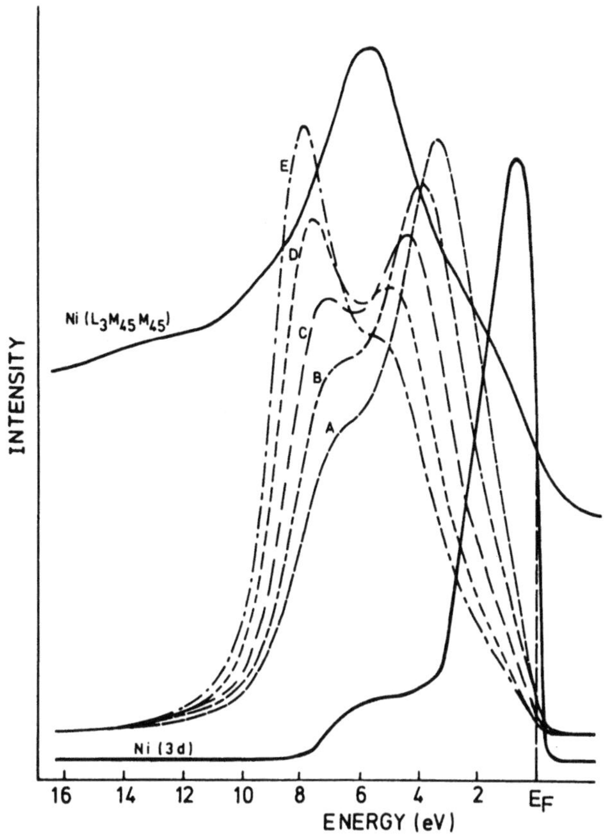

Figure 5. Calculated Auger line shape assuming the Ni(3d) XPS spectrum to be representative of the one-particle density of states, for various values of U: A: U = 1.0, B: U = 1.5, C: U = 2.0, D: U = 2.5, E: U = 3.0

and Im $g_o = \rho(\varepsilon)$ is the one hole density of states as can be determined by photoelectron spectroscopy. This all is fine and well defined for an initially full band. What about partially full bands like Ni? Can we use eq. (1) and if so, what should we take for g_o? If we use the $\rho(\varepsilon)$ as determined by photoemission it includes the satellite which however is there only because U is not negligible. But in eq. (1) g_{oo} was defined as the local two hole Greens function in the absence of U i.e. if U = 0. So it is obvious that if we follow this procedure we are taking U into account two times or maybe we should say three times once in eq. (1) explicitly and once each in g_o. The result of such a procedure for Ni is shown in Fig. 5 from which it is obvious that the experimental line shape cannot be reproduced [33]. This problem was previously pointed out by Bennet et al. [34] who showed that a better approach was to use a $\rho(\varepsilon)$ from XPS but with the satellite removed. This can easily be understood since the XPS spectrum already contains d^9 and d^8 states and a self-convolution effectively gets us to d^6, d^7 and d^8 states of which only the d^8 should be used.

Another perhaps better way is to follow a procedure consistent with the assumptions made in the theory in the first place. The assumptions are that the band is initially full and that we can treat the Auger emitting atom as an "impurity". Locally the band is effectively full if we consider that part of the Auger spectrum originating from the "main" Ni 2p core hole state which corresponds to the "screened" $2p^5 3d^{10}$ state. By g_o we then mean the "impurity" one hole Greens function as we would get it if U were zero on that site. We could then approximate g_o by a calculation of a Ni impurity with U = 0 embedded in a Ni host where the host density of states is given by the XPS spectrum including the satellite and the coupling of the impurity with the host is given by an Anderson Hamiltonian of the form

$$H = \sum_k \varepsilon_k c_k^+ c_k + \varepsilon_d d^+ d + \sum_k V(c_k^+ d + d^+ c_k) \qquad (3)$$

where the first term describes the host d density of states, the second an impurity state at energy ε_d and the third the hybridization. Now for $[\pi V^2 \rho(\varepsilon)]_{aver}$ = 1 eV and ε_d equal to the average host d band energy (E_H) the impurity density of states is equal to the host density of states. But for $\varepsilon_d = E_H - (10-n_d)\,U \simeq 0.65$ eV for $U \simeq 3$ eV and $n_d = 9.4$, as we would get from the sum rule if U were switched off, we get a local "impurity" one particle density of states as shown by the dashed curve in Fig. 6. We see that the satellite now has all but disappeared. Using this one particle density of states in eq. (2) but now also taking the d^8 multiplet structure into account we get the Auger spectrum shown in Fig. 7 which describes the experimental APECS spectrum extremely well for $U(^1G)$ = 3.0 eV. The APECS spectrum shown is that in coincidence with the 2p main ($2p^5 3d^{10}$) full screened line so that we are sure to reach a d^8 final state.[33,35]

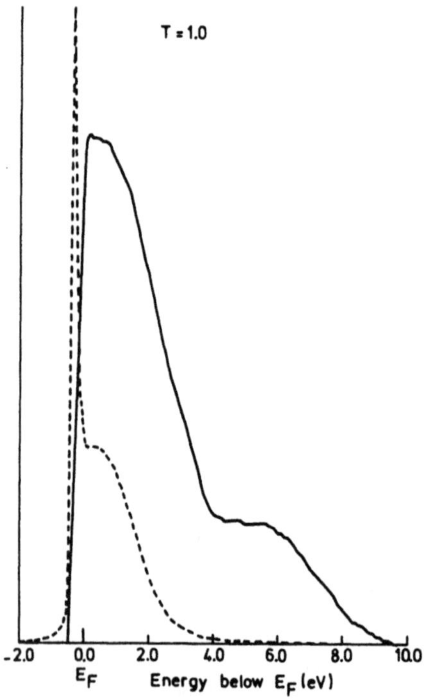

Figure 6. The solid line is the measured XPS spectrum of the valence band of Ni which is taken to be the host density of states for the calculation described in the text. The dashed line is the resulting impurity density of states as described in the text. The sharp delta function peak above E_F is an artifact of the approximations made but carried negligible spectral weight

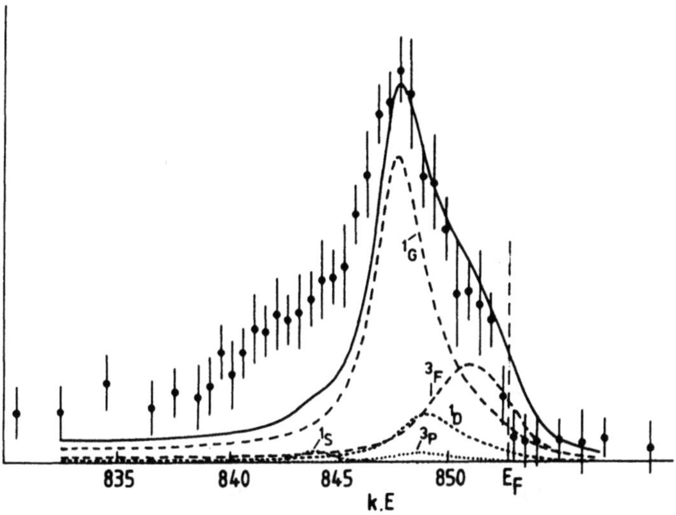

Figure 7. The dots show the experimental APECS spectrum of Ni in coincidence with the Ni $2p^{3/2}$ main line. The solid line is the theoretical Auger spectrum using the impurity density of states of Fig. 6 and the theory as described in the text. The dashed curves are the multiplet components

4. Mott-Hubbard-Anderson

Mott [2] and Hubbard [3] proposed that for materials containing atoms with open shell orbitals with a radial extent small compared to the interatomic spacing we could better start with localized electrons rather than putting the electrons in Bloch states. In this way we would quite naturally get the atomic limit in our previous example for a large interatomic spacing **a**. As an ansatz we consider the solid to consist of a lattice of atoms each with exactly n d or f electrons neglecting for the moment the possible presence of the more free electron like outer shell s and p electrons. This we recall is quite different from a band theory approach for which the probability that an atom has n electrons is given by the binomial distribution [3]

$$P(n) = \frac{z!}{(z-n)!n!} c^n(1-c)^{z-n} \tag{4}$$

in which z is the degeneracy i.e. 10 for d or 14 for f and zc is the electron concentration per atom. This ansatz therefore assumes that polarity fluctuations are completely suppressed. To determine the stability of our ansatz we consider the energy required to move an electron from atom i to atom j. This will turn out to be our definition of the electron-electron interaction U.

$$U = E(d^{n+1}) - E(d^n) + E(d^{n-1}) - E(d^n) \tag{5}$$

We see that U involves the energy difference between d^{n-1} and d^{n+1} which states differ by 2 electrons. Note that is a different definition from that used in Auger spectroscopy as described above for Cu and Ni.

$$U = (E(d^{n-2}) - E(d^n)) - 2(E(d^{n-1}) - E(d^n)). \tag{6}$$

Since all the one particle potential energy terms cancel in eq. (5) any

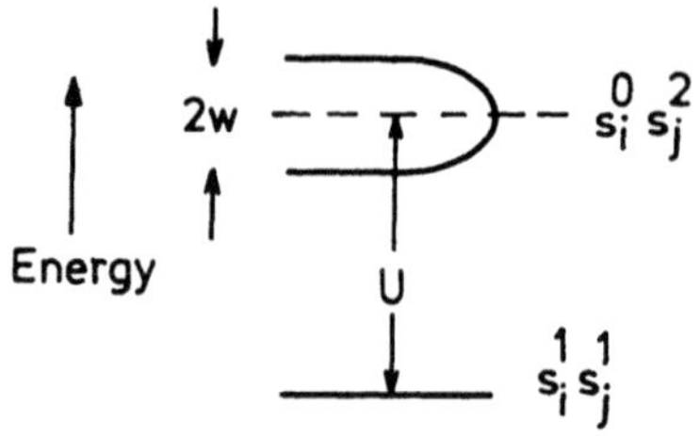

Figure 8. The total energy level diagram for a solid of hydrogen atoms with U >> W showing the stability of the ground state with strongly reduced polarity fluctuations. In fact the states $S_i^1 S_j^1$ and $S_i^0 S_j^2$ will mix for i and j nearest neighbors by an amount given approximately by $2(W/4)^2\ 1/U$ which mixes in some higher polarity states in the ground state and also causes an antiferromagnetic exchange between the unpaired spins on nearest neighbor atoms

theory including density functional which replaces exchange and correlation by a one particle potential in effect assumes that U=0. In this sense U is truly a measure of correlation effects and involves the difference in the ionization potential and the electron affinity.

The state with one extra electron on site j and a hole on site i will in fact form a band of width 2W because both the hole and the electron can now freely propagate. The total energy level diagram is shown in Fig. 8 from which we see that if U>W our ansatz is stable and we will have a Mott-Hubbard insulator with close to n electrons on each atom. As expected the explicit inclusion of correlation in the wave function reduces polarity fluctuations. More generally each atom has n, d or f electrons which will interact just as in the free atom yielding a lowest energy electronic configuration given by Hunds rule. Since Hunds rule states that the lowest energy state is that with a maximum spin and a maximum orbital angular momentum the materials with U>>W will have local magnetic moments. Hunds rule and more generally the multiplet splitting of a $d^n(f^n)$ electronic configuration strongly complicates Fig. 8 and also raises the question as to which states should we consider in determining U. Herring [5] has taught us that we should for each case in eq. (5) take the lowest (Hunds rule) state to define U. Before we discuss this in more detail, another brief intermezzo.

5. **Intermezzo:** Auger of Partially Filled Bands

I now use the above ansatz to demonstrate how misleading the low density approximation i.e. the Cini-Sawatzky theory can be. Consider a lattice of hydrogen atoms with a large and U>>W for the 1s electrons. Consider also a ferromagnetic state with one electron on each atom in the majority spin configuration and add now a small number, say 10%, of minority spin electrons as shown in Fig. 9. This now will be a metal with a density of states looking approximately like that shown in Fig. 9b. The photoelectron spectrum will look like Fig. 9c consisting of two peaks, one at an energy U below E_F and a much lower intensity peak (10%) at E_F. In Fig. 9d we show the self-convolution of the occupied part of the density of states which consists of three structures at 2U, U and 0 relative to E_F with relative intensities (areas) of 1:0.1:0.01 respectively so that the average energy of the self-convolution would be close to 2U below E_F. Now look at the Auger spectrum or rather the local two hole density of states. Two holes can only be produced in those initial states containing two electrons per

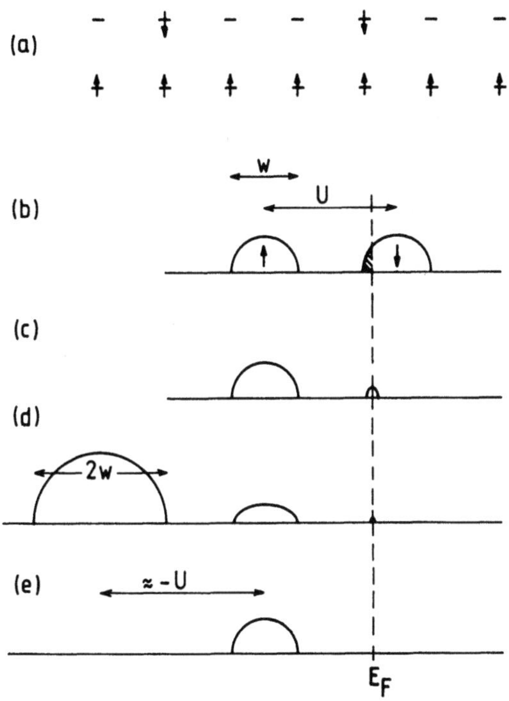

Figure 9. In 9a is shown a concept of a ferromagnetic lattice with a low density of minority spins leading to a metallic system. 9b shows the resulting density of states for U large and 9c shows the photoelectron spectrum. 9d is the self convolution of the photoelectron spectrum and Fig. 9e is the Auger spectrum which only has contributions from those states with two electrons per atom

atom. Therefore virtually all the Auger spectral weight will be at U below E_F as shown in Fig. 9e. The value of U determined by comparing the Auger spectrum to the self-convoluted photoelectron spectrum would be $U_{Auger} = -U$ or a large attractive interaction which obviously is quite ridiculous.

Now as the minority spin band fills up the peak at U in the photoelectron spectrum gradually decreases since its integrated intensity is equal to the number of singly occupied atoms, and the peak at E_F grows until for the filled band case only the peak at E_F remains and the self-convolution also has only the peak at E_F and comparison with the Auger spectrum now gives us the correct value of U, i.e. $U_{Auger} = U$.

From this rather extreme example, which I nonetheless think nicely demonstrates the physics involved, we see that the U value found using a straight forward application of the Cini-Sawatzky theory and a density of one particle states as given by the photoelectron spectrum (which is the low density approximation) will always result in a too small U value and perhaps even a negative U value. This approximation is only valid if the hole density is so low that the satellite structure in the photoelectron spectrum has negligible intensity. So for systems with a substantial hole density we will have to try to find other ways of determining U or we will have to wait for an improved theory.[36]

6. Back to Mott-Hubbard

We still had to deal with a number of complicating aspects like the importance of the multiplet structure and the problem of which states should be treated as atomic like or which band like.

It is clear that if the radius of an atomic orbital is small compared to the interatomic spacing in the solid we can to good approximation use atomic orbitals to describe these states. Although we will not be able to define sharp boundaries for states which can be treated as atomic like and states for which the wave function is strongly k dependent it is possible to at least discuss some extremes. In Fig. 10 is shown the radial extent of the charge distribution for the 3d, 4d, 5d and rare earth and actinide f states as compared to the Wigner Seitz radius of the elemental solid [37]. For the rare earths the f radial extent is so small compared to the interatomic distance that atomic behaviour is expected. Also the late actinides fall in this category. The 3d transition metals even for example Ni lie well above the rare earths as do the 4d, 5d and early actinides. Obviously the smaller the radial extent of the wave function, the smaller the band width and the larger is U. Harrison [38] has shown that for d bands $W \approx a^{-5}$.

Let's look at the extreme case of the rare earths. First from a straight forward application of a band theory approach we expect to see a broad s,p,d band and for the non-magnetic case a narrow 4f band (<1 eV wide [38]). Since the 4f band is not full the Fermi level must lie in it so we get the qualitative structure as shown in Fig. 11. What we really mean by the states above and below E_F drawn in Fig. 11 are the eigenstates of the system with one electron added (states above E_F) and those with one

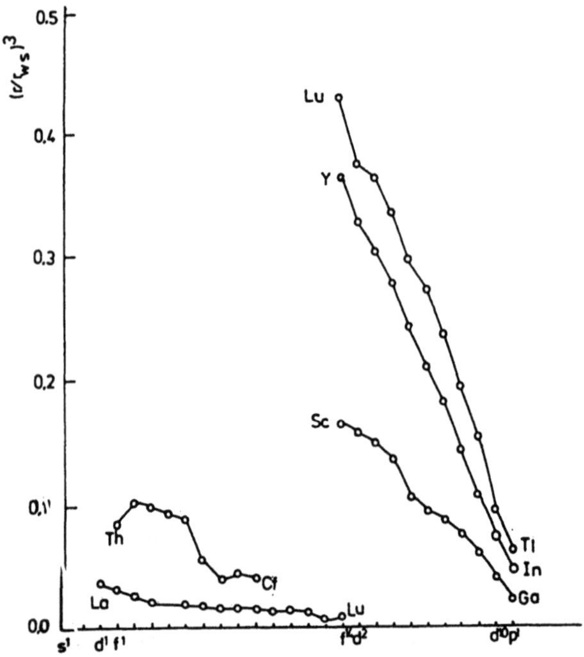

Figure 10. Ratio of l shell "volume" to Wigner-Seitz volume of the 3d, 4sd and 5d elements (l=2) and 5f elements (l=3) as a function of their position in the periodic system

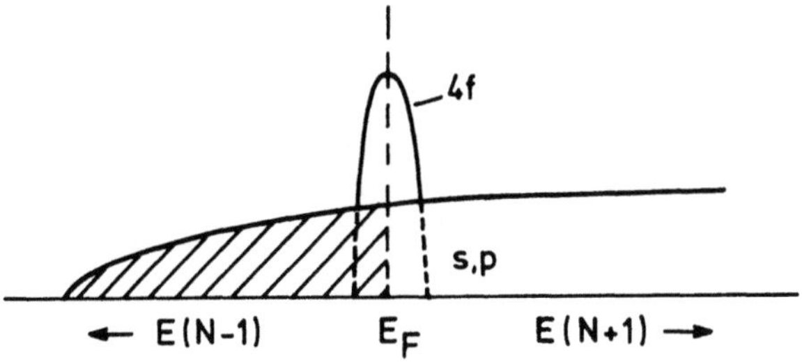

Figure 11. An artist's concept of the density of states from a one particle theory point of view for a rare earth metal with a narrow partly full 4f band and a wide partly full s,p,d band. Note that the states below E_F are the eigenstates of the system with one electron removed and those above E_F are the eigenstates with one electron added

electron removed (states below E_F). We see that this is exactly how we also defined U in eq. (5). Also we see that these states can be directly measured by a combination of photoelectron spectroscopy PS and inverse photoelectron spectroscopy (IPS). A classic and beautiful demonstration is the measurement by Lang and Baer [39] on the rare earth elemental metals shown in Fig. 12. The sharp structures seen result from the 4f states. We see energy spreads in the 4f states of up to 20 eV and usually a large gap in the PS-IPS 4f states - all quite different from our band theory prediction. The structure seen in the spectra is the multiplet structure of the $4f^{n-1}$(PS) and $4f^{n+1}$ (IPS) electronic configurations as demonstrated by the correspondence with the free atom calculations shown with the solid bars. The height of the solid bars are the free atom fractional parentages assuming the ground state to be that determined by Hunds rule for a $4f^n$ configuration. This really is a textbook example demonstrating

a) The importance of electron correlation
b) The stay at home principle for the ground state electronic configurations.
c) The power of PS-IPS to determine directly U
d) The importance of the multiplet structure which is apparently not screened by solid state effects.

In Fig. 13 are shown the energies of the lowest (closest to the Fermi level) $4f^{n-1}$ and $4f^{n+1}$ states as determined by PS-IPS. From this we can directly determine U which is the largest for a $4f^7$ ground state. We also see that for Ce, Eu and Yb the $4f^{n-1}$ states and for Sm and Tm the $4f^{n+1}$ states are quite close to E_F. By alloying or compound formation we can move the chemical potential (E_F) relative to the 4f states by about -2 to +1 eV relative to the elemental rare earths and we should for these elements be able to obtain other valence states in the ground state. The elements mentioned are indeed known to form mixed valent and valence fluctuation compounds.[40]

What about the multiplet structure and the apparent absence of screening? The coulomb interaction between d or f electrons is given by the Slater F^0, F^2, F^4, F^6 integrals which describe the monopole (F^0) dipole (F^2) etc. like contributions. The multipole contributions result in the multiplet structure and they are due to the dependence of the Coulomb integrals on the relative spin (exchange) as well as orbital occupations. The monopole part (F^0) is strongly screened in the solid since it is the direct interaction between two charges. Say it costs us $\varepsilon - E_{pol}$ to produce a local charge where E_{pol} is the reduction in the ionization potential due to polarization and screening. To create two such charges well spatially

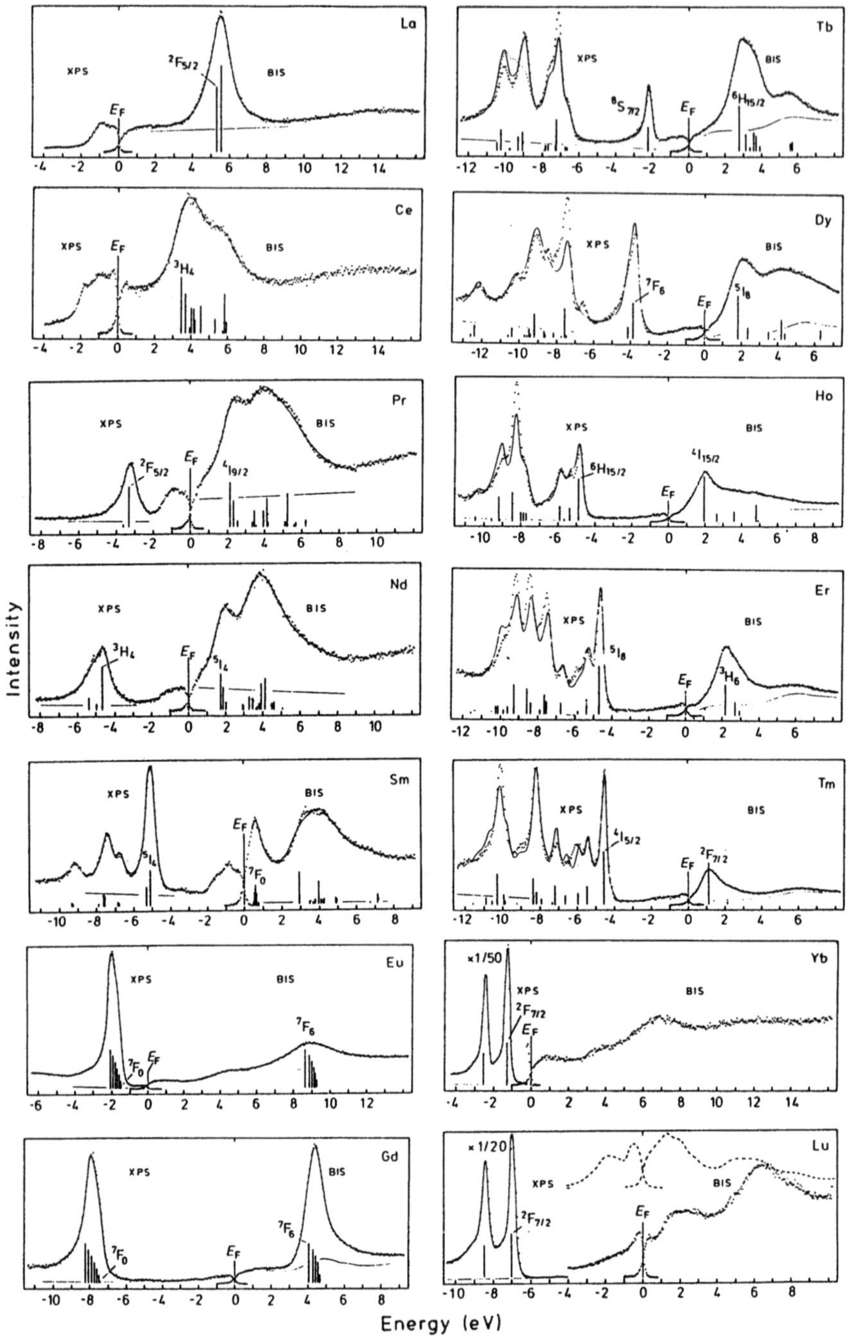

Figure 12. The photoemission and inverse photoemission spectra of the rare earths from Lang and Baer [39] showing the sharp contrast with the one particle picture of Fig. 11. The vertical bars are the calculated atomic multiplets with heights given by the coefficients of fractional parentage for electron removal ($E<E_F$) and electron addition ($E>E_F$) starting with the Hunds rule ground state

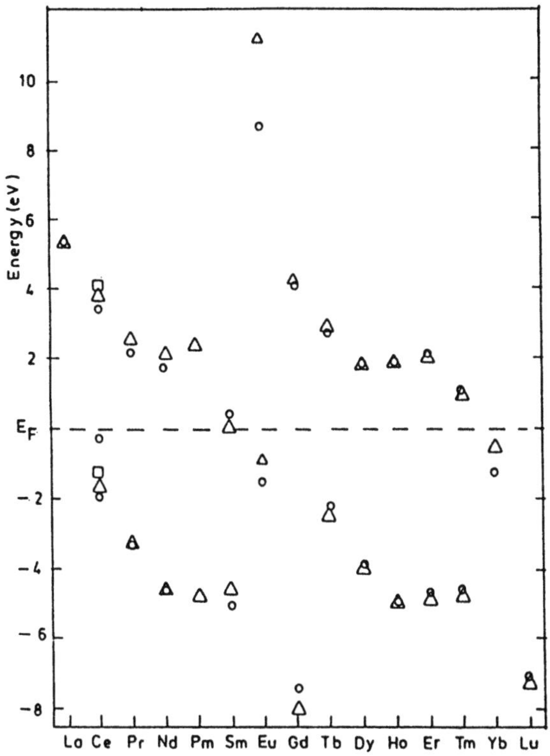

Figure 13. The lowest energy 4f electron removal and electron addition states for the rare earth metals taken from ref. 37. The open circles are from the data of Lang and Baer [39]. The triangles are from atomic theory with a screened monopole Coulomb interaction

separated would cost $2\varepsilon-2E_{pol}$. But to create two charges on the same atom would cost $2\varepsilon-4E_{pol}$ because the polarization or screening energy goes as Z^2 resulting in a screening of F^0 of $2E_{pol}$. However, the system really does not care much what the relative spin direction of the two holes (or electrons) or what their orbital (m_l) quantum number is if the radial extent is small compared to the interatomic distance. So for the states in materials which have radial extents much smaller than the interatomic spacing we have atomic wave functions with atomic multiplet structure but a strongly screened F^o Coulomb integral. This shows that we expect F^0 to be strongly screened but the F^2, F^4, F^6 are expected to be atomic like as found for the rare earths. This is a quite general and for magnetism an extremely important result which holds not only for the rare earths but we saw it above also for the Auger spectrum of Cu and has also been demonstrated for numerous 3d, 4d and 5d transition metals. If the screening of F^2 and F^4 were as large as that for F^0 we would have few, if any, magnetic 3d transition metal systems.

The multiplet structure is determined by F^2 and F^4 for the d electrons and F^2, F^4 and F^6 for the f electrons [41]. The multiplet spread is large for the rare earths as demonstrated in Fig. 12 but also very significant for the 3d elements. This is shown in Fig. 14 using F^2 and F^4 integrals from experiments on Mn impurities in Ag [42]. Fig. 14 clearly shows the above mentioned Hunds rule stabilization of the lowest energy highest spin state in each case. This stabilization tells us that F^2 and F^4 integrals

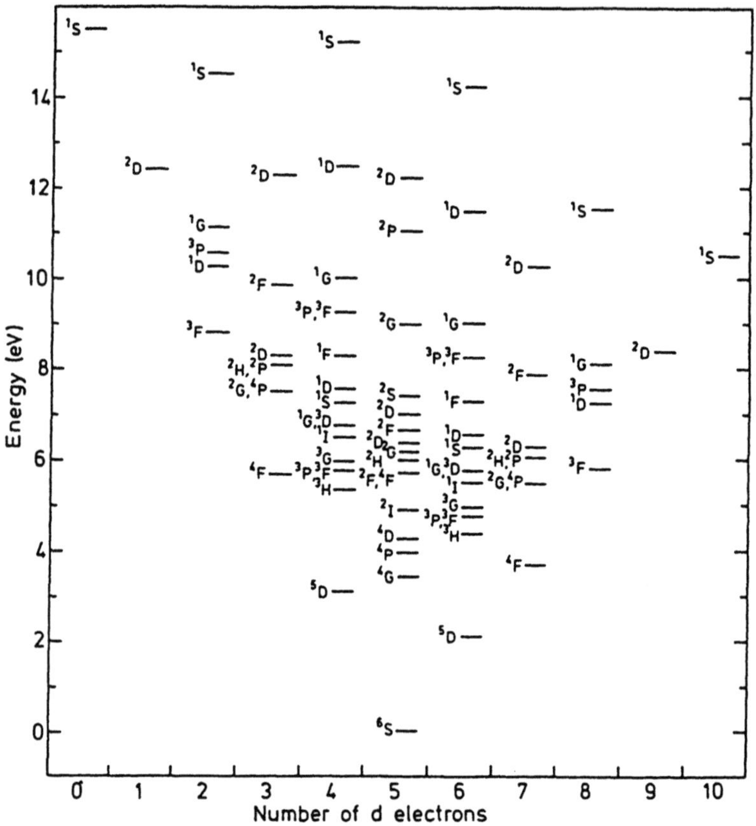

Figure 14. Shows the multiplet structure of 3d atoms with F^0, F^2 and F^4 integrals as determined for Mn impurities in Ag

can also have significant contributions to U. Using a notation so as to separate out the exchange part [37] given by

$$J(dd) = \frac{1}{14} (F^2 + F^4) \tag{7}$$

$$J(ff) = \frac{1}{6435} (286F^2 + 195F^4 + 250F^6) \tag{8}$$

and

$$C(dd) = \frac{1}{14} (\frac{9}{7} F^2 - \frac{5}{7} F^4) \tag{9}$$

$$C(ff) = \frac{1}{6435} (286F^2 + \frac{780}{11} F^4 - \frac{1750}{11} F^6). \tag{10}$$

In terms of these and F^0 we can obtain U as given in tables 1 and 2. For the 3d transition metals J ≃ 0.6 eV and C ≃ 0.3 eV. From tables 1 and 2 we see clearly why U for half-filled shells is so large. It is only for the half-filled shell that J adds to F^0 for the others J in fact reduces U. I think this is a very important observation which has not really been sufficiently appreciated up to recently. In fact since F^0 is strongly screened and J and C are not, it might be possible to get an attractive U say for d^6 of f^1 [43]?

Table 1. Parameters determining the energies of the Hund's rule ground states in LS coupling for all d occupations and the expression for the d-d gap for those occupations as described in reference 37

state	α_I	α_{F0}	α_J	α_C	U^{eff}
$d^0(^1S)$	0	0	0	0	
$d^1(^2D)$	1	0	0	0	F^0-J-C
$d^2(^3F)$	2	1	-1	-1	F^0-J+C
$d^3(^4F)$	3	3	-3	-1	F^0-J+C
$d^4(^5D)$	4	6	-6	0	F^0-J-C
$d^5(^6S)$	5	10	-10	0	F^0+4J
$d^6(^5D)$	6	15	-10	0	F^0-J-C
$d^7(^4F)$	7	21	-11	-1	F0-J+C
$d^8(^3F)$	8	28	-13	-1	F^0-J+C
$d^9(^2D)$	9	36	-16	0	F^0-J-C
$d^{10}(^1S)$	10	45	-20	0	

Table 2. Parameters determining the energies of the Hund's rule ground state in LS coupling for all f occupations and the expression for the f-f gap as described in reference 37

state	α_I	α_{F0}	α_J	α_C	U^{eff}
$F^0(^1S)$	0	0	0	0	
$F^1(^2F)$	1	0	0	0	F^0-J-3/2C
$F^2(^3H)$	2	1	-1	-3/2	F^0-J-1/2C
$F^3(^4I)$	3	3	-3	-7/2	F^0-J+2C
$F^4(^5I)$	4	6	-6	-7/2	F^0-J+2C
$F^5(^6H)$	5	10	-10	-3/2	F^0-J-1/2C
$F^6(^7F)$	6	15	-15	0	F^0-J-3/2C
$F^7(^8S)$	7	21	-21	0	F^0+6J
$F^8(^7F)$	8	28	-21	0	F^0-J-3/2C
$F^9(^6H)$	9	36	-22	-3/2	F^0-J-1/2C
$F^{10}(^5I)$	10	45	-24	-7/2	F^0-J+2C
$F^{11}(^4I)$	11	55	-27	-7/2	F^0-J+2C
$F^{12}(^3H)$	12	66	-31	-3/2	F^0-J-1/2C
$F^{13}(^2F)$	13	78	-36	0	F^0-J-3/2C
$F^{14}(^1S)$	14	91	-42	0	

For the rare earths we obviously are in the limit U >> W so the f electrons behave atomic like and in a Mott-Hubbard way. For the elemental transition metals, however, things are not so clear. PS and IPS measurements do not show the by correlation split band behaviour [44]. In fact we had just found above that for Ni U (^1G) ≃ 3 eV but using the lowest Hunds rule state we should look at U(^3F) = 1.6 eV which in fact is less than W but as pointed out above not negligible [45]. So for the 3d transition metals we are more in the band like regime but for at least Mn, Fe, Co, Ni the effects of correlation are certainly important. For the 4d and 5d transition metals we expect increasingly good agreement with band theory and a decreasing importance of correlation as evidenced also by the predominantly non-magnetic behaviour. For the actinides we expect as may be evident from Fig. 10 a behaviour intermediate between the 3d transition metals and the rare earths and perhaps even a change over as we progress from Pu to Am. If we dilute the 3d transition metal atoms or actinides so that the band width decreases the importance of electron correlation again emerges which brings us to the Anderson impurity, Kondo and heavy fermion systems.

7. Anderson-Kondo-Heavy Fermion

Going back to the rare earths we found that the 4f states behaved atomically and we could just as well have discussed them as impurity states in a metal where the metallic character is determined by the broad sp(d) bands. Also for impurities or diluted systems we define U by eq. (5) realizing that what is important is the energy required to remove or add one electron to system. In Fig. 15 we have sketched the electronic structure of an impurity with n-d electrons in a metallic host. The states below E_F (d^{n-1}) will have a finite lifetime because they are degenerate with a continuum of states (the sp band) with which they hybridize. This hybridization causes a broadening resulting in so-called virtual bound states [11,12]. If both d^{n-1} and d^{n+1} are far removed from E_F and if U>>Γ where Γ is the virtual bound state width, the lowest Hunds rule state of d^n will be the ground state and the impurity will have a local moment. This is a simplified description of Anderson's criterion for local moment formation showing again the competition of electron-electron interactions with the one electron terms leading to delocalization.

We could describe the rare earth elemental metals and alloys in this way but also now we can understand that transition metal impurities in non-magnetic metals can have large local moments. In Fig. 16 we show the local d spectral weight as determined from PS-IPS measurements of Mn impurities in Ag. Clearly U>Γ explaining the local moment (s=5/2)(d^5), for this system. From the d^{n-1} - d^{n+1} splitting we find

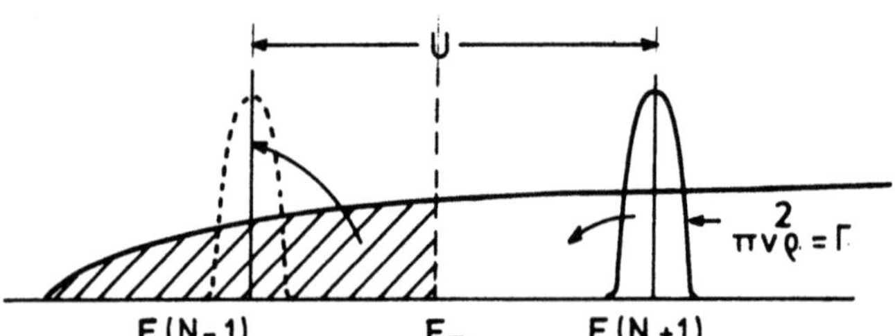

Figure 15. An artist's concept of correlation split impurity states hybridizing with a broad band. This results in a gap in the impurity states given by U and broadenings into virtual bound states of widths $\pi v^2 \rho$ where v is the impurity-host hybridization interaction and ρ is the host density of states

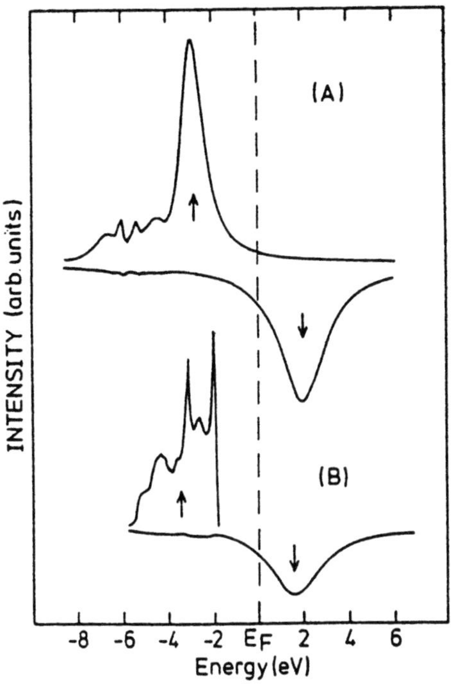

Figure 16. Calculated spin-projected local d density of states at a Mn site in a (A) Ag and (B) Cu surrounding

$$U = F^0 + 4J = 4.6 \text{ eV}. \tag{11}$$

Combining this with an Auger measurement leads to $F^0 = 1$ eV showing the strong screening from the atomic value of about 16 eV. [46] In a way Mn diluted in Ag looks similar to Gd metal showing nicely how dilution increases the relative importance of correlation.

There is another interesting phenomena which can occur for magnetic impurities in metals. The d^{n-1} state of Fig. 16 can be filled by an electron hopping into it from the s, p band states yielding the finite width. However this possibility abruptly stops when we reach the fermi level since at T=0 K the occupation abruptly cuts off at E_F. (An analogous thing happens for d^{n+1}.) This leads to the build up of a very sharp resonance at, or very close to, the fermi level resulting in the Kondo effect [9]. This resonance is so narrow for Mn in Ag that we could not observe it in PS-IPS but it is evident from the minimum in the low temperature resistivity of Mn Ag alloys [47]. Such a sharp resonance has been observed by Allen et al. [48] for the high Kondo temperature materials like Ce Ir_2 and Ce Ru_2 with PS-IPS and also by Baer et al. in a high resolution low temperature PS measurement of $CeSi_2$ [49].

A lot of attention has been placed on the experiments and theory of various Ce, Pr alloys with the result that we now have available a unified theory describing both the high energy spectroscopies, including core level spectroscopies, and the low energy properties like the Kondo effect and magnetic susceptibility with the same Hamiltonian and the same so-called 1/N many body theory developed by Gunnarsson and Schönhammer [50]. This is a really remarkable bit of physics in which we can so clearly see the use of high energy spectroscopies to determine model Hamiltonian parameters like U and the subsequent detailed predictions of the same theory for the physical properties.

8. Transition Metal Compounds

There are two reasons why a discussion of transition metal (TM) compounds follows logically. First of all the anions in the compound act as a diluent so just as above we expect the influence of electron correlation effects for 3d transition metal compounds to be considerably larger than in the pure metals. Secondly since the 3d band widths often turn out to be very small (<0.5 eV) the problem could be akin to the rare earth metals where we found that the 4f states could be treated in an impurity model to good approximation. The difference is that for TM compounds we will be dealing with 3d impurities in a semiconductor rather than a metallic host density of states. To see this let's take a very simplified ansatz for a TM compound like a halide or an oxide. At stoichiometry as for example in $NiCl_2$ or NiO, or CuO the anions are assumed to be closed shell (i.e. Cl^- or O^{2-}) and the TM cations have empty 4s, p states and n-d electrons where n is determined by charge neutrality. For the examples given Ni or Cu are 2+ and have n=8 for Ni and n=9 for Cu. In this ionic ansatz the Cl or O valence p band is full and the TM, 4s,p band is empty and the d "band" is partially full. In a band structure approach this d band would be in the gap formed by the O2p and TM 4s,p bands and at least for the non-magnetic states band theory would predict all of these to be d band metals whereas they all have substantial gaps (for example 4 eV for NiO). In Fig. 17 a band structure calculated density of states is shown for (non magnetic) $NiCl_2$ [51] displaying the large O2p - TM 4s,p gap, the very narrow (impurity like) d bands in this gap, and the Fermi level in one of the crystal field split d sub-bands.

Let us see what we would get from the ionic ansatz. The basic picture is shown in Fig. 18 with an ansatz for which each TM has a configuration d^n and the anions are in an inert gas configuration (i.e. $2p^6$ for O^{2-}). Similar to the discussion around Fig. 8 we consider the stability of this ansatz by looking at the possible charge fluctuations. These are:

$d_i^n d_j^n \rightarrow d_i^{n-1} d_j^{n+1}$ involving U as in eq. 5,

$d_i^n d_j^n \rightarrow d_i^{n+1} d_j^n \underline{L}$ where $\underline{L}$ is a "ligand" O 2p hole and this involves the charge transfer energy Δ,

$d_i^n d_j^n \rightarrow d_i^{n-1} d_j^n 4s_j$ which involves an energy $\Delta_{d,s}$.

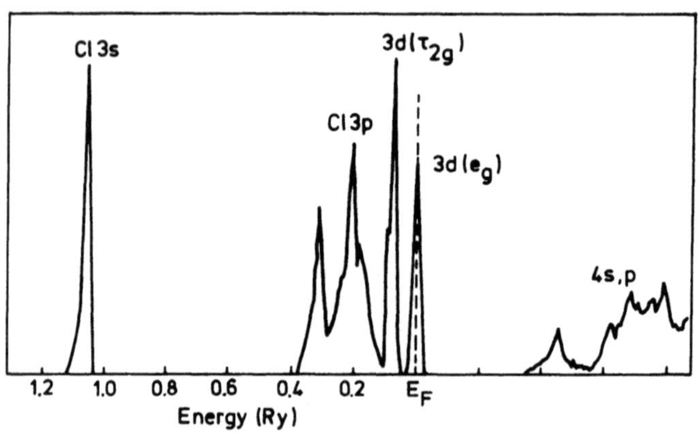

Figure 17. The calculated density of states of $NiCl_2$ showing the full Cl^- 3p band, the empty Ni 4s,4p band and in this gap the very narrow crystal field split Ni 3d bands

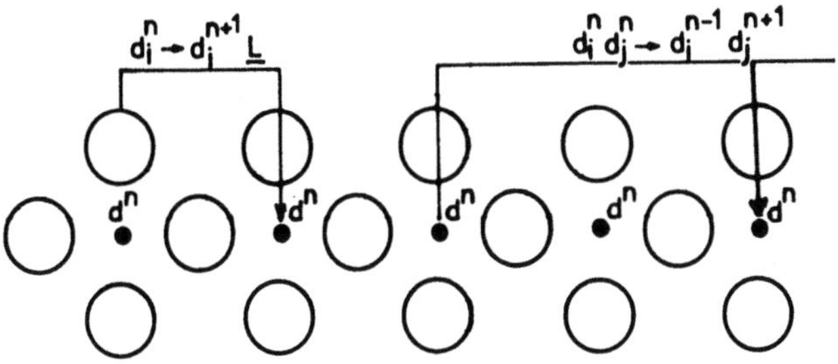

Figure 18. Representation of an ionic lattice consisting of TW ions (d^n) and closed shell anions. The most important charge fluctuation excitations are indicated

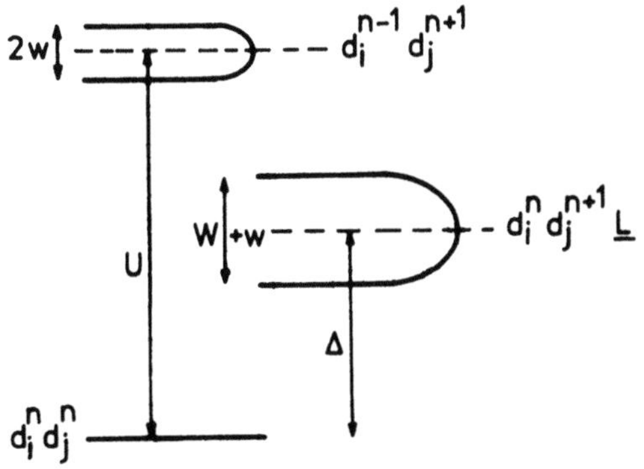

Figure 19. Total energy local diagram corresponding to an ionic ground state and excitations as shown in Fig. 18

Let us for the moment neglect the last possibility and look at the total energy level diagram in analogy to Fig. 8. This is shown in Fig. 19 in which also the band width W of the 2p hole is included. We see that for $U \ll \Delta$ the charge transfer state can be neglected and we have a Mott-Hubbard insulator or a d band metal for $U > w$ or $U < w$ respectively as for Fig. 8. However for $U \gg \Delta$ we have a charge transfer gap insulator for $\Delta > (W+w)/2$ or a hole like metal (holes in the 2p band) for $\Delta < (W+w)/2$. This leads us to a phase diagram shown in Fig. 20 in which the various possibilities are represented which is a simplified version of that presented recently by Zaanen et al. [52]. As shown first by Fujimori [53], Sawatzky and Allen [54] the late 3d TM compounds are in the charge transfer gap regime which has very important consequences also for the high T_c materials. This conclusion was basically reached by a detailed analysis of the PS, IPS spectra of NiO as shown in Fig. 21. This figure shows that the first ionization states are of primarily O 2p character rather than Ni 3d character. This in itself shows that $U > \Delta$ and we are in the charge transfer gap regime of Fig. 20.

Since these experiments and theoretical interpretations numerous other investigations have been done which support these conclusions. These ideas have rather interesting consequences for the CuO layers in high T_c materials. It is generally believed that the substitution of Sr^{2+} for La^{3+} in $La_{2-x}Sr_xCuO_4$ is charge compensated by holes in the CuO planes. For $U > \Delta$ these holes would be in the O 2p band as has in fact been confirmed by O_{1s} electron energy loss [55] as well as O_{1s} x-ray absorption spectroscopy [56]. As for Cu metal and Ni metal also for the compounds resonant photoemission and Auger spectroscopy have played an extremely important role in determining the importance of electron correlation.

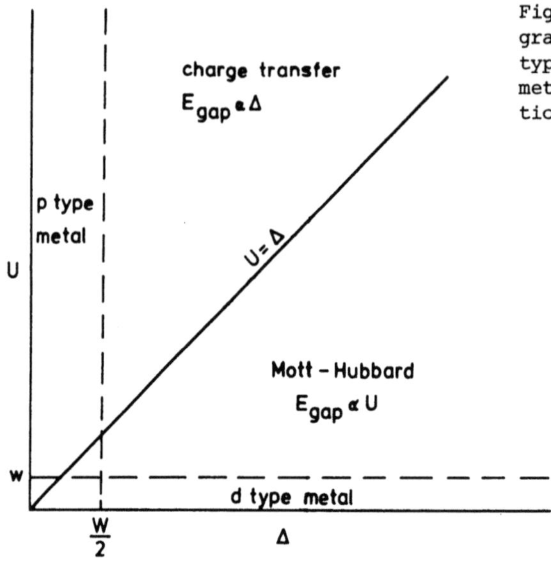

Figure 20. Simple phase diagram showing the various types of insulating and metallic states in transition metal compounds

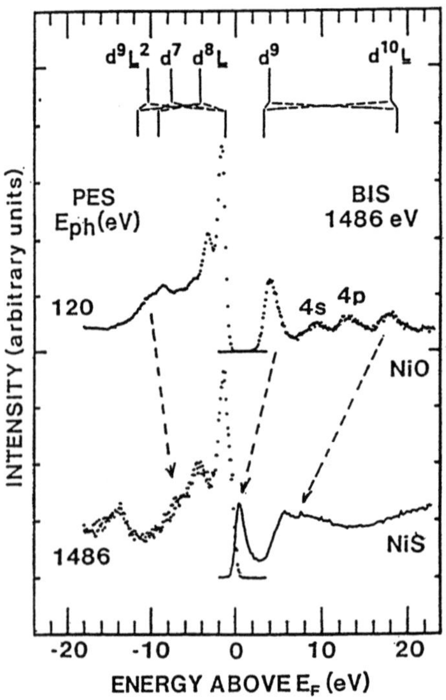

Figure 21. Top is the photoelectron - inverse photoelectron spectrum of NiO with the states labeled according to the ionic ansatz. The gap closes as we go from NiO to NiS as shown in the lower part

There are many aspects in addition to the gap of transition metal compounds for which detailed knowledge of the electron-electron interaction is of extreme importance like the superexchange [57] and the O 2p - TM 3d exchange [58] just to mention a few. A detailed discussion however would take too much space so I refer to the quoted literature.

It is rather interesting to put all of the above into one picture and note the similarities, differences and the classes of phenomena. This is shown in Fig. 22. We start at the top 22(a) which describes a magnetic impurity for $U \gg \pi V^2\rho$, a Kondo system for $E(d^{n-1}) - Ed(^{n})$ and $E(d^{n+1}) - E(d^{n}) \gg \pi V^2\rho$, a mixed valence or valence fluctuation system for $E(d^{n-1}) - Ed(^{n})$ or $E(d^{n+1}) - E(d^{n}) < \pi V^2\rho$ and $\pi V^2\rho \le kT$. In Fig. 22(b) is shown a 3d impurity in a semiconducting host for $U_{eff} < \Delta$ and both the d^{n-1} and d^{n+1} states in the gap. This could be Fe^{2+} in CaF_2 for example. For a lattice of such impurities and U > W where W is the d bandwidth this would represent a Mott Hubbard gap like for example in V_2O_3 below the semi-conductor metal transition. Fig. 22c shows a large U case with only one of the d states (d^{n-1}) in the gap and the other (d^{n+1}) in the conduction band. Such a material would have quite heavy localized d holes (d^{n-1}) and light free electron like electrons. Fig. 22d shows the case for $U \gg \Delta$ so that the gap is of a charge transfer type as in $NiCl_2$ for this case the electrons would be heavy and d like and the holes relatively light and anion p like.

Fig. 22e is rather special. Although basically the same as 22d the hybridization between the d_{n-1} state and the state with a hole in the valence band i.e. $d^n\underline{L}$) is strong enough to push a bound state out of the top of the valence band (dashed line) which now forms a state in the gap. This state has primarily ligand hole character but has the spin and symmetry of the d^{n-1} states. This is very similar to the Haldane Anderson [59] description of multiple charged impurity states in semiconductors. This state also has similarities to the Kondo state of Fig. 22a which is also a result of hybridization. In fact it has recently been proposed that this state in CuO is a local singlet because of a large O2p-Cu3d exchange and is a likely candidate for the quasi particle state providing the conductivity in the high T_c materials.[58]

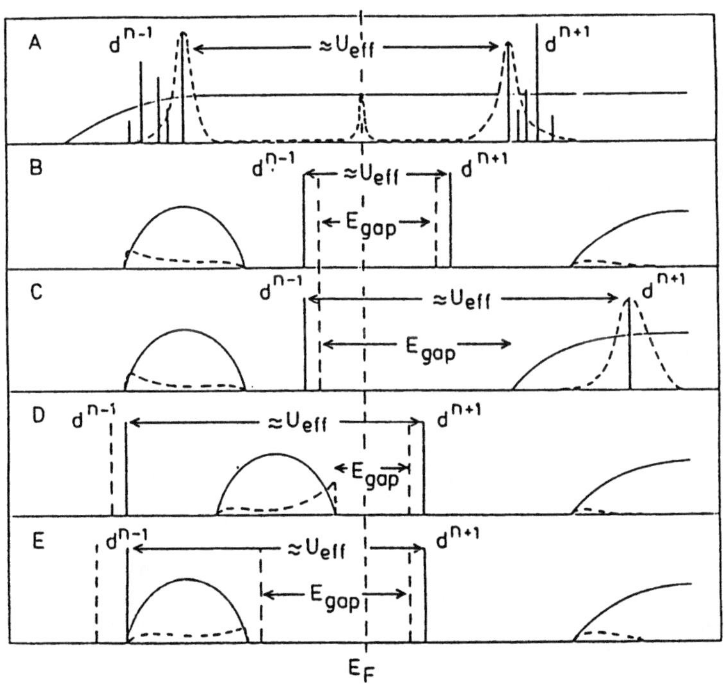

Fig. 22. An artist's concept of the possible situations encountered for strongly correlated impurities in solids. Dashed lines indicate the effect of hybridization

9. Conclusions

In this paper I have attempted to give a simple physical picture of why electron correlation effects are important and how we can measure the electron-electron interactions using examples out of some of the "hottest" topics in solid state physics and chemistry of the last few decades. I have especially emphasized Auger and combinations of PS and IPS as techniques which yield quite direct information about electron correlation effects. I have also indicated severe short comings of these techniques which we should all be aware of. More recent applications and more reviews of these techniques to the systems discussed can be found in recent conference proceedings [60].

10. Acknowledgements

This investigation was supported by the Netherlands Foundation for Chemical Research (SON) and the Foundation for Fundamental Research on Matter (FOM) with financial support from the Netherlands Organization for the Advancement of Pure Research (NWO).

REFERENCES

1. H.J. de Boer and E.J.W. Verweij, Proc. Phys. Soc. 48, 59 (1937).
2. N.F. Mott, Proc. Phys. Soc. Sect. A 62, 416 (1949).
3. J. Hubbard, Proc. Phys. Soc. Ser A, 277, 237 (1964).
4. N.F. Mott, in Metal Insulator Transitions, (Taylor & Francis, London 1974).
5. C. Herring, in Magnetism, Vol. IV, edited by T. Rado and H. Suhl (Academic, New York, 1964).
6. P.W. Anderson, Phys. Rev. 115, 2 (1959); Solid State Physics 14, 99 (1963).
7. J.B. Goodenough, Phys. Rev. 100, 564 (1955).
8. J. Kanamori, J. Phys. Chem. Solids, 10, 87 (1959).
9. J. Kondo, Prog. Theoret. Phys. (Kyoto) 28, 846 (1962); 32, 37 (1964); 34, 204 (1965).
10. M.B. Brodsky in Valence Instabilities and Related Narrow Band Phenenomena, edited by R.D. Parks (Plenum, New York, 1977), p. 351.
11. J. Friedel, J. Phys. Radium 19, 573 (1958).
12. P.W. Anderson, Phys. Rev. 124, 41 (1961).
13. J.R. Schrieffer and P.A. Wolff, Phys. Rev. 149, 491 (1966).
14. J.C. Bednorz and K.A. Müller, Z. Phys. B64, 189 (1986).
15. A.R. Williams and U. von Barth in Theory of the Inhomogeneous Electron Gas, edited by S. Lundqvist and N.H. March, Plenum, New York (1983) p.189.
16. P. Hohenberg and W. Kohn, Phys. Rev. 136, B864 (1964); W. Kohna and L.J. Sham, Phys. Rev. 140, A1133 (1964).
17. F.J. Himpsel, Adv. Phys. 32, 1 (1983).
18. P.Thiry, O. Chandesris, J. Lecant, C. Guillet, R. Pinchaux and Y. Petroff, Phys. Rev. Lett. 43, 82 (1979).
19. N.V. Smith, (Reports in Progress of Physics, in press)
20. D.P. Woodruff and N.V. Smith, Phys. Rev. Lett. 48, 283 (1982); W. Jacob, V. Dose and A. Goldman, Appl. Phys. A41, 145 (1986).
21. For energies larger than about 5 eV above E_F systematic deviations are found which are due to the exchange correlation hole (as used in DF theory) not being able to follow the added electron. W. Speier, R. Zeller and J.C. Fuggle, Phys. Rev. B32, 3597 (1985).
22. J.J. Lander, Phys. Rev. 91, 1382 (1953).
23. J.C. Riviére, AERE Harwell rep. no. AERE-R 10384 (1982).
24. H. Haak, G.A. Sawatzky and T.D. Thomas, Phys. Rev. Lett. 41, 1825 (1978).

25. M. Cini, Solid State Commun. 24, 681 (1977).
26. G.A. Sawatzky, Phys. Rev. Lett. 55, 418 (1985).
27. G.A. Sawatzky and A. Lenselink, Phys. Rev. B21, 1790 (1980).
28. See also the reviews, ref. 23, J.C. Fuggle "High Resolution Auger Spectroscopy" in "Electron Spectroscopy IV", ed. C.R. Brundle and A.D. Baker, p.85 (Academic, London) 1981.
P. Weightman, Rep. Prog. Phys. 45, 753 (1982);
D.E. Ramaker "Auger Spectroscopy as a Probe of Valence Bonds and Bands", Springer series in Chemical Physics, 1982.
29. L.C. Davis, J. Appl. Phys. 59, R25 (1986).
30. S. Hüfner and G.K. Wertheim, Phys. Lett. A51, 299 (1975).
31. C. Guillot, Y. Ballu, J. Paigné, J. Lecante, K.P. Jain, P. Thiry, R. Pinchaux, Y. Petroff and L.M. Falicov, Phys. Rev. Lett. 39, 1632 (1977).
32. L.M. Falicov in "X-Ray and Inner Shell Processes", edited by P. Lagarde, F.J. Wuilleumier and J.R. Briand (Les editions de Physique, France) p. C9-891 (1987).
33. H.W. Haak, Ph.D. thesis, University of Groningen,; 1983.
34. P.A. Bennett, J.C. Fuggle, F.U. Hillebrecht, A. Lenselink and G.A. Sawatzky, Phys. Rev. 27B, 2194 (1983).
35. H.W. Haak, G.A. Sawatzky, L. Ungier, J.K. Gimzewski and T.D. Thomsa, Rev. Sci. Instrum. 55, 696 (1984).
36. V. Drchal (to be published).
37. D. van der Marel and G.A. Sawatzky, Phys. Rev. 37, 10674 (1988).
38. W. Harrison in Electronic Structure and the Properties of Solids (Freeman, San Francisco, 1980).
39. J.K. Lang, Y. Baer and P.A. Cox, J. Phys. F11, 121 (1981).
40. Valence Instabilities and Related Narrow-Band Phenomena, edited by R.D. Parks (Plenum, New York, 1977);
Valence Fluctuations in Solids, edited by L.M. Falicov, W. Hanke, and M.B. Maple, (North-Holland, Amsterdam, 1981);
Proceedings of the 4th International Conference on Valence Fluctuations, Cologne, 1984, edited by E. Müller-Hartmann, B. Roden and D. Wohlleben, (North-Holland, Amsterdam, 1985);
Proceedings of the 5th International Conference on Crystalline Field and Anomalous Mixing Effects in f-electron Systems, Sendai 1985, edited by T. Kasuya, (North-Holland, Amsterdam, 1985).
41. J.S. Griffith in The Theory of Transition Metal Ions, (Cambridge University Press, 1961).
42. D. van der Marel, G.A. Sawatzky and F.U. Hillebrecht, Phys. Rev. Lett. 53, 206 (1984).
43. D. van der Marel and G.A. Sawatzky, Solid State Commun. 55, 937 (1985).
44. W. Speier, J.C. Fuggle, R. Zeller, B. Ackermann, K. Szat, F.U. Hillebrecht and M. Campagna, Phys. Rev. B30, 6921 (1984).
45. J.C. Fuggle, P. Bennet, F.U. Hillebrecht, A. Lenselink and G.A. Sawatzky, Phys. Rev. Lett. 49, 1787 (1982).
46. The atomic values can be obtained from atomic spectroscopic data. C.E. Moore, in Atomic Energy Levels, Natl. Bur. Stand. (USA), Circ. No. 467.
47. A.N. Gerritsen and J.O. Linde, Physica, 17, 573 (1951).
48. J.W. Allen, S.-J. Oh, O. Gunnarsson, K. Schönhammer, M.B. Maple, M.S. Torikachvilli and I. Lindau, Adv. in Phys. 35, 275.
49. F. Patthey, J.-M. Imer, W.-D. Schneider, Y. Baer, B. Delley and F. Hulliger, Phys. Rev. B36, 7697 (1987).
50. O. Gunnarsson and K.Schönhammer, Phys. Rev. Lett. 50, 604 (1983); Phys. Rev. B28, 4315 (1983). See also ref. 48.
51. R. Coehoorn, Thesis, University of Groningen, the Netherlands (1985).
52. J. Zaanen, G.A. Sawatzky and J.W. Allen, Phys. Rev. Lett. 55, 418 (1985).

53. A. Fujimori and F. Minami, Phys. Rev. B30, 957 (1984).
54. G.A. Sawatzky and J.W. Allen, Phys. Rev. Lett. 53, 2339 (1984).
55. J.A. Yarmoff, D.R. Clarke, W. Drube, U.O. Karlsson, A. Talab-Ibrahimi and F.J. Himpsel, Phys. Rev. B36, 3967 (1987).
56. N. Nücker, J. Fink, J.C. Fuggle, P.J. Durham, W.M. Timmerman, Phys. Rev. B37, 5158 (1988).
57. J. Zaanen and G.A. Sawatzky, Can. J. Phys. 65, 1262 (1987).
58. H. Eskes and G.A. Sawatzky, Phys. Rev. Lett. 61, 1415 (1988); F.C. Zhang and T.M. Rice, Phys. Rev. B37, 3759 (1988).
59. F.D.M. Haldane and P.W. Anderson, Phys. Rev. B13, 2553 (1976).
60. Narrow-Band Phenomena - Influence of Electrons with Both Band and Localized Character, edited by J.C. Fuggle, G.A. Sawatzky and J.W. Allen (Plenum Press, New York, 1988);
Proceedings 1988 Adriatico Research Conference, Trieste, Int. J. Mod. Phys. 1, (1988);
Proceedings NEC Symposium, Hakone, Japan, 1988 (to be published);
Core-Level Spectroscopy in Condensed Systems, edited by J. Kanamori and A. Kotani, (Springer-Verlag, 1988).

Part II

Theory of the Auger Process and Electronic Structure Determination

Theory of the Auger Spectra of Ca-Si Compounds

O. Bisi

Dipartimento di Fisica, Università di Modena,
Via Campi 213/A, I-41100 Modena, Italy

The theoretical analysis of the Si $L_{2,3}VV$ Auger lineshape of Ca-silicides is performed on the basis of the electronic properties of the valence states computed by the LMTO-ASA method. In $CaSi_2$ the spectrum is directly interpreted on the basis of the ground state valence density of states, while final-state hole-hole repulsion is not negligible ($U \sim 3$ eV) in Ca_2Si. This correlation effect, which is present only when a final state hole is found in an *s* state, is due to the quasi-atomic configuration of the Si 3*s* states in this compound.

1. Introduction

The Auger process is widely used not only as a probe of surface elemental composition, but of the electronic properties of the investigated system as well [1-7]. The latter information is provided by the analysis of the Auger lineshape. One of the most studied lineshape is that of the well-characterized Si $L_{2,3}VV$ transition [8-20]. In the present paper the electronic states of Ca-silicides ($CaSi_2$ and Ca_2Si) are investigated through a theoretical analysis of the Si $L_{2,3}VV$ Auger spectrum.

Ca-silicides belong to the family of earth-alkaline and rare-earth silicides where metal atoms exhibit a *d*-occupancy close to zero. The chemical bond in these compounds shows many features and peculiarities [21-24]. As far as silicon derived states are concerned we may have well mixed $s - p$ orbitals in the silicon rich phase, while in Ca_2Si the Si-*s* states are in a $quasi - atomic$ configuration, well separated from the *p* orbitals which are involved in the bond with calcium. The Ca *d* states unlike the case of pure metal, where only the low energy tail of the *d* band is occupied, are now forming a well defined structure below the Fermi level. This is due to the interaction with silicon atoms which bond with Ca $s - p - d$ orbitals.

In this paper, I present an investigation of the electronic properties of Ca-silicides through the analysis of the Si $L_{2,3}VV$ Auger lineshape. The properties of the valence states will be found to be directly related to the experimental data. In particular the different character of the Si *s* states in these two compounds will determine the nature of the Auger process: the final-state hole-hole repulsion is found significant only for the *s* states in Ca_2Si, where the *s* orbitals form an atomic-like narrow band.

The main features of the Auger process are described in Sect. 2, while the following section introduces the basic features of the chemical bond in Ca-silicides. Section 4 presents the theoretical analysis of the Auger spectrum in these compounds and the last section summarizes the main results of the present investigation.

Springer Series in Surface Sciences, Vol. 18 **Auger Spectroscopy and Electronic Structure**
Editors: G. Cubiotti · G. Mondio · K. Wandelt

2. The Auger CVV process

In the Auger process an initial hole created in a shell is transferred to an outer shell or orbital, and the energy involved in this transition results in the emission of a secondary electron. The kinetic energy of the Auger electron will be related to the final state energy of the atom, which is governed by the positioning of the double-hole formation and the interaction between remaining electrons. In core-valence-valence (CVV) Auger process the double-hole configuration resides in valence levels.

A draft of the Auger CVV process is shown in Fig. 1. The energies refer to the Auger electron final energy (E_f), the initial core-hole energy E_c and to the valence electron energies E_1, E_2. The indexes 1 and 2 represent the quantum numbers of the valence electrons, i.e. **k** vector, band (n) and spin (σ) labels. Figure 1 describes the simplest picture of the Auger process, where the energies are *single – particle* energies.

Many – body interactions may affect the CVV spectrum mainly through:

(i) *Initial-state core-hole screening.* Due to the presence of the core hole a screening relaxation takes place, which may change the number of local valence electrons and the relative number of their angular momentum weight. Therefore the Auger profile can be altered from that expected on the basis of ground-state configuration.

(ii) *Final-state hole-hole correlation.* If the final-state hole-hole Coulomb repulsion is not negligible it is necessary to introduce correlation effects. These may be expressed in terms of U, the fully screened hole-hole Coulomb interaction [25-27].

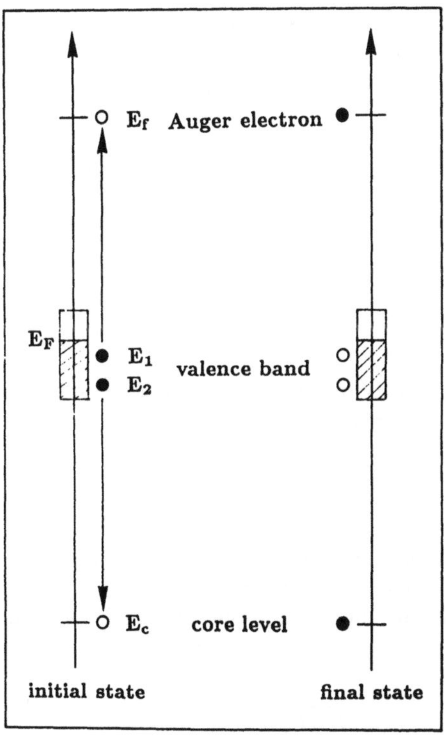

Fig. 1 - Sketch of the CVV Auger process

When these many-body effects are negligible, as for the case of the $L_{2,3}VV$ Auger spectrum of elemental Si, the transition is called *band-like*. In these conditions the energy conservation is simply

$$E_f = E_1 + E_2 - E_c, \tag{1}$$

and the Auger current for fixed $\mathbf{k}_f, E_f$ and σ_f is

$$I(\mathbf{k}_f, E_f, \sigma_f) = \frac{2\pi}{\hbar}\rho(\mathbf{k}_f, E_f, \sigma_f)\frac{1}{2l_c+1}\sum_{m_c=-l_c}^{+l_c}\frac{1}{2}\sum_{\mathbf{k}_1 n_1}\sum_{\mathbf{k}_2 n_2} < M >^2 \times \\ \delta(E_f - E_1 - E_2 + E_c), \tag{2}$$

where $\mathbf{k}_f = \sqrt{2mE_f}$, l_c and m_c indicate the angular momentum and magnetic quantum number of the core level, ρ is the density of final states and $< M >^2$ is the square of the matrix element of the transition.

For usual experiments the current is measured as a function of the energy only:

$$I(E_f) = \frac{1}{2\pi}\frac{m}{\hbar^3}\frac{\sqrt{2mE_f}}{2l_c+1}\sum_{m_c=-l_c}^{+l_c}\sum_{\mathbf{k}_1 n_1}\sum_{\mathbf{k}_2 n_2} < M >^2 \delta(E_f - E_1 - E_2 + E_c). \tag{3}$$

By using the identity:

$$\delta(E_f - E_1 - E_2 + E_c) = \int dE\ \delta(E_f - E - E_1)\delta(E_c - E_2 + E), \tag{4}$$

we get

$$I(E_f) = \frac{1}{2\pi}\frac{m}{\hbar^3}\frac{\sqrt{2mE_f}}{2l_c+1}\sum_{m_c=-l_c}^{+l_c}\int dE < M >^2 \times \\ \sum_{\mathbf{k}_1 n_1}\delta(E_f - E - E_1)\sum_{\mathbf{k}_2 n_2}\delta(E_c - E_2 + E). \tag{5}$$

A strong simplification takes place if the matrix elements of the transition M can be considered constant.

If M is constant:

$$I(E_f) = \frac{1}{2\pi}\frac{m}{\hbar^3}\sqrt{2mE_f}\ < M >^2 \int dE\ N(E_f - E)N(E_c + E), \tag{6}$$

where $N(E) = \sum_{\mathbf{k}n}\delta(E - E_{\mathbf{k}n})$ is the total density of valence states of the system. If Eq. 6 holds the Auger current is simply proportional to the self-fold D° of the valence density of states:

$$I(E_f) \propto N \otimes N = \int dE\ N(E_f - E)N(E_c + E), \tag{7}$$

If the final-state hole-hole repulsion is not negligible it is possible to take into account the Coulomb energy repulsion U between the two holes. According to the Cini-Sawatzky theory [28-32] this can be done by substituting the self-fold D° by the two-hole spectrum:

$$D(E_f) = \frac{D^\circ(E_f)}{[1 - F(E_f)U]^2 + [\pi U D^\circ(E_f)]^2}, \tag{8}$$

where $F(E)$ is the Hilbert transform of D°. Equation 8 has been derived in the case of a filled valence band.

For an unfilled band different treatments have been proposed. According to Tréglia et al. [33] Eq. 8 is still valid but now D° is the self-fold of the single particle spectrum, with self-energy effects included. In this light various Auger spectra have been investigated by computing D° from the measured photoemission spectrum [25]. More recently Cini and Verdozzi [34] suggested that D° should be computed from the undressed particle hole, without self-energy effects, as in the filled band case. The two holes tend to interact as bare holes essentially because the size of their screening clouds is larger than the range of hole-hole repulsion. This bare ladder approximation (BLA) is correct when the number of the holes in the valence band is low.

The main features of the $L_{2,3}VV$ lineshape in elemental Si can be interpreted without invoking many-body effects. However it is necessary to take into account matrix element effects, since $< M >^2$ cannot be considered constant and Eqs. 6-7 are not justified. Feibelman and McGuire [9] showed that the Auger current (Eq. 5) can be well approximated by the following combination of the self-folds of angular momentum density of states N_s and N_p

$$I(E_f) = \alpha_{ss} N_s \otimes N_s + \alpha_{sp} N_s \otimes N_p + \alpha_{pp} N_p \otimes N_p, \tag{9}$$

Assuming the matrix elements to be constant the $\alpha_{pp} : \alpha_{sp} : \alpha_{ss}$ ratio would be 1:2:1. Direct calculation of the matrix elements through atomic wave functions leads to a ratio of 1:0.9:0.1. It is evident that the main effect of the matrix element is to suppress the contribution of the self-fold of the s-like density of states and to reduce somewhat the contribution of the cross-fold of the s- and p-like densities of states.

The comparison with the experimental data shows that the computed α_{sp} is still too large with respect to α_{pp} by a factor $\geq$1.4. Therefore Eq. 9 can be used to compute the Auger current if the ratio is around 1:0.6:0.1. An equivalent treatment is to use Eq. 7 all the same, but substituting the valence density of states by the *transition density of states*:

$$N'(E) = \alpha N_s(E) + N_p(E), \tag{10}$$

Since $\alpha < 1$ the contribution of the s-like states is reduced. The results presented in this paper are computed in this way, with $\alpha = 0.3$. This choice is equivalent of using Eq. 9 with the $\alpha_{pp} : \alpha_{sp} : \alpha_{ss}$ ratio equal to 1:0.6:0.09.

3. The electron states in Ca-silicides

The electronic states of $CaSi_2$ and Ca_2Si are computed by the linear muffin-tin orbitals (LMTO) method in the atomic sphere approximation (ASA) [35-37]. The structure of these compounds is quite complex: $CaSi_2$ presents two molecules per unit cell with two non-equivalent Si atoms, while Ca_2Si has four molecules per unit cell with two non-equivalent Ca atoms [38]. For this reason no calculation of their electronic properties has been previously performed. The results of this calculation have been successfully compared with both the photoemission and inverse photoemission experimental data. Details of the calculation are out of the scope of the present work and are published elsewhere [23].

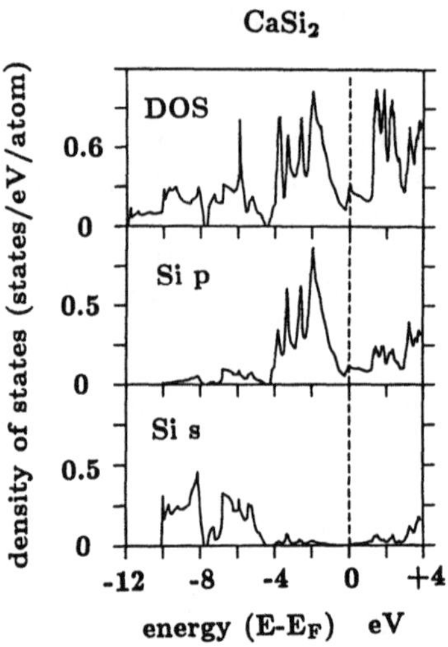

Fig. 2 - Total and Si-site projected DOS for $CaSi_2$

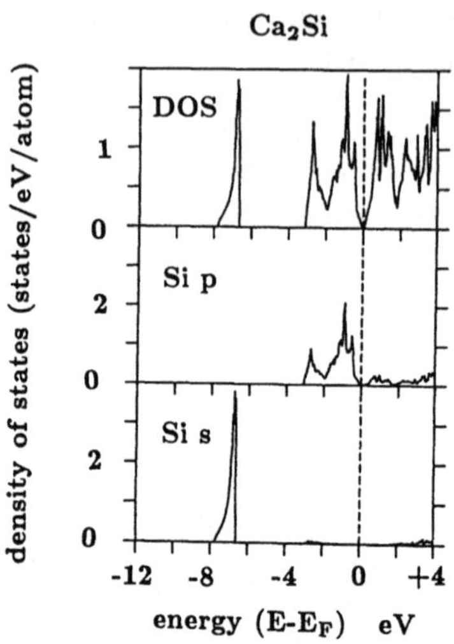

Fig. 3 - Total and Si-site projected DOS for Ca_2Si

For our present purpose is sufficient to illustrate the total density of states (DOS) and the Si-site projected DOS. These data are shown in Fig. 2 ($CaSi_2$) and in Fig. 3 (Ca_2Si).

In $CaSi_2$ the total DOS shows a unique valence band with Si *s*-derived states at higher binding energy and the Si *p*-like orbitals, interacting with the Ca $s-p-d$ states, above -4 eV. In Ca_2Si the *s*- and *p*-derived bands are completely separated by a gap of $\sim$ 4 eV. The *s*-like states are found below this gap, around -7 eV, forming an atomic-like state. The remaining part of the valence band, above -3 eV, is characterized by the Si *p*-like states strongly interacting with Ca orbitals. The density of states of Ca_2Si is zero at the Fermi level E_F, showing a semimetal behaviour.

The differences between these two compounds, as shown by the analysis of Fig. 2 and Fig. 3 are evident; we expect that such differences in the valence states reflect in the Auger spectrum.

4. The Auger spectrum of Ca-silicides

The experimental spectrum of $CaSi_2$ and Ca_2Si has been published by Sancrotti et al. [39]. Figure 4 shows the Si $L_{2,3}VV$ lineshape as computed by analytical integration of the raw spectra acquired in the derivative form. In $CaSi_2$ the main peak is located around 90 eV and only a shoulder is recognizable around 82-84 eV. On the other hand the Ca_2Si spectrum shows two well distinct structures: the main peak is still located at $\sim$ 90 eV and another feature is present around 82.5 eV. Furthermore the main peak is now narrower than that of $CaSi_2$. These differences can be related to the different features of the valence electron states, shown in Figs. 2-3.

The theoretical Si CVV lineshape, as computed from Eqs. 7 and 10 is shown in Fig. 5. It is evident that the experimental lineshape of $CaSi_2$ is well described by this theoretical calculation. The existing minor discrepancies have different origins like

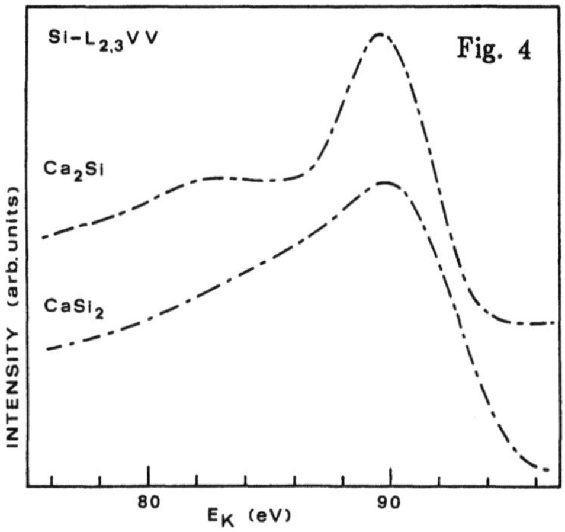

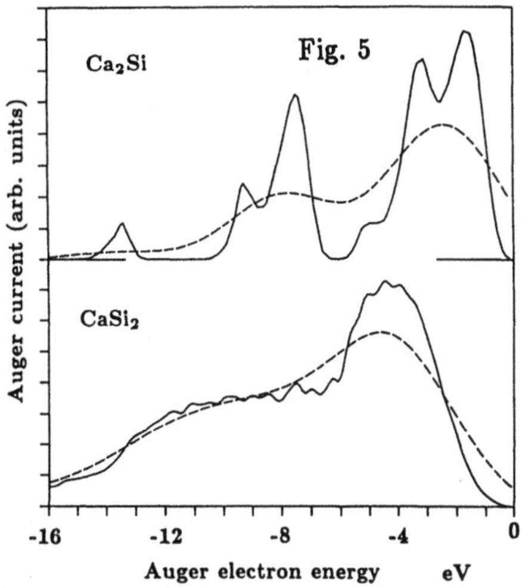

Fig. 4 - The experimental Auger spectrum for the Si $L_{2,3}VV$ transition in $CaSi_2$ and Ca_2Si

Fig. 5 - The theoretical Auger spectrum for the Si CVV transition in $CaSi_2$ and Ca_2Si. The full line presents the raw spectrum, while in the dashed curve a gaussian broadening ($\sigma = 1.5$ eV) to simulate the experimental resolution is added. The energy of the Auger electron is referred to the maximum kinetic energy, chosen as zero

energy losses, the choice of the background line for the analytical integration of the measured spectra, and the crude estimation of the matrix element effect. On the basis of the valence states calculation shown in Fig. 2 we can unambigously assign the main feature of the $CaSi_2$ Auger spectrum to the emission leaving two holes of p character on the Si site, while the second shoulder around 82-84 eV is related to an $s-p$ two-hole final-state contribution. According to the previous description of the matrix element effects (Sect. 2) the $s-s$ self-fold is greatly reduced.

In Ca_2Si the $p-p$, $s-p$ and $s-s$ structures are well distinct and separated. This result is a direct consequence of the wide gap present between the s and p states (Fig. 3). For this reason the main structure, which is still related to a $p-p$ final state contribution, is narrower than in $CaSi_2$. The second peak is assigned to the $s-p$ feature. The energy location of this structure in the measured and computed spectrum of Ca_2Si is different. While in the experimental lineshape the $s-p$ feature lies ~ 7.5 eV below the main peak, this distance is reduced to ~ 5.6 eV in the computed spectrum. A disagreement of the order of 2 eV cannot be explained in terms of the approximations of this calculation. It is directly related to the energy location of the single-particle states, which does not correspond to the measured Auger lineshape. Furthermore energy loss effects, related to the main peak $p-p$ of Fig. 4, may be ruled out as being responsible for the second structure on the basis of the electron energy loss spectrum directly acquired from Ca_2Si.

The disagreement between theory and experiment is interpreted as an evidence of final-state hole-hole correlation effects. The Coulomb repulsion U between the two valence holes may be introduced through Eq. 8 in the BLA approximation (see Sect. 2). This approximation does not hold rigorously in Ca_2Si, since the valence bands are separated from the empty states by a zero-gap. However it is interesting to study the effect of correlation even with an approximate treatment, which is the only available at present. The approximate treatment of correlation effects should be always kept in mind during the examination of the results, which cannot be used to interpret the fine details of the spectrum.

For the sake of completeness and in order to check that the inclusion of U is important only in Ca_2Si, similar calculations are presented also for $CaSi_2$. Various correlated spectra with different values of U, considered as a parameter, have been computed. Figure 6 shows the effect of correlation with $U = 1.5$ eV and $U = 3.0$ eV. It is evident that correlation effects are not present in $CaSi_2$. In this compound the agreement with the experimental data (Fig. 4) is better with the uncorrelated spectrum (Fig. 5) than with the correlated curve of Fig. 6. In Ca_2Si, where the uncorrelated spectrum does not agree with the measured one, the situation is more complex. The two main structures, i.e. the $p-p$ and $s-p$ features, are strongly distorted by the onset of correlation with a shift of their center of gravity towards lower kinetic energy. As a consequence the relative distance between the $p-p$ and $s-p$ features is still of the order of 5 eV, in contrast with the experimental value ($\sim$ 7.5 eV).

In order to solve this puzzle it is necessary to observe that correlation effects are important in the presence of strongly localized states, forming narrow bands. The density of states of Fig. 3 shows that the localization of s- and p-like valence states in Ca_2Si is quite different: only the former ones are found in an atomic-like configuration. Therefore processes involving an s hole have to be separately treated. This according to the suggestion of Liebsch [40-41] of allowing different Coulomb parameters for different angular momentum terms. In order to minimize the number of independent U's the choice $U_{sp}=U_{ss}$ and $U_{pp}=0$ has been made.

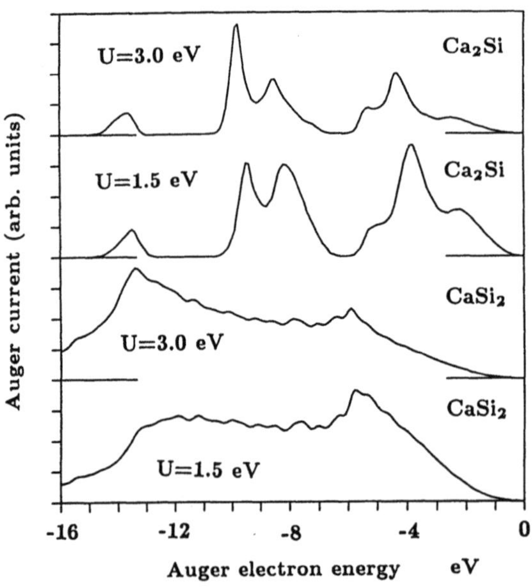

Fig. 6 - The theoretical Auger spectrum for the Si CVV transition in $CaSi_2$ and Ca_2Si, after inclusion of correlation effects. The energy of the Auger electron is referred to the maximum kinetic energy, chosen as zero

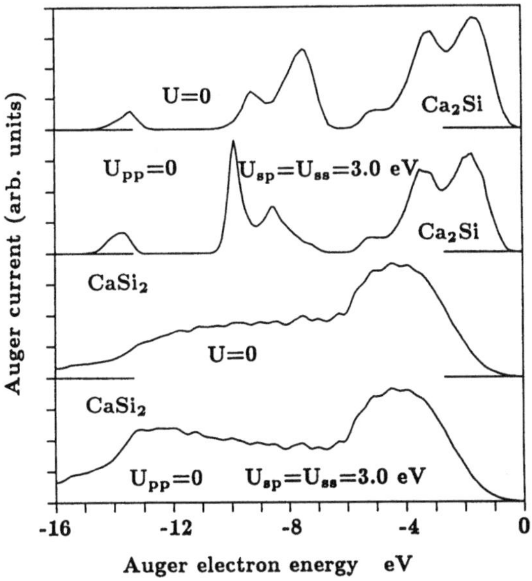

Fig. 7 - The theoretical Auger spectrum for the Si CVV transition in $CaSi_2$ and Ca_2Si, with $U=0$ and after inclusion of correlation effects for the s states only. The energy of the Auger electron is referred to the maximum kinetic energy, chosen as zero

In this case only the $s-p$ structure is distorted and the distance from the $p-p$ main peak is increased. While this effect does not improve the theoretical lineshape of $CaSi_2$ in the case of Ca_2Si a substantial improvement is achieved in the theoretical lineshape. In the data shown in Fig. 7, computed with $U_{sp}=U_{ss}=3$ eV, the distance between the centres of gravity of the $p-p$ and $s-p$ structures is ~7 eV, to the compared with the measured value of 7.5 eV.

As already pointed out a detailed agreement with the experimental spectrum is beyond the aim of this calculation. The main purpose of this work is to show that correlation effects are present in the Si $L_{2,3}VV$ spectrum of Ca_2Si, and that only s states are involved, the value of the fully screened Coulomb integral being of the order of 3 eV. This effect is directly related to the atomic-like nature of the s states in Ca_2Si, not present in $CaSi_2$ where correlation effects are found to be not significant. These results show that in Ca-silicides, on varying the Ca content, different Auger processes take place. A summary of this work, with a comparison with the experimental data of Fig. 4 is presented in Fig. 8.

4. Conclusions

The availability of accurate electronic density of states for $CaSi_2$ and Ca_2Si allows to investigate the basic mechanisms governing the lineshape of the Si $L_{2,3}VV$ Auger spectrum.

In $CaSi_2$ the process is simply related to the self-fold of the density of states, with the matrix elements weighting differently the angular momentum terms. A different mechanism is found in Ca_2Si, where the final-state hole-hole repulsion is important

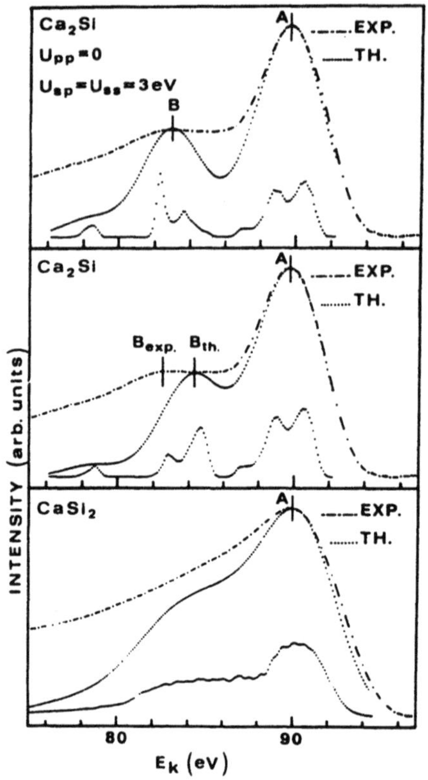

Fig. 8 - Comparison between the measured and computed Si $L_{2,3}VV$ Auger spectra for $CaSi_2$ (bottom panel: no correlation) and Ca_2Si (central panel: no correlation; top panel: correlation for transitions involving s states). The theoretical results are also shown after Gaussian broadening ($\sigma = 1.5$ eV)

(U ~3 eV), but only for processes involving Si 3s orbitals. The onset of correlation effects is related to the atomic-like nature of the Si 3s states in Ca_2Si, as opposed to the delocalized s states found in $CaSi_2$.

The author is indebted to M.Sancrotti, F.Marchetti, L.Braicovich F.Manghi and C.Calandra for fruitful discussions and suggestions. This work was partially supported by Consiglio Nazionale delle Ricerche (CNR) and Ministero Pubblica Istruzione (MPI). Calculations have been performed at the Computer Center of the University of Modena.

5. References

1. M.Thompson, M.D.Baker, A.Christie and J.F.Tyson: **Auger Electron Spectroscopy** (John Wiley & Sons, New York, 1985).
2. D.R.Jennison: J.Vac.Sci Technol. **20**, 548 (1982).
3. D.R.Jennison, J.A.Kelber and R.R.Rye: Phys.Rev. B **25**, 1384 (1982).
4. D.E.Ramaker: Phys.Rev. B **25**, 7341 (1982).
5. M.Vos, D. v.d. Marel and G.A.Sawatzky: Phys.Rev. B **29**, 3073 (1984).
6. M.Davies, D.R.Jennison and P.Weightman: Phys.Rev. B **29**, 5313 (1984).
7. G.Bruno, G.Cubiotti, B.Ginatempo and E.S.Giuliano: Physica Scripta **35**, 699 (1987).
8. P.J.Feibelman, E.J.McGuire and K.C.Pandey: Phys.Rev. B **15**, 2202 (1977).
9. P.J.Feibelman and E.J.McGuire: Phys.Rev. B **17**, 690 (1978).

10. J.A.Roth and C.R.Crowell: J.Vac.Sci Technol. **15**, 1317 (1978).
11. D.R.Jennison: Phys.Rev. B **18**, 6865 (1978).
12. D.E.Ramaker, J.S.Murday, N.H.Turner, G.Moore, M.G.Lagally and J. Houston: Phys.Rev. B **19**, 5375 (1979).
13. P.S. Ho, G.W.Rubloff, J.E.Lewis, V.L.Moruzzi and A.R.Williams: Phys. Rev. B **22**, 44784 (1980).
14. S.D.Bader, L.Richter, M.B.Brodsky, W.E.Brower and G.V.Smith: Solid St. Commun. **37**, 729 (1981).
15. G.Zajac and S.D.Bader: Phys.Rev. B **26**, 5688 (1982).
16. U. del Pennino, P.Sassaroli, S.Valeri, C.M.Bertoni, O.Bisi and C. Calandra: J.Phys. C **16**, 6309 (1983).
17. G.Zajac, J.Zak and D.S.Bader: Phys.Rev. B **27**, 6649 (1983).
18. G.Rossi and I.Lindau: Phys.Rev. B **28**, 3597 (1983).
19. L.Calliari, M.Sancrotti and L.Braicovich: Phys.Rev. B **30**, 4885 (1984).
20. O.Bisi: In **Thin-Films - Interfaces and Phenomena**, ed. by R.J. Nemanich, P.S.Ho and S.S.Lau (Materials Research Society, Pittsburg, Pennsylvania, 1983) p.13.
21. C.Calandra, O.Bisi and G.Ottaviani: Surf.Sci.Rep. **4**, 271 (1985).
22. D.D.Sarma, W.Speier, L.Kumar, C.Carbone, A.Spinsanti, O.Bisi, A.Iandelli, G.L. Olcese and A.Palenzona: Z.Phys. B **71**, 69 (1988).
23. O.Bisi, L.Braicovich, C.Carbone, I.Lindau, A.Iandelli, G.L.Olcese and A. Palenzona: to be published.
24. M.Sancrotti, I.Abbati, A.Rizzi, L.Calliari, F.Marchetti and O.Bisi: Surf. Sci. **189/190**, 300 (1987).
25. P.A.Bennett, J.C.Fuggle, F.U.Hillebrecht, A.Lenselink and G.A.Sawatzki: Phys. Rev. B **27**, 2194 (1983).
26. O.Bisi, C.Calandra, U. del Pennino, P.Sassaroli and S.Valeri: Phys.Rev. B **30**, 5696 (1984).
27. C.Calandra and O.Bisi: Phys.Rev. B **31**, 8288 (1985).
28. M.Cini: Solid St. Commun. **20**, 605 (1976).
29. M.Cini: Solid St. Commun. **24**, 681 (1977).
30. G.A.Sawatzky: Phys.Rev.Lett. **39**, 504 (1977).
31. G.A.Sawatzky and A.Lenselink: Phys.Rev. B **21**, 1790 (1980).
32. M.Cini: Surf.Sci. **87**, 483 (1979).
33. G.Tréglia, M.C.Desjonquères, F.Ducastelle and D.Spanjaard: J.Phys. C **14**, 4347 (1981).
34. M.Cini and C.Verdozzi: Nuovo Cimento D **9**, 1 (1987).
35. O.K.Andersen: Phys.Rev. B **12**, 3060 (1975).
36. H.L.Skriver: **The LMTO Method** (Springer, Berlin, 1984).
37. O.Bisi, O.Jepsen and O.K.Andersen: Phys.Rev. B **36**, 9439 (1987).
38. W.B.Pearson: **Handbook of lattice spacing and structures of metals and alloys** (Pergamon Press, New York, 1958).
39. M.Sancrotti, I.Abbati, L.Calliari, F.Marchetti, O.Bisi, A.Iandelli, G.L. Olcese and A.Palenzona: Phys.Rev. B **37**, 4805 (1988).
40. A.Liebsch: Phys.Rev. B **23**, 5203 (1981).
41. D.E.Ramaker, F.L.Hutson, N.H.Turner and W.N.Mei: Phys.Rev. B **33**, 2574 (1986).

Surface Effects and Core-Hole Effects in Auger Spectra of Metals and Semiconductors

C.-O. Almbladh and A.L. Morales

Department of Theoretical Physics, Lund University,
Sölvegatan 14A, S-23362 Lund, Sweden

Auger core-valence-valence (CVV) spectra of simple metals and semiconductors have been rather thoroughly investigated both experimentally and theoretically during the last decade. The first detailed calculation on Si by Feibelman and McGuire[1] demonstrated the importance of matrix element effects and showed that more simple-minded interpretations in terms of self-convoluted state densities are hardly even qualitatively correct. Later works by Jennison and collaborators[2-4] confirm these conclusions, and the picture that has emerged from several studies is that one-electron-like bulk calculations which account for matrix elements and in some cases also static core-hole screening describe experimental CVV spectra rather well in sp-bonded materials. However, recent works on CVV spectra by Almbladh, Morales, and Grossmann[5-7] suggest that the corrections to the one-electron and bulk approximations may not be as small as believed earlier. We will here summarize these new results, and the reader is referred to Refs. 5-7 for more details.

As has been demonstrated by previous workers a detailed knowledge of the one-electron structure is necessary in order to understand Auger CVV spectra in metals and semiconductors. We have performed accurate one-electron calculations for all the second and third-row simple metals using self-consistent orbitals and proper matrix elements[5,6]. In all cases investigated the spectra from ground-state orbitals peak at <u>lower</u> energy than do the experimental results. We are rather confident that our results are inherent to the one-electron and bulk approximations and not due to insufficient numerical accuracy. To explain the data we therefore go beyond the one-electron and bulk approximations and study the effects of the surface and mean free path, and the effects of the core hole present in the initial state of the Auger process. As we will describe, these effects give important modifications and a much better agreement with experiments. Interestingly enough, however, in each individual case studied here <u>either</u> the surface effects are important <u>or</u> the core-hole effects. Thus, the explanation is different in each case.

In order to study core-hole effects we have performed selfconsistent calculations with a core-hole impurity in addition to the ground-state calculations[5]. The spectra may be decomposed according to the angular momentum character ℓ of the valence holes in the final state. As shown by von Barth and Grossmann[8] the <u>shape</u> of a (subchannel) spectrum generally agrees well with the one-electron spectrum obtained from orbitals solved in the <u>final-state</u> potential without a core hole. The total (Γ) and partial ($\Gamma_{\ell\ell'}$) yields, on the other hand, are determined by the valence-electron wavefunction in the initial core-hole state (see e.g. Refs. 5 and 9). Combining these two results we obtain a final-state approximation used earlier by Ramaker[10] and by Jennison[3],

Springer Series in Surface Sciences, Vol. 18 **Auger Spectroscopy and Electronic Structure**
Editors: G. Cubiotti · G. Mondio · K. Wandelt

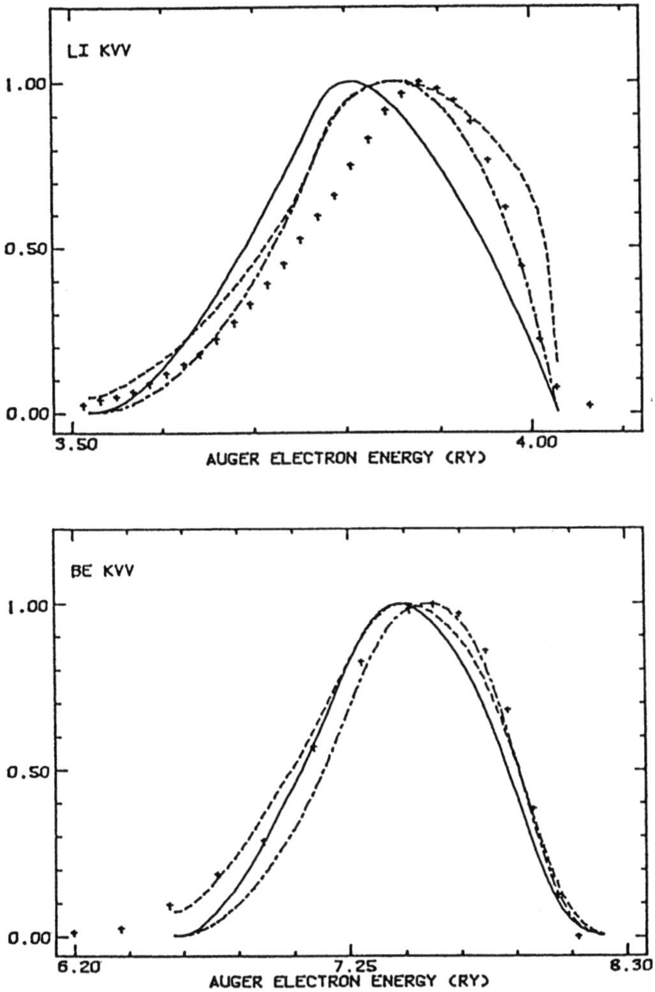

Fig. 1. KVV spectra of Li and Be. Solid curve: One-electron bulk theory; Dashed-dotted curve: Final state rule, Eq. (1); Short-dashed curve: Full MND results. Theoretical results from Ref. 6, and experimental results (crosses) from Ref. 11 (Li) and Ref. 3 (Be).

$$D(\varepsilon_A) = \sum_{\ell \geq \ell'} \frac{\Gamma^{*}_{\ell\ell'}}{\Gamma^{o}_{\ell\ell'}} D^{o}_{\ell\ell'}(\varepsilon_A) \quad . \tag{1}$$

Here the superscripts o and * refer to one-electron results without and with a static core hole, respectively. Our calculations show that there are important modifications of the relative subchannel yields only for Li and Be. In these cases the core hole increases the relative contributions involving p-electrons, and Eq. (1) then gives a much better agreement with experiment (Fig. 1).

The weakness of static core-hole screening effect for the third-row metals may seem surprising in view of the rather strong effects seen in the corresponding KLV spectra[12]. These spectra have an anomalous shape, first explained by von Barth and Grossmann[13], and show a pronounced distorsion of the s-DOS near the core hole. Because there is a core hole also in the final state, the s-DOS distorsion is reflected is the KLV spectra, as opposed to CVV spectra which according to final state rule mainly reflect the local DOS of the no-hole state. We have used our relaxed orbitals to calculate the KLV spectra of Mg and Al[14], and we have obtained a similar good agreement with experiment as reported for Na by von Barth and Grossmann[13]. In the case of KLV spectra there are no dynamical core-hole effects and, owing to the longer Auger-electron mean free path, corrections from the surface are most probably small. These results therefore support the overall reliability of our methods.

Our selfconsistent results with a core hole also allow us to obtain the lifetime widths of the uppermost core level in simple metals. These widths are given by the total yields which, as discussed above, are determined by the completely relaxed wavefunction. The core-hole screening increases the widths by 2-4 times compared to the results from ground-state orbitals but has never been properly included before. In the case of Li, the 1s hole lifetime monitors large effects of incomplete lattice relaxation and is rather accurately known. In Ref. 5 we obtain a width of 17 meV, in excellent agreement with the value 16 meV deduced[15] from experimental x-ray emission data by Arakawa et al.

In order to treat core-hole effects of dynamical nature we adopt the independent fermion model due to Mahan, Nozières, and DeDominicis (MND)[16]. We thus represent the system with an effective one-electron Hamiltonian H_* in the initial core-hole state and a different one-electron Hamiltonian H_0 in the final state without a core hole. To determine H_0 and H_* we utilize our self-consistent ab initio results with and without a core-hole impurity. We thus require the model to reproduce the correct one-electron spectra with and without a static core hole, and we do the fit in each angular momentum and spin subchannel separately. To obtain a fully dynamical spectrum we then solve the model numerically[6]. Owing to the well-known MND edge enhancement the peak moves to higher energies for both Li (Fig. 1) and Na (Fig. 2), and for Li the leading edge is actually too

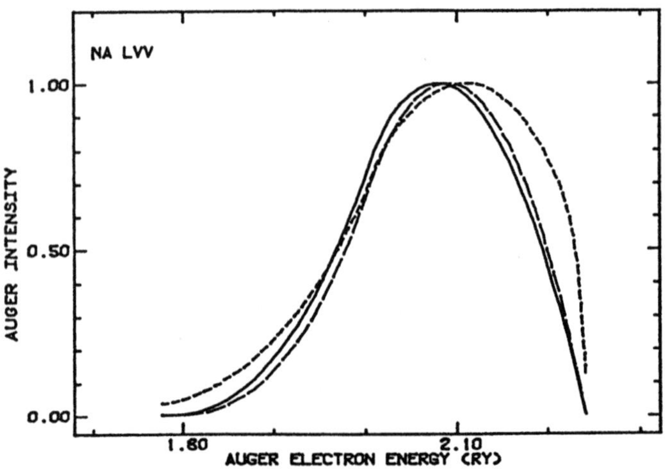

Fig. 2. Theoretical Auger $L_{2,3}VV$ spectra of Na. Solid curve: One-electron bulk theory; Short-dashed curve: Full MND results; Long-dashed curve: Results for the surface model.

steep. In the case of Li, however, one encounters difficulties when trying to reconcile the singularity exponent observed in x-ray photoemission with results of inelastic electron scattering which show no trace of s-wave enhancement (see e.g. Ref. 9). It has been suggested that spin-flip scattering mediated by the core-valence exchange interaction is a likely reason for this shortcoming in the case of Li[9,19]. Our calculations suggest that the MND model overestimates the dynamical effects also for the Li Auger spectrum. In the case of Na on the other hand, both x-ray emission and absorption spectra show no doubt strong singularity effects, and we thus believe the dynamical effects predicted by our MND calculations to be genuine. There are no significant redistributions among subchannels in the case of Na, and the entire effect is truly dynamical.

We have found no evidence that the core-hole effects are important for the $L_{2,3}$ spectra of Mg and Al. In order to explain these spectra we have studied the effects of the surface and finite Auger-electron mean free path. In the case of Al we have modeled an (100) surface by selfconsistent slab calculations. The slab spectra agree very well with the corresponding results from a much simpler infinite-barrier model, which we use for the remaining systems. In the simplified model we extend an idea by Gadzuk[20] and represent a valence orbital near the surface by a Bloch wave specularly reflected in a surface plane about one atomic radius outside the outermost nuclei. In the calculations a mean free path of about 4 Å was assumed. In the case of Al and Mg the surface effects move the intensity maximum to higher energies and give results in rather good agreement with the data, but the surface had little effect on the other simple-metal spectra investigated (Fig. 3).

Finally, we have also made calculations for the $L_{2,3}VV$ spectrum of Si[7]. Our results on the third-row simple metals show that core-hole effects are important for mono-valent systems and that surface effects are important for polyvalent ones. Thus one would expect important surface effects in the Si case as well, and our calculations confirm this expectation. Our results correspond to a mean free path of 7 Å and are based on self-consistent slab calculations on an ideal (111) surface and on ideal and reconstructed (100) surfaces. We consider the 2x1 reconstruction and adopt Chadi's asymmetric dimer model[21]. Our bulk spectrum, which is quite close to the earlier results by Jennison[2], peaks at too high an energy and exhibits a shoulder from sp channels which is too strong. The surface effects move the peak to lower energies and weaken the shoulder, thus improving the agreement with experiment. Our calculations show that spectra from different surfaces are quite similar. Thus, it would be rather difficult to isolate the surface effects experimentally.

The spectra discussed here are all rather structureless and easy to fit by parametrized theories. Thus, in order to investigate the importance of various many-body and surface effects, detailed calculations are needed. We have in Refs. 5-7 presented direct parameter-free calculations of the magnitude of such effects, which allows us to draw rather firm conclusions about their importance. The corrections to the one-electron bulk approximation furthermore greatly improve the agreement with experiment.

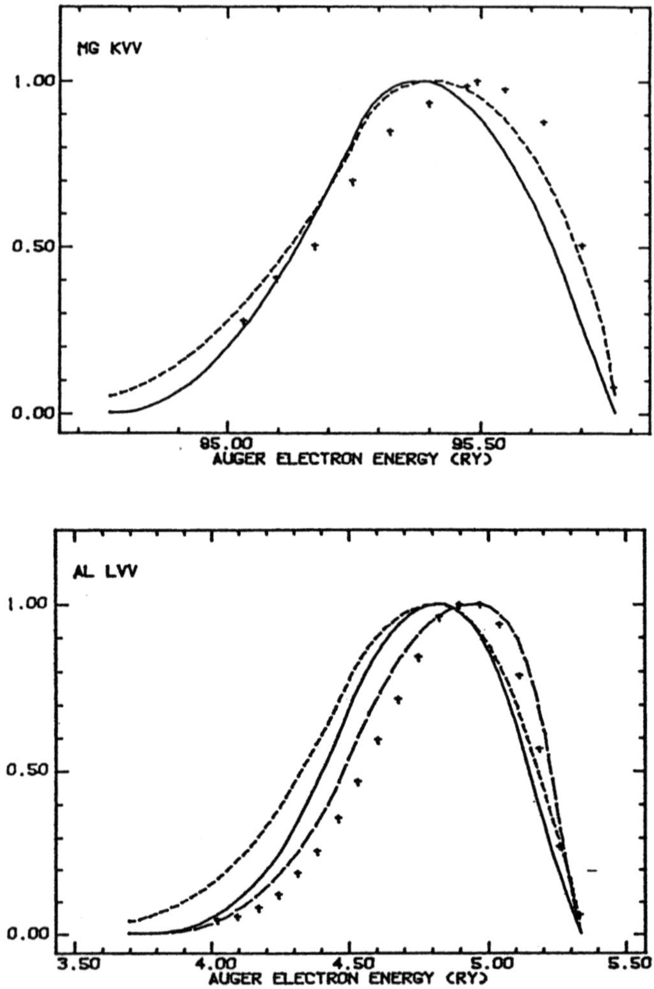

Fig. 3. Auger $L_{2,3}VV$ spectra from Mg and Al. Solid line: One-electron bulk theory; Short-dashed curve: Full MND results; Long-dashed curve: Theoretical surface results. Experimental results (crosses) from Ref. 17 (Mg), and Ref. 18 (Al).

ACKNOWLEDGMENTS

The present work was supported in part by the Swedish Natural Science Research Council.

REFERENCES

1. P.J. Feibelman and E.J. McGuire, Phys. Rev. B 17, 690 (1978).
2. D.R. Jennison, Phys. Rev. B 18, 6865 (1978);
3. D.R. Jennison, H.H. Madden, and D.M. Zehner, Phys. Rev. B 21, 430 (1980).
4. M. Davies, D.R. Jennison, and P. Weightman, Phys. Rev. B 29, 5313 (1984).

5. C.-O. Almbladh, A.L. Morales, and G. Grossmann, submitted to Phys. Rev. B.
6. C.-O. Almbladh and A.L. Morales, submitted to Phys. Rev. B.
7. C.-O. Almbladh and A.L. Morales, Paper Presented at the International Conference on X-Ray and Inner-Shell Processes in Atoms, Molecules, and Solids, Paris, September 1987, J. Physique Colloque C9, no. 12, 879 (1987).
8. U. von Barth and G. Grossmann, Phys. Rev. B 25, 5150 (1982).
9. C.-O. Almbladh and L. Hedin, in "Handbook on Synchrotron Radiation" vol. 1b, edited by E.E. Koch (North-Holland, Amsterdam, 1983) pp 607-904.
10. D.E. Ramaker, Phys. Rev. B 25, 7341 (1982).
11. H.H. Madden and J.E. Houston, Solid State Commun. 21, 1081 (1977).
12. A. Barrie and F.J. Street, J. Electr. Spectr. Rel. Phenom. 7, 1 (1975); R. Lässer and J.C. Fuggle, Phys. Rev. B 22, 2637 (1980).
13. U. von Barth and G. Grossmann, Physica Scripta 28, 107 (1983); U. von Barth and G. Grossmann, Proceedings of the 8th Annual Sympsium on Electronic Structure of Metals and Alloys, edited by P Ziesche, Dresden 1978, p. 76.
14. C.-O. Almbladh and A.L. Morales, (unpublished).
15. C.-O. Almbladh, Phys. Rev. B 16, 4343-57 (1977).
16. G.D. Mahan, Phys. Rev. 163, 612, (1967); P. Nozières and C.T. DeDominicis, Phys. Rev. 178, 1097 (1969).
17. A.M. Baró and J.A. Tagle, J. Phys. F 8, 563 (1978).
18. J.E. Houston, J. Vac. Sci. Techn. 12, 255 (1975).
19. C.-O. Almbladh and U. von Barth, Calculation of exchange effects in x-ray spectra, Extended abstract, Vth Int. Conf. on Vacuum-Ultraviolet Radiation Physics, Monpellier (1977).
20. J.W. Gadzuk, Phys. Rev. B 9, 1978 (1974).
21. D.J. Chadi, Phys. Rev. Lett. 43, 43 (1979); J. Vac. Sci. Technol. 16, 1290 (1979).

Auger Transition Rate Calculations for Systems with Open Subshells: Insight into Core Hole Screening in Simple Metals

A.D. Laine [1], *G. Cubiotti* [2], *and P. Weightman* [1]

[1]Department of Physics and Interdisciplinary Research Center in Surface Science, University of Liverpool, P.O. Box 147, Liverpool, L69 3BX, UK

1 Introduction

The screening of core holes following inner shell ionization in metals is not well understood. It is, in principle, possible to obtain screening information such as the particular orbital character of the valence configuration screening the core hole or holes from Auger profiles of transitions involving valence levels. These profiles are complex, however, and often contain contributions from unresolved multiplet structure. A crucial preliminary step in analysing such spectra is to determine the relative intensity of the various components of the multiplet structure. We address this problem in this work and in particular consider the influence of different screening configurations on Auger transition rate calculations. As an example, we consider in detail the $KL_{2,3}$-$L_{2,3}L_{2,3}V$ transitions of Mg and Al, the spectra of which have been recently measured [1,2].

2 Theory

McGuire has given general expressions [3] which can be used to evaluate Auger transition rates in multiply ionised systems. In this formalism, the transition rate of a system with an initial state described by jj coupling and a final state described in the LS coupling scheme can be computed providing the matrix elements between the initial and final states are known. These latter have been calculated for a large number of atoms by McGuire from the results of Herman and Skillman's Hartree-Fock-Slater atomic structure calculations [4]. The calculation of Auger transition rates using McGuire's formalism is most appropriate for free atoms and such calculations are generally in good agreement with experimental spectra for free atoms. Since the wavefunctions of atomic core electrons are expected to be almost independent of phase, we expect, and find, that transition rates calculated according to McGuire's scheme are also in good agreement with the experimental spectra of core-core-core Auger spectra observed from solids. Attempts to use McGuire's formalism to calculate transition rates for Auger processes involving the valence levels of metals are open to a number of objections. The most obvious one is that the valence wavefunctions in the metal will be very different to those of a free atom. Due to the presence of core electron wavefunctions in the Auger matrix element, however, Auger transitions are a very local

Springer Series in Surface Sciences, Vol. 18 Auger Spectroscopy and Electronic Structure
Editors: G. Cubiotti · G. Mondio · K. Wandelt

probe of electronic structure. In the metal the wavefunctions of valence electrons must be orthogonal to the core electron wavefunctions and, since the core wavefunctions should be independent of phase, this constraint should mean that the valence wavefunctions do not deviate strongly from atomic valence electron wavefunctions in the core region. Thus Auger matrix elements deduced from atomic structure calculations may be a quite reasonable approximation to the matrix elements of the corresponding transition in the metal. In this work we also make use of the equivalent cores approximation, by using aluminium matrix elements to describe Auger processes involving the 3p electrons assumed to screen core holes in magnesium. An unavoidable assumption in our approach is that of an integer screening charge in our atomic-like calculation: it is impossible to perform the Clebsch-Gordan algebra with fractions of electrons.

The final state rule [5,7] tells us that the intensities of the various components of an Auger profile depend on the initial state configuration, and that the lineshapes depend on the final state configuration. Calculating the intensities therefore opens up the possibility of investigating the character of the initial state screening configuration. A full analysis of the Auger profiles, however, also requires final state density of states calculations.

3 Results

The results of transition rate calculations for Mg and for Al with different possible screening configurations are given in tables 1 and 2. These were obtained by evaluating an expression derived from McGuire's treatment of systems with open subshells [8].

All the numbers given are in atomic units and should be multiplied by 10^{-4}.

Table 1

Auger transition rate calculations for the magnesium $KL_{2,3}$-$L_{2,3}^{2}V$ satellite lines for different initial state valence configurations

Initial Valence Configuration	$KL_{2,3}$-$L_{2,3}^{2}V_{s}$ ^1S	^1D	^3P	Total
$3s^1 3p^1$	0.0625	0.312	0.562	0.937
$3s^2 3p^0$	0.125	0.625	1.125	1.875
$3s^2 3p^1$	0.125	0.625	1.125	1.875
$3s^2 3p^2$	0.375	1.875	3.375	5.625
$3s^2 3p^3$	0.375	1.875	3.375	5.625
Initial Valence Configuration	**$KL_{2,3}$-$L_{2,3}^{2}V_{p}$ ^1S**	**^1D**	**^3P**	**Total**
$3s^1 3p^1$	0.098	0.490	0.883	1.472
$3s^2 3p^0$	0.0	0.0	0.0	0.0
$3s^2 3p^1$	0.098	0.490	0.883	1.472
$3s^2 3p^2$	0.591	2.94	5.301	8.83
$3s^2 3p^3$	1.556	7.78	14.0	23.36

Table 2

Auger transition rate calculations for the aluminium $KL_{2,3}-L_{2,3}^{2}V$ satellite lines for different initial state valence configurations

Initial Valence Configuration	$KL_{2,3}-L_{2,3}^{2}V_{s}$ ^1S	^1D	^3P	Total
$3s^1 3p^2$	0.216	1.08	1.95	3.25
$3s^1 3p^3$	0.216	1.08	1.95	3.25
$3s^2 3p^1$	0.145	0.72	1.3	2.17
$3s^2 3p^2$	0.435	2.17	3.9	6.62
$3s^2 3p^3$	0.435	2.17	3.9	6.62

Initial Valence Configuration	$KL_{2,3}-L_{2,3}^{2}V_{p}$ ^1S	^1D	^3P	Total
$3s^1 3p^2$	0.59	2.94	5.3	8.83
$3s^1 3p^3$	1.56	7.8	14.04	23.36
$3s^2 3p^1$	0.098	0.49	0.883	1.472
$3s^2 3p^2$	0.59	2.94	5.3	8.83
$3s^2 3p^3$	1.56	7.8	14.04	23.36

In all cases, the ^1S:^1D:^3P final state intensity ratio is 1:5:9 in agreement with a statistical weight argument. Each ^1S and ^1D final state is reached by a decay from a ^1P initial state and each ^3P final state from a ^3P initial state. The term splitting due to the coupling of valence l and s angular momenta has been disregarded as this is much smaller than the initial state ^1P, ^3P and final state ^1S, ^1D and ^3P splittings.

4 Discussion

The experimental magnesium and aluminium $KL_{2,3}-L_{2,3}^{2}V$ Auger satellite spectra have been measured recently [1,2]. The spectra show unresolved multiplet structure arising from ^1P and ^3P terms produced by coupling between the 1s and 2p holes in the initial state and ^1D, ^1S and ^3P terms resulting from the coupling of the $2p^2$ holes in the final state. In the absence of transition rate calculations, the $KL-L^2V$ spectra of Mg and Al were analysed in earlier work by attributing the $KL_{2,3}V$ lineshape to each component of the multiplet structure and varying the relative intensity of each component. In Mg it was found that the observed profile could be reproduced by assigning intensities in good agreement with statistical 2 weights, while this method failed for Al. For Mg, it was assumed that the ^1P and ^3P initial states could each decay into ^1S, ^1D and ^3P final states. The ^1P:^3P intensity ratio was given as 1:2.3 as opposed to the supposed ideal ratio of 1:3 and the ^1S:^1D:^3P final state intensity ratio was given by 1:6.2:9 as opposed to the ideal 1:5:9. The initial state splitting was calculated from self consistent field atomic structure calculations, and the absolute energies were shifted empirically.

Our calculations show that this modelling in the earlier work is invalid because the ^1P initial state cannot decay to a ^3P final state, and a ^3P initial state can only decay to a ^3P final state. The experimental spectrum is thus composed of three features as opposed to six. This suggests that the apparent agreement for the Mg satellites was fortuitous, and explains the lack of success in accounting for the Al data.

In seeking to understand the spectral profile of these transitions, it is also important to consider the work of VON BARTH and GROSSMAN [9], who showed that the s and p densities in the vicinity of a core hole are very different. Thus it may be possible, by looking at the differences in the relative s and p contributions for different configurations to deduce the actual screening configuration. In addition, we found in our calculations that the s:p ratio depends more on the particular valence configuration than on the use of Mg or Al matrix elements, validating the use of the equivalent cores approximation and thus the use of Al matrix elements for the Mg V_p calculations. The dependence is such that combinations of given screening configurations with the appropriate calculated density of states surrounding the core holes would be expected to yield manifestly different calculated spectral profiles. The large differences in both the s:p transition rates and the s and p densities of states should make it possible to determine the correct screening configuration around double core ionised sites in these simple metals.

Earlier work based on an analysis of the energies of single and double core hole states suggests the screening configuration for a two hole state would be $3s^2 3p^2$ for Mg and $3s^2 3p^3$ for Al. It should be possible to determine the actual screening configuration from the Auger profile by combining the present transition rate calculations with accurate density of states calculations for the two core hole state.

5 Conclusions

Although statistical weight arguments have been previously used to fit the $KL_{2,3}-L_{2,3}^{2}V$ Auger satellite spectrum of magnesium, this procedure has been shown to be invalid because the number of spectral features is halved when a rigorous calculation is performed.

The biggest assumptions used in the calculations, those of the validity of atomic wave-functions and of the equivalent cores approximation are shown to contribute less to differences in the calculated rates than does the particular configuration studied.

Actual fitting of the experimental spectra requires the use of accurate densities of states surrounding the two core holes: such calculations are currently underway [10]

References

1 M. Davies, D.R. Jennison, P. Weightman: J. Phys. C: Solid State Phys. 17 L107 (1984)
2 P.H. Hannah, P. Weightman: J. Phys. C: Solid State Phys. 18 L239 (1985)
3 E.J. McGuire: In Atomic Inner Shell Processes (Academic, N.Y. 1975)
4 F. Herman, S. Skillman: Atomic Structure Calculations (Prentice Hall, N.J. 1963)
5 D.R. Jennison: Phys. Rev. Lett. 40 807 (1978)
6 D.R. Jennison, H.H. Madden, D.M. Zehner: Phys. Rev. B21 430 (1980)
7 D.E. Ramaker: Phys. Rev. B25 7341 (1982)
8 G. Cubiotti, A. Laine, P. Weightman: (in press)
9 U. von Barth, G. Grossman: Phys. Scr. 28 107 (1983)
10 P.S. Fowles, J.E. Inglesfield: (private communication)

Comments on the Relationship Between Auger Lineshapes and Local Electronic Structure

G. Cubiotti

Istituto di Fisica Teorica, Università di Messina,
P.O. Box 50, I-98166 S. Agata di Messina, Italy

Since the discussion is intended to be devoted to the perspectives of the subject we are dealing with, the present contribution will attempt to focalize two points, which in my opinion deserve further work, both experimental and theoretical.

The subjects I would like to talk biefly about are a) the effect of using realistic solid state matrix elements in the calculations of Auger spectral intensities and b) the possibility of computing and detecting effects due to the screening charge around the atom on which the Auger process takes place. On both subjects I am going to present recent and preliminary results of investigations which are still under way.

1 The Effect of Solid State Matrix Elements

Does it exist? If one looks at the excellent review paper by RAMAKER /1/ a definite conclusion cannot be reached. On one hand, it is striking that the atomic matrix elements, when scaled in a suitable way, seem not only to work well enough for the condensed matter but also to present a rather small variation with the atomic number. On the other hand, for some systems large differences can be found between the atomic and the condensed phases.

In more recent times some papers have pointed out that the question is still open. As an example, our work on the AgPd alloy /2/ was able to link the form of the low energy side of the XVV Auger spectra

/3/ with this effect. Previous authors /3/ have only been able to get agreement between the XVV spectra of AgPd alloys and the results of calculations of the local density of states by arbitrarily reducing the intensity of the contribution from M_4VV transitions. We have shown that this reduction is unnecessary if one takes into account solid state matrix elements.

Preliminary results on the magnetic PdNi alloy /4/ also show that considering solid state matrix elements can lead to sensible effects. The results for the $Pd_{0.95}Ni_{0.05}$ alloy are displayed in Fig. 1 for the Ni site and for both spin up and spin down contributions.

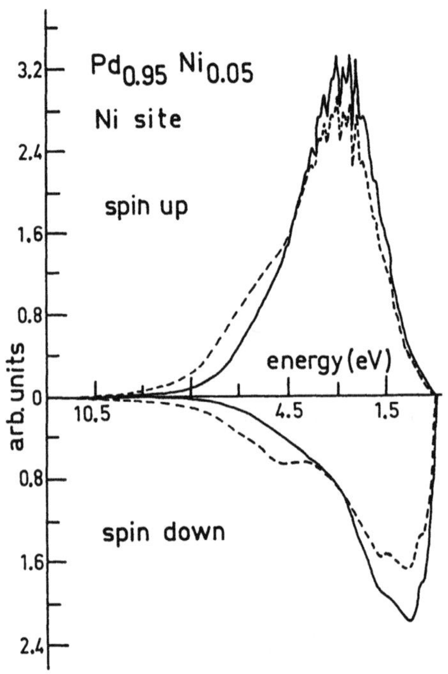

Fig. 1. Calculated (——) spin up and spin down contributions to the $M_{23}VV$ Auger spectra for the $Pd_{0.95}Ni_{0.05}$ alloy, compared with the self-convolution (---) of the valence density of states. The Fermi energy is taken as zero

It can be seen how, especially at the low energy side, the differences can be important.

On the other hand, the above quoted works /2,4/ do not include any calculation of the multiplet structure arising from the effect of the Coulomb correlation between the two holes in the valence band, which could also lead to a large deformation of the spectrum. The question of the need to include multiplet structure in such an analysis is among the problems we would like to present for discussion. In fact, in our opinion some further theoretical work would be necessary in order to clarify if the Cini-Sawatzky theory /5,6/ for treating the Coulomb correlation can be used in conjunction with a solid state calculation like the one of /2,4/, which originates from potentials computed in a local density approximation scheme, in which exchange and correlation are already included. Thus it would seem that in putting the calculated electron density of states in the Cini-Sawatzky formulae one would screen twice the hole-hole interaction. We shall not go into the details of the problems, leaving it to the following discussion.

2 Screening of the core hole

It is well known that, by means of the Final State Rule /7-10/, it is possible to state that in an Auger transition the shape of the spectrum is determined by the final state density of states, while the intensities are determined by the electronic structure in the initial state. Thus, the core-valence-valence spectra probe an unscreened valence density of states and the core-core-valence spectra probe a screened density of states. It is then clear that an Auger spectrum may give informations on the way the initial core hole is screened in a solid, particularly in the case of a metal.

We shall present in the discussion some preliminary results /11/, obtained by applying those considerations, in an attempt to gain some insight into the way the screening takes place in simple metals, like Mg and Al.

The experimental results /12,13/ on the KLV transitions have contributions from s and p components. The determination of the separate s and p contributions to the spectra is difficult but can be accomplished by careful comparison of the KL_1V and $KL_{2,3}V$ profiles.

We have calculated the KL_1V_s/KL_1V_p and the $KL_{2,3}V_s/KL_{2,3}V_p$ ratios as functions of different local valence charge configurations, in order to see which of them would better reproduce the experimental data. The calculations have been made by using the McGUIRE formalism /14,15/ and the atomic $D(K,l_2)$ and $E(K,l_2)$ integrals, as obtained by McGUIRE /16/. The results for the various configurations we have tried are listed in Table 1.

Table 1. Comparison of the ratios between the different components of the KLV spectra, as calculated for different electronic configurations, with the experimental ones. Units are 10^{-4} atomic units.

	KL_1V_s/KL_1V_p		$KL_{2,3}V_s/KL_{2,3}V_p$	
config.	Mg	Al	Mg	Al
$3s^1$ $3p^1$	2.53	=	0.64	=
$3s^1$ $3p^2$	1.26	1.47	0.32	0.37
$3s^2$ $3p^1$	5.06	=	1.27	=
$3s^2$ $3p^2$	2.53	2.94	0.64	0.74
$3s^2$ $3p^3$	=	1.14	=	0.28
Experim.	1.0	0.9	0.25	0.25

It can be seen that the theory overestimates the ratios, because it seems to give too much weight to the s contribution. We cannot perform the atomic-like calculation with fractional charges, but by extrapolating the results we can show that the theory would reproduce the experimental data by using the configurations $3s^{0.9}$ $3p^{2.1}$ for Mg and $3s^{1.6}$ $3p^{2.4}$ for Al. Thus, within our approximations, this analysis show that p-type screening is important for these metals.

It would be quite interesting to compare the results of this kind of computations with the ones of solid state calculations of the s and p densities of states around the atom with the core hole. The only meaningful check of the present results can be made at the moment by comparing them to those by ALMBLADH et al. /17/, obtained by self-consistent solid state calculations. The agreement between the two sets of results is not very close for Mg, while it is very good for Al (the ALMBLADH et al. figures are $3s^{1.29}$ $3p^{1.56}$ for Mg and $3s^{1.46}$ $3p^{2.14}$ for Al) and the latter calculations do confirm a major role of p electrons in screening the core hole. That seems to be encouraging in going firther to study more complex metals, both by solid state techniques and by our approach, in an attempt to describe in a quantitative manner the screening of the core hole. It would gain us an important step forward in understanding completely the Auger processes.

References

1. D.E. Ramaker: in "Chemistry and Physics of Solid Surfaces IV", ed. R. Vonselow and R. Howe, Springer-Verlag, Berlin 1982
2. G. Bruno, G. Cubiotti, B. Ginatempo, E.S. Giuliano: Phys. Scr. 35, 699 (1987)

3. M. Davies, P. Weightman: J. Phys. C: Solid State Phys. 17, L1015 (1984)
4. G. Bruno, G. Cubiotti, B.Ginatempo, to be published
5. M. Cini: Solid State Commun. 24, 681 (1977)
6. G.A. Sawatzky: Phys. Rev. Lett. 39, 504 (1977)
7. U. von Barth, G. Grossman: Solid State Commun. 32, 645 (1979); Phys. Scr. 21, 580 (1980)
8. G.D. Mahan: Phys. Rev. B21, 1422 (1980)
9. L.C. Davis, L.A. Feldkamp: Phys. Rev. B23, 4269 (1981)
10. D.E. Ramaker: Phys. Rev. B25, 7341 (1982)
11. G. Cubiotti, A. Laine, P. Weightman, to be published
12. M. Davies, D.R. Jennison, P. Weightman: Phys. Rev. B29, 5313 (1984)
13. S. Waddington, Ph.D. Thesis, University of Liverpool (unpublished)
14. E.J. McGuire: in "Atomic Inner Shell Processes", ed.B. Crasemann, Academic, New York 1975
15. G. Cubiotti, A. Laine, P. Weightman: to be published
16. E.J. McGuire: Phys. Rev. 185, 1 (1969)
17. C.O. Almbladh, A.L. Morales, G. Grossmann: to be published

The Local Densities of States of 4d Transition Metals

P.J. Durham [1], *W.M. Temmerman* [1], *and A.M. Begley* [2]

[1] Daresbury Laboratory, Keckwick Lane, Daresbury, Cheshire, UK
[2] Birmingham University, Bristol Road South, Birmingham, UK

We present a set of calculations of XPS and soft x-ray emission (SXE) spectra for all the 4d elements, based on self-consistent band structures. We comment on the correspondence with available data, and suggest that while photoemission can be accurately described, SXE spectra are less directly related to the local densities of states.

In multicomponent systems the local densities of states (LDOS) are basic, indeed indispensable, elements in the description of the electronic structure. In substitutionally disordered transition metal alloys, for example, the LDOS provides the most direct way of distinguishing between the limits of rigid band and split band behaviour. In ordered compounds the LDOS is a valuable tool for interpreting the stucture and bonding. For these reasons electronic structure theory for ordered and disordered systems has always made calculation of the LDOS one of its major aims, and has paid a good deal of attention to the relevant experimental probes. Of the many experiments giving LDOS information we concentrate in this Note on x-ray photoemission (XPS) and soft x-ray emission (SXE).

The 1-electron theory for XPS and SXE (and their inverses BIS and SXA - since we are interested here in the occupied states, we shall not consider BIS or SXA any further) are rather fully worked out; the formulae for the respective transition rates can be written in terms of the LDOS, $n^{\alpha}{}_{l}(E)$ for atom α, angular momentum l, as follows

SXE:

$$W_c^{\alpha} = \sum_{l} \left| m_{ll_c}^{\alpha} \right|^2 n_l^{\alpha}(E) \tag{1}$$

XPS:

$$I(E) = \sum_{\alpha} \sum_{l_1 l_2} \left| M_{l_1 l_2}^{\alpha} \right|^2 n_{l_2}^{\alpha}(E) \tag{2}$$

where

$$m_{ll_c}^{\alpha} = \frac{e\hbar}{mc} \int d\underline{r} (\phi_c^{\alpha}(\underline{r}))^* \underline{A}.\nabla \psi_l^{\alpha}(\underline{r}, E)$$

$$M_{l_1 l_2}^{\alpha} = \frac{e\hbar}{mc} \int d\underline{r} \left(\chi_{l_1}^{\alpha}(\underline{r}, E + h\nu) \right)^* \underline{A}.\nabla \psi_{l_2}^{\alpha}(\underline{r}, E)$$

Here $\phi^{\alpha}{}_{c}$is the wavefunction for core level c on atom α, $\chi^{\alpha}{}_{l}$ is a single-scatterer solution [1] for atom α, and $\psi^{\alpha}{}_{l}$ is the same single-scatterer solution but normalised to unity within the muffin-tin sphere. This XPS result represents a local approximation shown by Winter et al [1] to be valid in the limit of high photon energy. It can be seen that the XPS spectrum sums over all l-components of the LDOS on each type of site, weighted by a matrix element, while the SXE spectrum, because it involves a core excitation, is site-specific and contains at most two l-components of the LDOS (ie. l_c-1 and l_c+1). The price to be paid for this is that the short core hole lifetimes broaden SXE spectra to the extent that only fairly gross features of the LDOS can be discerned. Application of these expressions has given valuable insight into the electronic structure of quite complex transition metal systems. For example, the split band nature of the disordered binary alloys Cu-Ni [2] and Ag-Pd [3] has been convincingly demonstrated by spectroscopic studies, the formation of Ni and Pd virtual bound states at low concentrations being particularly clear in SXE data. Perhaps surprisingly, the behaviour of disordered and ordered Cu-Pd alloys seems to be rather more complicated [4].

Irregardless of such successes, one must question the extent to which a one-electron interpretation is modified by many-electron processes. Focussing still on the transition metals, it is clear that both XPS and SXE spectra should contain d-band self-energy effects, and that SXE will in addition be influenced by dynamic core hole effects. (Those static effects which are represented by the final state rule can be ignored in the sense that there is no hole in the final state for SXE). Such problems are best appreciated in systematic surveys of related systems. For example, Fuggle et al [5] have suggested that multiplet effects are important in the x-ray absorption spectra of the 3d elements, and have examined core hole effects by comparing many EXAFS and BIS spectra. Here we report preliminary results of a theoretical study of XPS and SXE spectra for all the 4d elements.

Our calculations are based on a set of self-consistent field band structures determined within the framework of the local density approximation (using the Hedin-Lundqvist parameterisation of the exchange-correlation potential) by the ASA-LMTO method. The correct crystal structures were used, but the calculated lattice parameters were within 2% of the experimental values. Relativistic effects at the level of the Pauli equation (ie. the Darwin, mass velocity and spin-orbit terms were included but the full Dirac equation was not solved). The l-resolved local densities of states were used in equations 1 and 2, the matrix elements being calculated from the solutions for the single self-consistent atomic potentials.

Figure 1 shows our XPS (and BIS) calculations for the 4d series (assuming Al Kα incident radiation). The major trends across the series are obviously the band-filling and the crystal structure; note, for example, the similarity between the two HCP elements Y and Zr, compared with the two BCC elements Nb and Mo. These spectra contain structure in quite good agreement with high- resolution data, but there is generally too much weight at high binding energies, as noted by Smith et al [6]. These authors speculated that the density of states at the bottom of the d-band might be suppressed by matrix element effects, and our calculations do indeed show this to be true. The size of the effect is, however, too small to bring the calculations into agreement with experiment. On the other hand, the results in figure 1 were obtained using a broadening of 0.3 eV. But the Auger lifetime broadening of d-band holes should actually increase with binding energy due to the larger phase space available, the variation being quadratic at least for free-electron models [7]. The data of Smith et al for Rh, Pd and Ag suggests that the broadening rises from about 0.3 eV at the top of the d-band to about 1.5, 1.0 and 0.5 eV (for Rh, Pd and Ag respectively) at the bottom. Using a quadratic interpolation between these limiting values, we obtain results, shown in figure 2, in very good agreement with the data of Smith et al. This broadening recipe is, of course, a crude approximation to the true self-energy correction. (Note that in the case of Ag a rigid shift of 1 eV to higher binding energies has also been applied; the largest contribution to this 'real part of the self-energy' is probably the self-interaction correction to which local density eigenvalues are subject [8].) However, it is a manageable correction which still leaves the one- particle spectrum dominant.

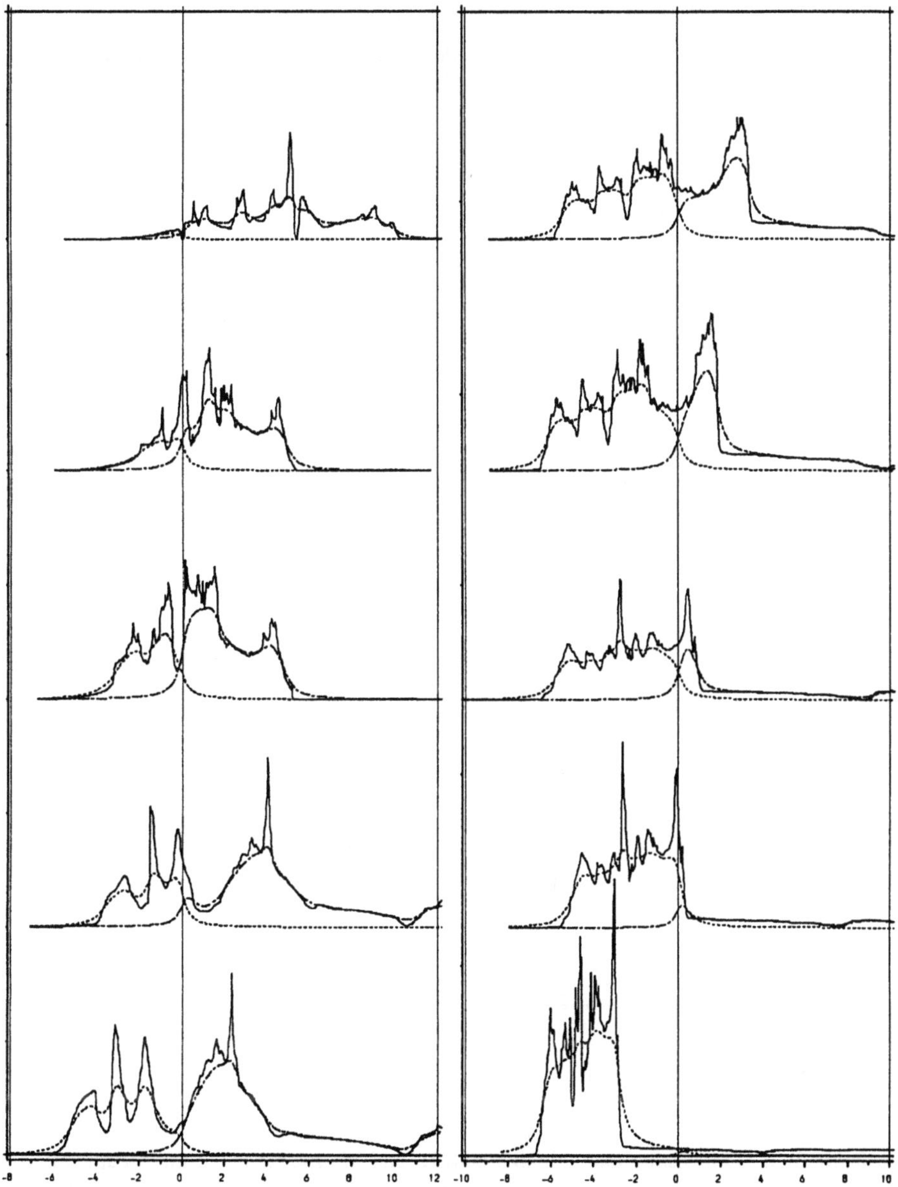

Fig. 1 : XPS (dashed curves) and BIS (chain curves) calculations for Sr (upper left panel) to Ag (lower right panel). The full curves are the bare (unbroadened) spectra, including matrix elements.

Figure 3 shows our calculations of the L_2 and M_2 soft x-ray emission spectra for the 4d series. Here we used tabulated values [9] for the width of the core holes, plus a constant broadening (assumed rectangular) of 0.5 eV to represent experimental resolution. Despite the consequent loss of detail, similar trends to those seen for the XPS spectra are visible

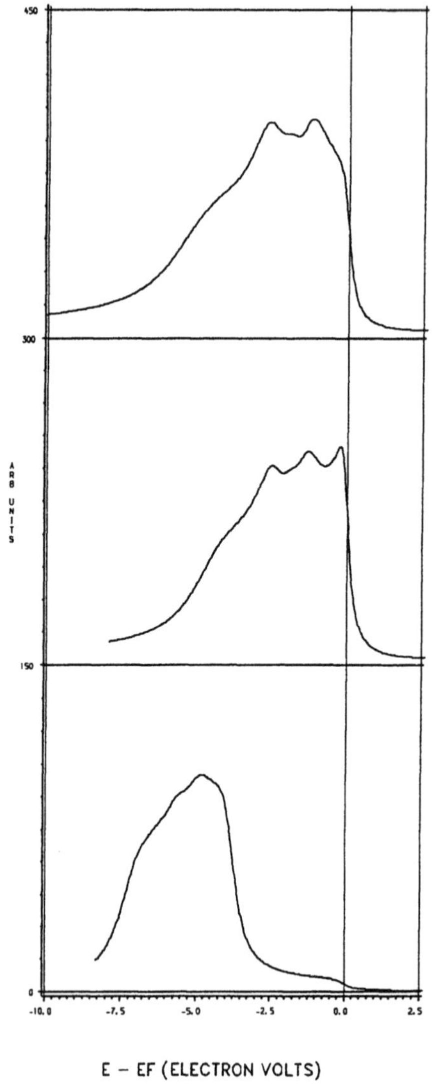

Fig. 2 : XPS calculations, using energy-dependent broadening parameters, for Rh (top), Pd and Ag (bottom).

in figure 3. It is interesting that although both L_2 and M_2 spectra probe the occupied part of the d-band, the shapes of the two emission bands can be rather different, particularly for the earlier elements of the series. In our calculations this is a matrix element effect. In the energy range of the emission band, the dipole coupling of the 2p core level with the valence d-states is always more than twice as great as that with the s-states, for all the 4d elements. This is not so for the 3p core level - except for Ag the ratio of the s to d matrix elements is always greater than 1/2, and at the bottom of the Sr d-band the ratio is about 2. It appears to be incorrect to assume that the $M_{2/3}$ spectra reflect only the local density of d-states. Although the trend in the widths of the calculated emission bands across the series seems to be in reasonable agreement with available data [3,10], their magnitudes are systematically too great by up to 20%. It could be argued that the core level widths we have assumed are too large, but the problem seems to be deeper than that. The spectra in figure 3 actually

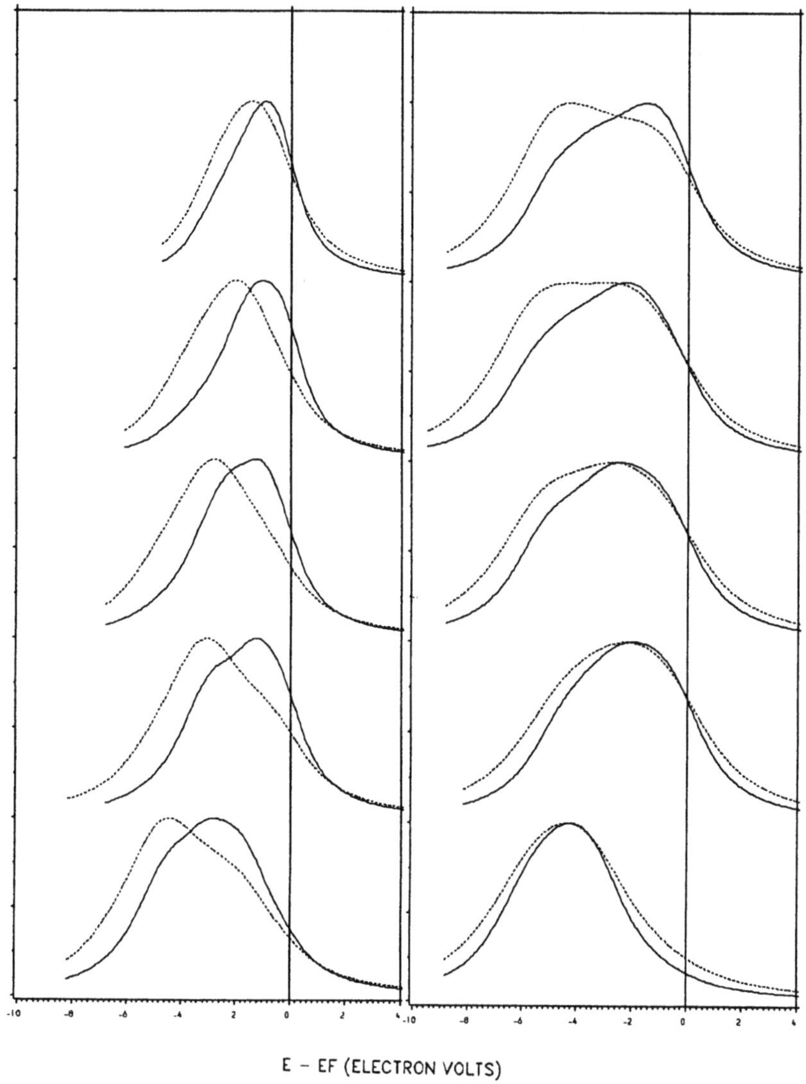

Fig. 3 : L_2 (full curves) and M_2 (dashed curves) SXE calculations for Sr (upper left panel) to Ag (lower right panel).

contain more structure (related to the local densities of states) than the experimental results - if our broadening parameters were reduced, more structure would inevitably be revealed.

These preliminary calculations therefore reveal systematic discrepancies between theoretical and experimental SXE spectra, even for these relatively simple systems. In view of the comparative success of the XPS theory (if the d-band self-energy corrections are reasonably modelled), it is tempting to attribute the SXE problems to core hole effects. Within the framework of the Nozieres-de Dominicis model, these effects are to be included by folding the one-particle spectrum with a broadening function proportional to both the real core level photoemission spectrum and a factor describing the excitonic enhancement of the spectral weight [11]. Substantial asymmetry parameters of perhaps 0.25 have been reported [12] for the photoemission lines of the 3p levels. Clearly, further systematic studies of such effects (together with the influence of d-band correlations) are needed to refine the analysis of SXE data to the point where more quantitative information about local densities of states can be obtained.

References.

1. H. Winter, P. J. Durham and G. M. Stocks, J Phys F: Metal Phys *14*, 1047 (1984).
2. P. J. Durham, D. Ghaleb, B. L. Gyorffy, C. F. Hague, J.-M. Mariot, G. M. Stocks and W. M. Temmerman, J Phys F: Metal Phys *9*, 1719 (1979).
3. P. J. Durham, B. L. Gyorffy, C. F. Hague, J.-M. Mariot, A. J. Pindor and W. M. Temmerman, Inst. Phys. Conf. Ser. No. 55, 131 (1980).
4. H. Winter, P. J. Durham, W. M. Temmerman and G. M. Stocks, Phys. Rev. *B33*, 2370 (1986).
5. J. C. Fuggle, preprint (1987); W. Speier, R. Zeller and J. C. Fuggle, Phys. Rev. *B32*, 3597 (1985).
6. N. V. Smith, G. K. Wertheim, S. Hufner and M. M. Traum, Phys. Rev. *B10*, 3197 (1974).
7. M. A. Blokhin and V. P. Satchenko, Izv. Akad. Nauk. *24*, 397 (1960).
8. eg. A. R. Williams and U. von Barth in "Theory of the Inhomogeneous Electron Gas", eds. S. Lundqvist and N. H. March, Plenum (1983).
9. O. Keski-Rahkonen and M. O. Krause, Atomic Data and Nuclear Data Tables *14*, 139 (1974).
10. I. Ya. Nikiforov, I. I. Gegusin and G. I. Alperovich, Proceedings of the International Symposium on X-Ray Spectra and Electronic structure, vol. II, eds. A. Faessler and G. Weich, Munchen, 122 (1973).
11. eg. G. D. Mahan, Phys. Rev. *B11*, 4814 (1975).
12. R. Nyholm and N. Martensson, Solid St. Commun. *40*, 311 (1981).

A Multiple Scattering Approach to the Interpretation of Interlayer and Intralayer Interactions in $TiSe_2$ Clusters

P. Pavone[1], *R. Girlanda*[2], *G. Martino*[2], *and E.S. Giuliano*[3]

[1]S.I.S.S.A. Strada Costiera 11, I-34014 Trieste, Italy
[2]Istituto di Struttura della Materia, Università degli Studi di Messina, I-98100 Messina, Italy
[3]Istituto di Fisica Teorica, Università degli Studi di Messina, I-98100 Messina, Italy

Recently the interest in layer materials has greatly increased due to the amount of physical phenomena which have been discovered in several of these compounds. In this field special efforts have been made to describe the electronic and structural properties of transition metals dichalcogenides. However, up until now there is still not a complete understanding of such properties. Among the most interesting features of these compounds we can on the whole underline the possibility of intercalating, into the Van der Waals gaps between the layers, a great variety of atoms, ions, molecules or halide compounds. A drastic change in the properties of the host material is usually obtained by intercalation. A useful application of this particular feature is the construction of cathodes of batteries based on Li^+ transport. Furthermore, more recently it has been found that some of these compounds present superconducting behavior.

Our attention has been especially attracted by titanium dichalcogenides and in particular by the selenide which still presents some problems as to its behavior as a semimetal or a semiconductor [1]. The electronic properties of titanium dichalcogenides has been studied by many authors in terms of band–structure theory [2],[3],[4], but our purpose is to give a description of these materials from a local point of view. For this reason we have explicitly considered the projected density of states (pDOS's) which we have obtained for different $TiSe_2$ clusters in the formalism of multiple scattering theory. More deeply we can observe that the pDOS's of given angular momentum above the Fermi energy can be compared, apart from a matrix element, to the absorption spectra from a core level; on the other hand the valence band pDOS's can be used to make a comparison with the photoemission spectra.

In the next section we will show some basic ideas about the multiple scattering theory, then we will consider the application of this theory to $TiSe_2$ clusters emphasizing the relevance of the size and geometry of the clusters for the resulting pDOS's, and finally we will present and discuss the obtained results.

1. Scattering from Muffin–Tin Potentials

1.1 Muffin–Tin Potentials

In these sections we will explicitly obtain an expression for the pDOS's of a cluster of non–overlapping muffin–tin (MT) potentials. A single MT potential $v(\underline{r})$ is defined as $v(\underline{r}) = v(|\underline{r}|)$ if $|\underline{r}| < r_0$ and $v(\underline{r}) = 0$ if $|\underline{r}| \geq r_0$. A first question has to be answered: why is it possible to consider such a kind of potential as an approximation of the real one?

In fact, in solids we can assume that far from the ionic or atomic cores the real potential is quite flat, whereas in the vicinity of the lattice sites it is strongly variable and weakly dependent on the direction; this justifies our assumption. Muffin–tin potentials can be obtained in a simple way, as follows. In one–electron approximation, the behavior of electrons in solids can be determined by solving the following Schrödinger equation:

$$[-\nabla^2 + V(\underline{r})] \; \psi_{\underline{k}}(\underline{r}) = \epsilon_{\underline{k}} \; \psi_{\underline{k}}(\underline{r}) \tag{1.1}$$

where $\psi_{\underline{k}}(\underline{r})$ is the one–electron eigenfunction and $\epsilon_{\underline{k}}$ the corresponding eigenvalue. The function $V(\underline{r})$ which appears in (1.1) is the crystal potential and can be generally written as a sum of site centered atomic–like potentials:

$$V(\underline{r}) = \sum_i v_i(\underline{r} - \underline{R}_i) \tag{1.2}$$

where $\underline{R}_i$ is the position in the lattice of the generic site. A MT potential is now obtained by choosing on each site a sphere of the radius r_{MT} which is the muffin–tin radius, and spherically averaging $V(\underline{r})$ inside this sphere. Outside these regions the potential $V(\underline{r})$ is averaged in the three dimensional space to a value called the muffin–tin zero.

1.2 Single Scatterer Theory

Let's consider a single MT potential $v(|\underline{r}|)$. As it is spherically symmetrical the solution of the Schrödinger equation can be written as:

$$\psi(\underline{r};\epsilon) = \sum_L a_L(\epsilon) \; R_l(r;\epsilon) \; Y_L(\hat{r}) \tag{1.3}$$

where the $Y_L(\hat{r})$ are the spherical harmonics with $L \equiv l, m$, and $R_l(r;\epsilon)$ is the solution of the radial equation:

$$[-\frac{1}{r}\frac{d^2}{dr^2}r + \frac{l(l+1)}{r^2} + v(r)] \; R_l(r;\epsilon) = \epsilon \; R_l(r;\epsilon) \tag{1.4}$$

The scattering properties of the muffin–tin potential are described by the corresponding phase shifts:

$$\delta_l(\epsilon) = \cot^{-1}\left[\frac{\sqrt{\epsilon}\, n_l'(\sqrt{\epsilon}\, r_{\mathrm{MT}}) - \gamma_l(\epsilon)\, n_l(\sqrt{\epsilon}\, r_{\mathrm{MT}})}{\sqrt{\epsilon}\, j_l'(\sqrt{\epsilon}\, r_{\mathrm{MT}}) - \gamma_l(\epsilon)\, j_l(\sqrt{\epsilon}\, r_{\mathrm{MT}})}\right] \tag{1.5}$$

where n_l and j_l are, respectively, the spherical Neumann and Bessel functions, and:

$$\gamma_l(\epsilon) = \frac{1}{R_l}\frac{dR_l}{dr}|_{r=r_{\mathrm{MT}}} \tag{1.6}$$

The density of scattering states in the formalism of Green functions is given by [5],[6]:

$$n(\epsilon) = -\frac{1}{\pi}\int d^3r \ \mathrm{Im}\ G(\underline{r},\underline{r};\epsilon) \tag{1.7}$$

where the imaginary part of the Green function (GF) is obtained as [6]:

$$\mathrm{Im}\ G(\underline{r},\underline{r}';\epsilon) = -\sqrt{\epsilon}\sum_L R_l(r,\epsilon)\, R_l(r',\epsilon)\, Y_L^*(\hat{r})\, Y_L(\hat{r}') \tag{1.8}$$

In our case the function G can be alternatively achieved introducing the t–matrix through the following expression (for details see [6]):

$$G(\underline{r},\underline{r}';\epsilon) = G_0(\underline{r}-\underline{r}';\epsilon) + \int d^3r_1 \int d^3r_2\ G_0(\underline{r}-\underline{r}_1;\epsilon)\ t(\underline{r}_1,\underline{r}_2;\epsilon)\ G_0(\underline{r}_2-\underline{r}';\epsilon) \tag{1.9}$$

where G_0 is the GF for identically zero potential, and then solving the integral equation:

$$t(\underline{r},\underline{r}';\epsilon) = v(\underline{r})\ \delta(\underline{r}-\underline{r}') + \int d^3r_1\ v(\underline{r})\ G_0(\underline{r}-\underline{r}_1;\epsilon)\ t(\underline{r}_1,\underline{r}';\epsilon). \tag{1.10}$$

The relevant quantities of interest to us are the "on the energy shell" angular momentum components of the t–matrix defined by:

$$t_{LL'}(\epsilon) = \int d^3r_1 \int d^3r_2\ Y_L^*(\hat{r}_1)\ j_l(\sqrt{\epsilon}\, r_1)\ t(\underline{r}_1,\underline{r}_2;\epsilon)\ j_{l'}(\sqrt{\epsilon}\, r_2)\ Y_{L'}(\hat{r}_2) \tag{1.11}$$

It can be shown that for spherically symmetrical potentials only the diagonal components can be different from zero, so we can write [5]:

$$t_{LL}(\epsilon) = -\frac{1}{\sqrt{\epsilon}}\sin\delta_l\ e^{i\delta_l} \tag{1.12}$$

In this way we finally obtain an expression for the density of states:

$$n(\epsilon) = -\frac{\epsilon}{\pi}\sum_L \int_0^{r_{\mathrm{MT}}} dr\ r^2 |R_l(r,\epsilon)|^2 \frac{\mathrm{Im}\ t_{LL}(\epsilon)}{\sin^2\delta_l(\epsilon)}\ . \tag{1.13}$$

1.3 Multiple Scattering Theory

As for the single MT potential the scattering properties of a cluster of scatterers can be expressed in terms of the phase shifts and the "on the energy shell" components. We can consider the following equation which is the expression of the analogue of (1.10) for multiple scattering in terms of operators (for a more complete description see [6]):

$$\tau^{ij}(\epsilon) = v_i\ \delta_{ij} + \sum_k v_i\ G_0(\epsilon)\ \tau^{kj}(\epsilon) \tag{1.14}$$

where, for example, $G_0(\underline{r}-\underline{r}';\epsilon) \equiv <\underline{r}|G_0(\epsilon)|\underline{r}'>$ etc. The "on the energy shell" components are now given by:

$$\tau^{ij}_{LL'}(\epsilon) = \int d^3r_i \int d^3r'_j\ Y^*_L(\hat{r}_i)\ j_l(\sqrt{\epsilon}\ r_i)\ \tau^{ij}(\underline{r}_i,\underline{r}'_j;\epsilon)\ j_{l'}(\sqrt{\epsilon}\ r'_j)\ Y_{L'}(\hat{r}'_j) \tag{1.15}$$

where L and L' are respectively the angular momentum corresponding to the sites $\underline{R}_i$ and $\underline{R}_j$, $\underline{r}_i = \underline{r} - \underline{R}_i$ and $\underline{r}'_j = \underline{r}' - \underline{R}_j$. It is possible to achieve an expression of the $\tau^{ij}_{LL'}(\epsilon)$ in terms of the equivalent single scattering quantity $t^i_{LL}(\epsilon)$ and of factors depending only on the structural properties of the assembly of MT potentials. In such a way we can write:

$$\tau^{ij}_{LL'}(\epsilon) = t^i_{LL}(\epsilon)\ \delta_{ij}\ \delta_{LL'} + \sum_{k\neq i}\ \sum_{L''}\ t^i_{LL}(\epsilon)\ G_{LL'}(\underline{R}_i - \underline{R}_j;\epsilon)\ \tau^{kj}_{L''L'}(\epsilon) \tag{1.16}$$

where the structure constants can be obtained from:

$$G_{LL'}(\underline{R}_i - \underline{R}_j;\epsilon) = 4\pi\sqrt{\epsilon} \sum_{L''} C^{L''}_{LL'}\ i^{l''}\ h^+_{l''}(\sqrt{\epsilon}\ |\underline{R}_i - \underline{R}_j|)\ Y_{L''}(\Omega) \tag{1.17}$$

$h^+_{l''}$ being the Hankel function, the $C^{L''}_{LL'}$ the Gaunt numbers [7] and Ω the angle between $\underline{R}_i$ and $\underline{R}_j$. Finally the expression of the density of the states projected on the $\underline{R}_i$ site is similar to that shown in (1.13) where the $t_{LL}(\epsilon)$ are replaced by the $\tau^{ii}_{LL}(\epsilon)$:

$$n^i(\epsilon) = -\frac{\epsilon}{\pi} \sum_L \int_0^{r_{MT}} dr\ r^2 |R^i_l(r,\epsilon)|^2 \frac{\mathrm{Im}\ \tau^{ii}_{LL}(\epsilon)}{\sin^2 \delta_l(\epsilon)}\ . \tag{1.18}$$

2. Application to Titanium Diselenide Clusters and Discussion of Obtained Results

Following the theory previously expressed, we have calculated the pDOS's of $TiSe_2$ for finite clusters of atoms the positions of which have been chosen according to the corresponding crystal structure. Usually, significant results on infinite solids cannot be obtained from the study of finite clusters, but in this work we are only looking at local properties of solids such as the pDOS's that can be successfully described in this context. Although in principle the theory of multiple scattering allows us to achieve all the angular momentum components of pDOS's we have retained in the present calculation only those values of the quantum number L which correspond to occupied states in the atomic elements.

Among these values particular attention has been directed to the p–like states ($l = 1$), the reason for this special interest is the possibility of obtaining information about states with axial symmetry with respect to the c–axis of the unitary cell (p_z–like), and about states which extend in the layer in the xy plane ($p_{x,y}$–like). Obviously the p_z–like pDOS's depend more deeply on the interaction between two contiguous layers than $p_{x,y}$–like ones, whereas these latter ones are principally due to the interactions with atoms in the same layer. We have considered clusters centered both on titanium and on selenium, but only Ti–centered clusters can be chosen invariant under inversion. All the Se–centered clusters do not have inversion symmetry, thus splittings appear which are not expected to be present in the infinite crystals which are the final objects of our analysis. For this reason we will only present results concerning clusters centered on Ti in this work.

We have considered three different kinds of clusters. Firstly we have calculated the pDOS's of clusters the shape of whichis taken to be as spherical as possible, and we have chosen those clusters so that the concentration of each element was approximately stoichiometric. In this way we have tried to obtain a better approximation of pDOS's in solids. Next, to point out the interlayer interaction, we have considered groups of atoms arranged in adjacent layers in a manner that we have called "cylindrical"; we have considered only atoms within a fixed distance from the c–axis and with the value of the axial coordinate included in a symmetrical interval around the origin. The third kind of cluster contains only atoms in the same layer within a fixed distance so we can call it a "layer" type; with such a choice we want to point out the effect of intralayer interactions on pDOS's.

The obtained results for spherical clusters are shown in Fig. 1 and Fig. 2 for $p_{x,y}$–like and p_z–like states respectively, whereas those corresponding to the "layer" and "cylindrical" ones can be found in Fig. 3 and Fig. 4. One of the most interesting features we can observe in these figures is that almost independently of the size and the shape of the clusters the p_z–like states are more affected by the arrangement of the atoms than the $p_{x,y}$–like ones. For example, though in Fig. 1 we note the growth of a new peak when we increase the number

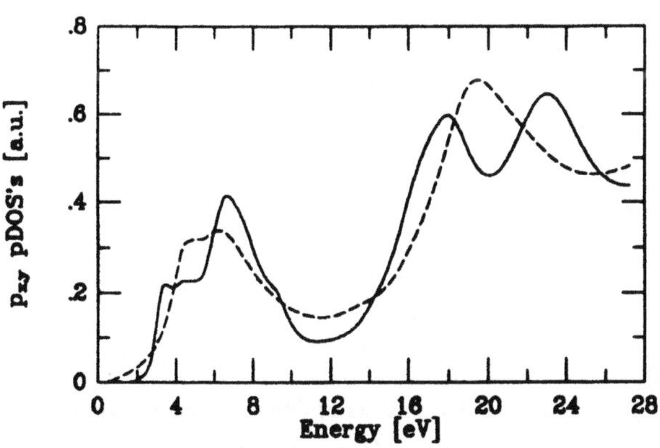

Fig. 1 Density of $p_{x,y}$–like states projected on Ti for spherical clusters of different size. Solid line: 51 atoms. Dashed line: 19 atoms

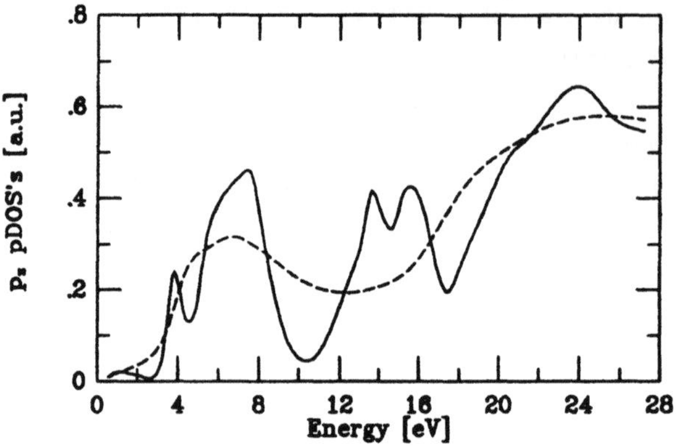

Fig. 2 The same as Fig. 1 for p_z–like states

of atoms, the analogue pDOS's in Fig. 2 are much more structured and the change from the same curve in a cluster with 19 atoms is more marked. This result is not unexpected due to the fact that the principal features corresponding to intralayer interactions are already contained in the smallest cluster (19 atoms), whereas those connected to the axial dimension are pointed out only in the bigger clusters consisting of atoms out of the layer too.

Another interesting result, especially for the $p_{x,y}$ states, is the strong influence of the symmetry of the spatial arrangement of the atoms on states with symmetry which is completely different. As evidence this feature we can observe that some structure in Fig. 1 such as the peak at about 23 eV seems to be due entirely to the presence of different layers, actually, as can be shown in Fig. 3, it is also present in all the "cylindrical" clusters whereas it is absent in the "layer". Therefore, the influence of effects dependent on interactions between atoms not in the same layer, also in states with layer symmetry, shows that for this compound any two–dimensional approximation can be considered only as the first step to a deeper

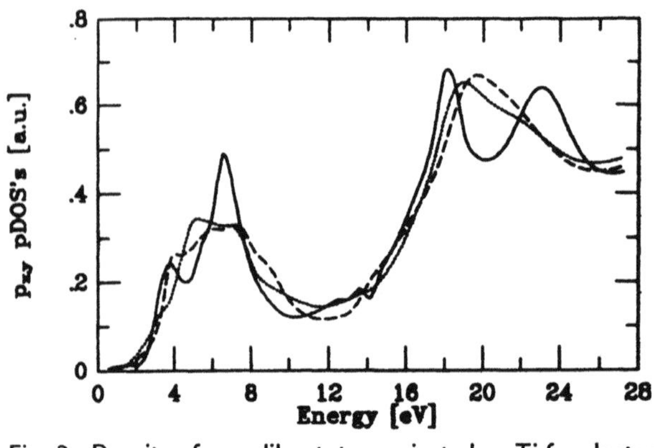

Fig. 3 Density of $p_{x,y}$–like states projected on Ti for clusters of different size and geometry. Solid line: 57 atoms in a "cylindrical" cluster (see text for explanation on the terminology). Dashed line: 37 atoms in "layer" cluster. Dotted line: 27 atoms in a "cylindrical" cluster

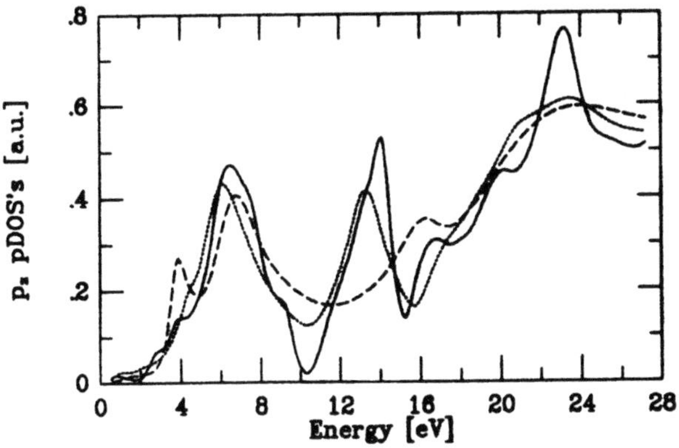

Fig. 4 The same as Fig. 3 for p_z–like states

approach. If the $p_{x,y}$–like pDOS's are greatly affected by the third dimension the p_z–like ones are obviously dependent in a critical way on this feature as can be seen in Fig. 4. This figure can also provide an explanation of the origin of the two peaks at about 13.5 eV and 15.5 eV in Fig. 2 , the former being clearly associated to the interlayer interactions, the latter, on the contrary, has comparable contributions from both kinds of interactions. We want now to conclude with some observations about the possibility of extrapolating our conclusions to infinite solids. It has to be pointed out that band gaps cannot be obtained in our formalism, the reason for this being found in the particular type of electron states we have considered. We have dealt with scattering states and for such states the density of states can vanish only when all the interferences are considered, and such a situation is possible only in the infinite solid. However, we can note in our results that in the region between 8 eV and 12 eV, where previous calculations [2] localize the gap, the density of states tends to go down as the size of the cluster is increased (see for example Fig. 2 and Fig. 4). Finally we remark that we have also calculated the total DOS's and we have found a good agreement with the total DOS's referred to in the literature [2].

It has to be mentioned that at present a collaboration is in progress with experimentalists from the Frascati Laboratory in order to compare our results on p-like pDOS's with measurements of polarized X–ray absorption spectra from K threshold of titanium.

References

1. M.M. Traum, G. Margaritondo, N.V. Smith, J.E. Rowe and F.J. Di Salvo: Phys. Rev. B17, 1836 (1978)
2. E. Doni, R. Girlanda: In Electronic Structure and Electronic Transitions in Layered Materials: Recent Developements, Vol. 7 of Physics and Chemistry of Materials with Low–Dimensional Structures, Series A: Layered Structures, ed. by V. Grasso (Reidel, Dordrecht, 1986)

3. A. Zunger and A.J. Freeman: Phys. Rew. B14, 1839 (1978)
4. J.A. Wilson and A.D. Yoffe: Adv. Phys. 18, 193 (1969)
5. P. Lloyd and P.V. Smith: Adv. Phys. 21, 69 (1972)
6. B.L. Gyorffy and G.M. Stocks: In Electrons in Finite and Infinite Structures, ed. by P. Phariseau (Plenum, New York, 1977)
7. E. Condon and G. Shortley: The Theory of Atomic Spectra (Cambridge University Press, Cambridge, England, 1967)

On the Theory of Auger Induced Desorption

M. Cini

Dipartimento di Fisica, Università di Roma II,
Via O. Raimondo, I-00173 Rome, Italy

Abstract. *Recent progress in the theory of induced desorption is briefly reviewed. By the exact solution of an interacting, time-dependent Anderson model we can take into account the non-adiabatic character of the electronic motion and the effects of on-site electron correlation. Thus, we obtain a quantum theory of the Knotek-Feibelman mechanism and also a better understanding of the Menzel-Gomer-Redhead process. Moreover, several new physical situations are analyzed and possible experimental verifications are suggested. On the other hand, the main limitations and/or shortcomings of the present state of the art are discussed. Attention is paid to the effects of orbital shrinkage on Auger line shapes and ion yields. An attempt is made to evaluate such progress in the light of existing work in this and related theoretical problems.*

1 Introduction

From a theoretical viewpoint, the problem of induced desorption mixes up several ideas originating in the field of electron spectroscopy with others that developed in the study of atom-surface interactions. Like all the experiments that simultaneously involve electronic excitations and ionic degrees of freedom, it poses very difficult problems that require a rather heavy and sophisticated mathematical treatment, and every advance in physical understanding corresponds to a marked increase in complexity. Thus, it is very encouraging to note that at least, some of the features can be understood quite simply. Consider for example the ion yields from chemisorbed atoms like O or H on several transition metal surfaces. The semiclassical theory initiated by MENZEL, GOMER and REDHEAD [1] ascribed its origin to bonding-antibonding one-electron transitions of the atom-surface complex (MGR mechanism). This explained why the threshold energy for inducing desorption corresponds to valence excitations, as observed. The fact that so few of the ionized species reach the detector as ions (typically, 10^{-5} or less, depending on several factors) can be understood, at least qualitatively, by the presence of very efficient bond-healing processes. Since desorbing ions are quite slow compared with electronic motions, there is "time" to reneutralize or recapture them before they get far. A semiclassical estimate showed that the ion escape probability must be extremely sensitive to ion velocity, leading to the prediction of an isotopic effect, which in fact is observed experimentally. The energy distribution curve of the desorbing ions was predicted to be roughly Gaussian, reflecting the oscillator-like nature of the initial state and an approximately uniform surface electric field experienced at the start of the desorption process. All these features are beautifully shown by experimental data. As an example, the reader might read the paper by MING L. YU [2] on the desorption of O ions from Mo surfaces.

The discovery of the KNOTEK-FEIBELMAN (KF) mechanism [3,4] showed that the independent-particles description does not apply to all situations. The desorption of O^+ from oxides and, more generally, of positive ions of electronegative species from ionic crystals have thresholds that often correspond to core levels in the system. Thus, desorption must be initiated by an Auger transition. Systems like TiO_2, where the core level involved belongs to Ti, actually behave as if the Auger process were interatomic. Moreover the ion yield is comparatively very high, suggesting that an important fraction of the ions which are created at the surface actually avoid reneutralization [5]. The existence of a "reneutralization bottleneck" can be understood [4] by the concept of two-holes resonances, that was introduced [6-8] in the context of Auger line shape theory. Actually, a two holes state will change F^- into F^+ and lead to F^+ desorption, e.g. from LiF surfaces [9], but three holes are needed to desorb O^+ from oxides; at any rate, the correlation-induced resonances are localized in nature, and therefore bond healing is much less probable than in a MGR process. Since the resonances can be monitored as quasi-atomic peaks in the Auger spectrum, we already know the physics behind them and the conditions under which the KF mechanism is going to operate. Thus, in quasi-atomic situations, one can understand qualitatively many interesting phenomena without having to resort to any sophisticated formalism. Even in this case, simplicity arises from a quite complex situation because a few of the many factors of the involved problem become dominant, and affect the ion yield qualitatively.

However, a much more detailed description is needed if one wants a microscopic, quantitative theory, which is essential in order to understand cases that do not fall into the above special categories. For instance, one would like to know what happens when the hole-hole correlation U is not large enough to produce truly localized states, but significantly distorts the Auger line shape. Moreover one expects that the ion yield cannot be properly understood in terms of the electronic structure of the initial state alone: the way the system evolves during the desorption process must enter the description, and in this way, the ion escape probability must depend on kinematics, too. Finally, to allow comparisons, it is desirable that one theory describe MGR and KF processes on equal footing, based on a quantum mechanical description of the electronic states. Thus, we need to upgrade also the previous semiclassical theories of the one-electron process. Such a theory will also make contact with the subject of atom and ion-surface scattering, where considerable work has already been done.

In this short note I cannot obviously review all aspects of the problem. Much experimental and theoretical work on induced desorption up to 1985 is reported in previous books of this series [10,11]. I wish to give a summary of the dynamical theory that I have published recently

[12-14], with emphasis on its physical contents. Then I try to give a critical assessment of the current degree of understanding. Unfortunately, several of the features that emerge from the theory still lack an experimental verification. However, by studying the inherent strengths and limitations of the approach we may perhaps understand whence further complications might come, and we can also anticipate future directions of the theory.

2 Overview of the Time-Dependent Theory [12,13,14]

In the present work, we consider the desorption of monovalent species, like F^+ from Fluorides, because then we are dispensed from considering more than two holes at one time. Like in the Newns theory of chemisorption [15] and in previous work on the Auger line shapes [6,7], we describe the adatom-surface complex by an interacting Anderson model

$$H = \varepsilon(t)[\, n_{0+}+n_{0-}] + U(t)\, n_{0+}\, n_{0-} + \Sigma_{k,\sigma}\, \varepsilon_k n_{k\sigma+}$$

$$+ \Sigma_{k,\sigma} [\, W_k(t)\, a^+_{0\sigma}\, a_{k\sigma} + h.c.]. \qquad (1)$$

Here, a and n are hole destruction and number operators, the label 0 refers to the orbital of the desorbing ion, while k denotes the band states of the solid,as usual. The desorbing ion energy level ε, the hole-hole repulsion U and the hopping parameters W_k are all allowed to vary with time in an arbitrary way, in order to mimic the variation of such parameters along the ion's trajectory. The Anderson model was also used in other work on the desorption theory [5,16], although the time dependence was not allowed. The time-dependent Anderson model was also used in much work on ion-surface scattering (see [17] and References therein). We differ from most workers in that we concentrate on the case when all the relevant electronic states are below the Fermi level. We do so because this is a quite important special case for the problem at hand, and it allows solving the interacting problem exactly and analytically. I believe that the use of analytical methods, which is a major difference between the present approach and [16], gives us a deeper understanding of the problem and an absolute control of the mathematical machinery. Important physical effects are usually associated to mathematical pathologies. In the present problem several kinds of them turn out to be important, including mild Van Hove singularities, ordinary poles and essential singularities. All kinds deserve and require an analytical treatment.The solution of the closed-band problem is also expected to give us important hints about the more difficult problem of incompletely filled bands.

We want to use (1) as a model for MGR-like processes, induced by one initial hole, as well as for the Auger induced ones, when the initial holes are two. It is clearly essential to us that both mechanisms are described on equal footing. In order to calculate the ion escape probabilities we need to solve for the amplitudes. To describe the MGR process we only need the one-hole two-times correlation function

$$g(t,\tau) = -i < a_{0+}(t)\, a^+_{0+}(\tau)> . \qquad (2)$$

To describe Auger induced processes we still need (2), but we also want the two-holes amplitudes, both site-diagonal,

$$G(t,\tau) = -i < a_{0-}(t)\, a_{0+}(\tau)\, a^+_{0+}(0)\, a^+_{0-}(0) > \tag{3a}$$

and off-diagonal, like

$$G_k(t,\tau) = -i < a_{0-}(t)\, a_{k+}(\tau)\, a^+_{0+}(0)\, a^+_{0-}(0) > \quad . \tag{3b}$$

It is also convenient to define

$$N(t) = i\, G(t,t)\, \theta(t)\, . \tag{4}$$

Although the exact solution is obtained in any case, the equations turn out to be simpler to handle and easier to interpret physically if one assumes that all the hopping matrix elements vary in time according to the same, arbitrary law, that is, $W_k(t) = V_k u(t)$. This unessential assumption is adopted in the present work for the sake of simplicity. The one-hole amplitude then can be expressed in terms of the local self-energy Σ calculated in equilibrium, before desorption, by solving

$$i\frac{\partial}{\partial t} g(t,\tau) = \varepsilon(t)\, g(t,\tau) + u(t) \int_{\tau}^{t} dt'\, u(t')\, \Sigma(t-t') g(t',\tau) \quad . \tag{5}$$

Equation (5) agrees with the results of studies of the time-dependent Anderson model in the context of ion surface scattering [18]. However in previous analytic treatments the energy dependence of the self-energy of the virtual level was neglected, and this drastic oversimplification prevented an analysis of the dependence on the electronic structure. The closed band limit of the theory of EASA and MODINOS [17] of atom-surface scattering is also essentially equivalent to (5). Their analysis is of interest for us because they did not start with the Anderson model, but justified the results from a more general viewpoint. Thus, they showed explicitly how the parameters in the Hamiltonian (except U) can be calculated, and demonstrated their dependence on the detailed band structure of the solid. For the interacting two-holes amplitude we have no previous analysis to compare with. I find

$$i\frac{\partial}{\partial \tau} G(t,\tau) = \varepsilon(\tau)\, G(t,\tau) + i\, U(\tau)\, g(t,\tau)\, G(\tau,\tau) + u(\tau) \int_{0}^{\tau} d\tau'\, u(\tau') \Sigma(\tau-\tau') G(t,\tau') \quad , \tag{6}$$

which is rather difficult to use directly because of the dependence on two time arguments. Fortunately the equal-times limit of (6) exists, and reads

$$N(t) = -\, g^2(t,0)\Theta(t) + i \int_{0}^{t} dt'\, g^2(t,t') U(t') N(t') \quad . \tag{7}$$

The probability of ionic desorption in the MGR process is $P_{1h} = | g(\infty,0)|^2$ and the ion escape probability $P^{\uparrow\downarrow}$ in the Auger induced process is just the long time limit of $| N(t)|^2$. The probability that a doubly ionized atom, produced by an Auger transition, suffers partial reneutralization is more complex to obtain. It can be expressed in the following form:

$$P^{\uparrow} = -\frac{1}{\pi}\int_{-\infty}^{\infty} d\omega \Sigma_2(\omega) \,|\int_0^{\infty} d\tau\; u(\tau)G(\infty,\tau)\exp(i\omega\tau)|^2 , \qquad (8)$$

where Σ_2 stands for Im(Σ). To solve (5,6,7) and evaluate of (8) we need to specify the time dependences in the Hamiltonian. Although, in a preliminary study, we may be willing to ignore the time dependences of ε and U, we cannot obviously do the same with u(t). It is natural to consider, as a simple model, $u(t) = \exp(-\lambda t)$, where the frequency λ is the inverse of the characteristic desorption time t_0 and is of the order of the normal component of velocity divided by a bond length. This model may be appropriate when the desorbing ion moves with roughly constant velocity, as is referred to as the "gradual bond-breaking" model. On the other hand, when considering desorption from a highly ionic surface under the influence of huge electric fields one can justify a "sudden bond-breaking" model $u(t) = \theta(t_0 - t)$. I have derived a number of analytic solutions, but the reader is referred to the original papers for the mathematical techniques that were developed to achieve them. Here, I focus on the main physical ideas.

The ion yield turns out to depend very much on t_0 and appreciably also on the form of u(t); knowing only the electronic structure of the adatom-surface complex before excitation does not even enable us to predict orders of magnitude,in general.The influence of the detailed time evolution shows up dramatically when the system possesses localized states. This is the case when the initial state is a "surface molecule" [15,19], characterized by a strong chemisorption bond on a narrow-band solid. The ion yield is then predicted to undergo oscillations as a function of the velocity parameter λ due to resonant charge exchange between the adatom level and the narrow band. Similar phenomena are already known in ion-surface scattering, as discussed by TULLY [20], but apparently they have not yet been discovered in desorption experiments. The theory also enables us to calculate the yield of doubly charged ions that have been partly reneutralized, and the yield of singly charged ions, produced by a MGR type process. Both kinds of ions obviously carry one less electron than they have in the initial state. It may be interesting that their yields are predicted to vary with velocity in very different ways, so that we could distinguish the two processes experimentally.

It was shown by explicit examples that the existence of a localized state which is split off a continuous band does not necessarily imply that a reneutralization bottleneck is going to operate. The ion yield is again determined by the evolution of the system. Indeed, the hopping terms of the Hamiltonian may play a role in producing localized states; during desorption, these terms vanish and the split-off state may be adsorbed by the continuum. In such cases, the ion escape probability is very small for slow ions. Correlation-induced two-holes resonances, instead, form quasi-atomic states that typically remain off the continuum all the time. The KF mechanism then neatly results. For slow and gradual bond breaking, $P^{\uparrow\downarrow}$ is exactly equal to the relative weight of the quasi-atomic peak in the Auger spectrum. In this case, the ion yield is independent of the details of the band structure and the λ dependence is also unimportant. In other terms, a "universal" behavior is approached with increasing U.

The KF mechanism arises in the theory as a special case, and actually, the most interesting situation occurs when the Auger spectrum is band-like. The general theory has been analyzed in some detail for sudden bond-breaking situations, when N(t) is just the Fourier transform of the two-holes density of states. $P^{\uparrow\downarrow}$ is determined by $N(t_0)$, and since the ions are slow, this means that we have to study the long-time, asymptotic behavior. Aside from the split-off state, there are contributions to $P^{\uparrow\downarrow}$ arising from all structures in the continuum, including Van Hove and band-edge singularities and, in particular, the ion yield is sensitive to the hole effective mass. Since hole-hole correlation is also capable of producing narrow structures in band-like situations [7], $P^{\uparrow\downarrow}$ may be enhanced by many orders of magnitudes even in the absence of quasi-atomic states. At the same time, $P^{\uparrow\downarrow}$ remains quite sensitive to ion velocity. It should be remarked that a strong dependence on λ also implies the existence of an isotope effect, and a strong angular dependence (since λ is related to the normal component of velocity). The extreme sensitivity of $P^{\uparrow\downarrow}$ to U suggests that experiments on thin films of ionic crystals grown on a metal surfaces should show an enhancement of the ion yield with film thickness, because U is also expected to increase.

Having summarized the model and its predictions, I wish to draw attention on several complications that we have neglected up to now.

3 Two-Holes Dispersion

In the original interpretation by FEIBELMAN [4], the ion escape probability in the KF mechanism depended primarily on the small width of the two-holes state, which implied a long enough lifetime for desorption to occur. In turn, the width was determined by two-holes dispersion [8], that is due to the propagation of the two-holes bound state in a periodic solid. An order-of-magnitude calculation showed that the width was too large to explain the observed magnitude of $P^{\uparrow\downarrow}$. However, FEIBELMAN suggested that a further enhancement of $P^{\uparrow\downarrow}$ by a factor of 50 or so could easily result from orbital shrinkage effects. If the "radius" of the ion bearing the two holes is substantially reduced compared to the negative ion, hopping in the final state must be much slower than could be inferred from photoemission measurements.

However, at the time of Ref. [4] the dispersion width was overestimated by about an order of magnitude, and was later shown [21] to be usually of the order of 0.1 eV or less, even for perfect solids. In addition, at the surface we must consider the effects of reduced coordination and band bending,that will also decrease the width and hinder two-holes hopping.The surface induced change in the polarization contribution to U should also impede delocalization into the bulk in a similar way. Therefore, it is most likely that two-holes hopping is essentially limited to the surface layer. Even if the surface were perfect enough to allow such a delicate delocalization, it appears to me that a strictly surface dispersion would hardly have any effect on the ion yield. Therefore the neglect of two-holes hopping may be expected to be of no consequence for us, although it could turn out to be important in the case of partially filled bands.

4 Orbital Shrinkage Effects

Even if we now discard the original motivation for including such effects, nevertheless we must worry about their possible relevance, since the orbital radii of the electronegative ions do depend on charge . First of all, let us investigate how orbital shrinkage effects could manifest themselves in the Auger line shape. Accordingly, we resume the Auger line shape model in its simplest form [6], which applies to nondegenerate orbitals and bands, and seek a suitable generalization. Assume then that the hopping matrix element V between site 0, where the primary core hole is produced, and its Z nearest neighbors in a tight-binding solid is changed into $V+\eta$ if a hole of opposite spin is also present at site 0. This is naturally modeled if the interaction term in the Hamiltonian is modified to read

$$H_I = U\, n_{0+}\, n_{0-} + \eta\, \Sigma_i \{ n_{0-}\, a^+_{0+}\, a_{i+} + n_{0+}\, a^+_{0-}\, a_{i-} + \text{h.c.} \} \quad , \tag{9}$$

where the sum includes the nearest neighbors and h.c. stands for the Hermitean conjugate. A closed form exact solution for the Auger line shape can be obtained as follows. Let the Latin indices denote sites,

$$G_{ijkl} = < a_{i-}\, a_{j+}\, (\omega - H)^{-1}\, a^+_{k+}\, a^+_{l-} > \tag{10}$$

the two-holes Green's function and G^0_{ijkl} its limit for $U = \eta = 0$. We readily obtain

$$G_{0000} = G^0_{0000} + U\, G^0_{0000}\, G_{0000} + \eta\, \Sigma_i \{ G^0_{0000}\, G_{i000} + G^0_{00i0}\, G_{0000} +$$

$$+ G^0_{0000}\, G_{0i00} + G^0_{000i}\, G_{0000} \} \tag{11}$$

and similar results for the off-diagonal terms G_{j000}. Since different neighbors are only mixed through the unperturbed Green's functions we achieve a closed result by summing the off-diagonal equations over neighbors. From now on we write the site-diagonal Green's functions simply as G and G^0. We use obvious symmetry properties of G_{ijkl} and the equations of motion, that yield

$$\Sigma_i\, G^0{}_{00i0} = \frac{(\omega\, G^0(\omega) - 1)}{(2V)} \tag{12}$$

and

$$\Sigma_{ij} \{ G^0_{j0i0} + G^0_{j00i} \} = \frac{\omega(\omega\, G^0(\omega) - 1)}{2V^2} \quad . \tag{13}$$

Hence we find the generalized expression

$$G = \frac{G^0}{(1-\frac{\eta}{V}(\omega G^0-1))^2 - (U + \frac{\eta^2\omega(\omega G^0-1)}{V^2})\, G^0} , \qquad (14)$$

that reduces to the familiar result $G = \frac{G^0}{1-UG^0}$ when η vanishes. The Auger line shapes of negative ions should be characterized by negative values of η, that generally entail a reduction of the solid induced features. Band-like spectra become narrower than could be expected from the self-convolution of the one-hole density of states, and tend to peak around the atomic energy level ε. For U such that the usual theory predicts strongly distorted band-like spectra the effect of η produces a more symmetric line shape, where the band edges are less evident. Atomic-like spectra are also modified, because the peak is shifted to lower binding energies, contrary perhaps to simple minded expectations. Orbital shrinkage effects might well be present in the measured spectra of Oxides and Halides; in the writer's opinion, they have never been discovered because no adequately accurate analyses of such line shapes have been published to date. After all, the modifications due to η in (14) are often qualitatively similar to those of a reduced apparent band width.

As noted above, orbital shrinkage effects are not included in the present form of the desorption theory; however, when they are significant ,they should show up already in the Auger line shape, essentially as a disagreement with the expectations based on the underlying band structure as determined by one-electron measurements, like photoemission. It seems most likely that the ion yield would then be understandable in terms of the width of the features of the Auger spectrum and the above theory should then be still a useful approximation. If such cases were seen experimentally, one would be motivated to extend the time-dependent approach, by modifying the interaction term like in (9). Generalized laws of desorption could be deduced, using very much the same mathematical techniques. Such developments would probably be premature at the present stage.

5 The Classical Trajectory

The aim of the present dynamical theory is to predict the yields of the various desorbing species as functions of the electronic parameters and of trajectory. Although here we wanted just to demonstrate how the theory works qualitatively, for quantitative work we could compute the electronic parameters as functions of distance, very much along similar lines as in [17]. One could then obtain the potential energy surfaces, and calculate trajectories according to Newton's laws. While such a computational program looks rather demanding, from a theoretical point of view the assumption of a classical trajectory is the most serious limitation of our approach. Treating the nuclear motion classically is not without justification, particularly for heavy atoms, but it must be kept in mind that the vibrational motion of the ions before the primary excitation can only be described quantum mechanically. A complete, first-principles theory should be based on a time-independent Hamiltonian and should treat the electronic and nuclear degrees of freedom according to quantum mechanics. Moreover, a theoretical scheme for doing that already exists. Several years ago, FREED [22] has published a quantum theory of electron stimulated desorption that translates the whole problem into an elegant Green's function language. The resulting formalism is very similar to the theory of Raman scattering processes, and is rigorous within the Born-Oppenheimer approximation. The trouble is that the practical calculation of the Green's functions appears to be intractable, except perhaps in lowest order in the hopping parameters. The approach has been useful, since FREED was able to show formally how the general scheme reduces to previous semiclassical theories at the proper limits. However, it is remarkable that he did so without even writing down a concrete model for the electronic Hamiltonian and for the forces acting on the ion core. No way has been found to implement the theory and make contact with realistic models of the electronic structure.

Since the mathematical difficulties of the first principles approach look prohibitive, to make progress we must aid ourselves with our physical understanding. Assuming that the classical trajectory approach (Trajectory Approximation) is not completely wrong, we may look for corrections. As a first improvement, we should like to make the trajectory selfconsistent, allowing it to be influenced by the time-dependent charge of the ion and by the response of the solid. In a further step, we should like to replace the fixed trajectory by a weighted average of many possible paths. This should take proper care of the "initial conditions" and should lead to a semiclassical approximation in the first place. There is little doubt that a mature form of the theory will be based on Feynman's sum-over-path approach to quantum mechanics, and will necessarily turn out to be quite involved mathematically. Many of these ideas were explored recently in a rather difficult, but important paper by NEWNS, in the context of a theory of atom-surface scattering [23]. The application of his scheme to the desorption problem is far from obvious, and among other things one should find ways to include a description of the electronic states, but the general relevance of the formalism is evident. Remarkably, his final results do lend support to the applicability of the Trajectory Approximation in many cases, although the inclusion of selfconsistency and of a recoil term is recommended. So it appears that a gradual, physically oriented approach to the solution of the full problem should be rewarding, and the time-dependent theory should be regarded as a step in that direction.

6 Conclusions

The dynamical theory provides a fully quantum description of the electronic states and their evolution during the bond-breaking process, including correlation effects. Several mathematical peculiarities of the integro-differential equations have been understood, and this knowledge should be valuable in developing numerical solutions in a large variety of cases. In the present scheme, we can also extend the formalism to include degenerate orbitals and bands, at the cost of an increased computational complexity. However I feel that, before dwelling with the details, we should now verify the general conclusions of Sect.2 with experiments. We need measurements of energy and angle-resolved yields of ions and neutrals from a variety of semiconductors and ionic crystals. The above results should also be useful in the case of metal surfaces covered with ionic chemisorbates or thin films. The desorption of monovalent species from solids with (at least, locally) filled bands should be privileged in the interest of fundamental understanding. However, this kind of studies should also be important for application-oriented Laboratories, since stimulated desorption is developing into a powerful local spectroscopy of surfaces. The threshold energy for desorption tells us which species is bonded to which in the topmost layer, and the energy distribution curves can tell us much about the nature of local bonds and local fields. Needless to say, the desorption data should be regularly supplemented by the Auger line shapes.

In the present paper, I have considered several complications that in principle make the theoretical problem quite formidable. We are still far from a fully comprehensive theory, and we cannot even be sure that all the main ingredients of a realistic physical picture have already been included. A rigorous description of the motion of the ion is likely to bring to the forefront new physical situations and fresh ideas, but appears to be beyond our reach at this time. Nevertheless, the overall results of our analysis are encouraging. The neglect of two-holes hopping should be of no concern to us. It should be possible to account for the main effects of orbital shrinkage properly in the present context, since they are observable already in the Auger line shape. The Trajectory Approximation remains our main concern, but there are good reasons to consider it a useful starting point at least. Above all, most of our conclusions involve very strong dependences of the ion yield on several parameters, affecting the qualitative behavior; as discussed in the Introduction, this is the kind of situations where analyses tend to simplify. Therefore, experimenters can accept the theoretical predictions with some confidence and, in the writer's opinion, most of the above results should survive the test. Time is ripe for systematic experimental work.

REFERENCES

1 D.Menzel and R.Gomer, J.Chem.Phys. 41,3311 (1964);

P.A.Redhead,Can.J.Phys. 42,886 (1964)

2 Ming L.Yu, Phys.Rev B 19,5595 (1979)

3 M.L.Knotek and P. J. Feibelman, Phys. Rev. B 18, 6531 (1978);

Surface Sci. 90, 78 (1979)

4 P.J.Feibelman, Surface Sci. 102, L51,(1981)

5 D.E.Ramaker, C.T.White and T.E.Murday, J. Vac. Sci. Technol. 18, 748 (1981); Phys. Lett. A 89, 211 (1982)

6 M. Cini, Solid State Commun. 20, 605 (1976)

7 M. Cini, Solid State Commun. 24, 681 (1977)

8 G.A.Sawatzky, Phys. Rev. Lett. 39, 504 (1977)

9 T. Yasue, A. Ichimiya and S. Ohtani,Surface Sci. 186, 191,(1987)

10 Desorption Induced by Electronic Transitions DIET I, Edited by N.H.Tolk, M.M.Traum,J.C.Tully and T.E.Madey (Springer-Verlag,Berlin, 1983)

11 Desorption Induced by Electronic Transitions DIET II, Edited by W.Brenig and D.Menzel (Springer-Verlag,Berlin, 1985)

12 M Cini, Phys. Rev. B 32,1945 (1985)

13 M.Cini, Nuovo Cimento D 8, 333 (1986)

14 M.Cini, Nuovo Cimento D 9, 1515 (1987)

15 D.M.Newns, Phys. Rev. 178, 1123 (1969)

16 D.R.Jennison,E.B.Stechel and J.A.Kelber, in [10]

17 S.I.Easa and A. Modinos, Surface Sci. 183, 531 (1987)

18 A.Blandin,A.Nourtier and W.Hone, J.Phys. (Paris) 37, 369 (1976); W.Bloss and D.Hone, Surf. Sci. 72, 277 (1978)

19 G. Tibbetts, J.M.Burkstrand and J.C.Tracy,Phys. Rev.B 15,3652,(1977);

J.W.Gadzuk, Surf. Sci. 43,44,(1974); Phys. Rev. B 10,5030,(1974)

20 J.C.Tully,Phys. Rev. B 16, 4324,(1977)

21 M.Cini and A.D'Andrea, J.Phys. C 16, 4469,(1983)

22 K. F. Freed, Surface Sci. 122, 317 (1982)

23 D.M.Newns, Surface Sci. 154, 658 (1985)

Part III

Auger Spectra from Solid and Gaseous Samples

The Local Density of States in Simple Metals and Their Alloys as Revealed by Auger Spectroscopy

P. Weightman

Physics Department and Surface Science Research Centre, University of Liverpool, Oxford Street, P.O. Box 147, Liverpool L69 3BX, UK

An account is given of recent work on the spectra of Auger transitions which involve final states in the valence levels of simple metals. Attention is focused on the information that Auger spectroscopy can provide on the local electron configuration, the local density of states and the screening of core-ionised and impurity sites in simple metals and their alloys.

1 Introduction

In the independent electron model of atomic structure we can describe the Auger process as a radiationless transition in which an atom initially ionised in a level a, decays to a doubly ionised final state creating holes in levels b and c. The energy for such a process can be written as the difference in binding energy of the initial and final states;

$$KE(abc) = E_B(a) - E_B(bc) \tag{1}$$

It is sometimes useful to rewrite the expression for the Auger energy in terms of the binding energies of the individual final state holes since these can often be determined from x-ray photoelectron spectroscopy, XPS. This requires the introduction of the term U(b,c) to represent the interaction between the final state holes.

$$KE(abc) = E_B(a) - E_B(b) - E_B(c) - U(b,c) \tag{2}$$

If the initial state is long lived then we can obtain the wavefunction of this N-1 electron ion by deleting the single electron orbital ϕ_a and electron i from the Slater determinant of the N electron atom: a wavefunction we denote by $\Psi(N-1, \phi_a, i)$. The final state wavefunction is $\Psi(N-1, \phi_b, (\phi_c \rightarrow \chi_d), j)$ where the term in brackets indicates that orbital ϕ_c has been replaced by the continuum orbital, χ_d of the ejected electron. The non-relativistic transition rate is then given by;

$$A_{ABC} = \frac{2\pi}{\hbar} \left| \langle \Psi(N-1, \phi_a, i) \left| \sum_{i>j} \frac{e^2}{r_{ij}} \right| \Psi(N-1), \phi_b, (\phi_c \rightarrow \chi_d), j) \rangle \right|^2 \tag{3}$$

where the sum over the interelectronic distances, r_{ij}, runs over all electrons i and j such that i > j. If we assume a central field and adopt the frozen orbital approximation then the normalisation and orthogonality properties of the ϕ make it possible to reduce (3) to the square of the difference of direct and exchange integrals each involving only those electrons directly involved in the transition;

$$A_{ABC} = \frac{2\pi e^2}{\hbar} \left| \langle \phi_a(1)\chi_d(2) \left| \frac{1}{r_{12}} \right| \phi_b(1)\phi_c(2) \rangle - \langle \phi_a(1)\chi_d(2) \left| \frac{1}{r_{12}} \right| \phi_b(2)\phi_c(1) \rangle \right|^2 \tag{4}$$

Springer Series in Surface Sciences, Vol. 18 Auger Spectroscopy and Electronic Structure
Editors: G. Cubiotti · G. Mondio · K. Wandelt

The expressions given above, when combined with the results of the appropriate atomic structure calculations, are found to give a good description of the Auger spectra of free atoms [1,2]. In this paper we are concerned with the spectra of Auger transitions in which one or both of the final state holes occur in the valence levels of atoms in simple metals or their alloys. We refer to these processes as core-core-valence, CCV, or core-valence-valence, CVV, transitions respectively. In order to describe Auger processes occurring in a metallic environment it is necessary to modify the expressions given above for the energies and transition rates of free atom Auger processes. To begin with the screening that accompanies any change in the local potential in a metal will change the Auger energies, the single electron binding energies and the interaction energy between the two final state holes. In addition the metallic environment produces changes in the electron configuration of the atom and in the wavefunctions of the valence electrons that might be expected to change Auger transition rates. It is the sensitivity of Auger processes to these features of the metallic environment that makes this field of Auger spectroscopy both difficult and interesting. It also means that the technique is a useful probe of the electronic structure of such materials, particularly with respect to the nature of electron screening.

2 Energetics

In the metallic state the direct observation of the Fermi level in XPS provides an accurate reference energy for the measurement of Auger and single electron binding energies. A comparison of the results of XPS and Auger measurements shows that for simple metals and their alloys the interaction energy between the two final state holes created in CCV and CVV processes, U(b,c), is negligibly small. The kinetic energy of the Auger spectrum is thus given by the differences in binding energy of the single hole states a, b and c and, since one or both of the latter come from the valence levels, the spectral profile has a width comparable to that of the single electron bandwidth, W, of the metal. The observation that $U(b,c) \simeq 0$ is important since it is well known that Auger profiles can be strongly distorted by correlation effects when U(b,c) is comparable to W[1,3,4]. For simple metals $U/W \ll 1$ and we can ignore such effects. Expressions (1) and (2) can be combined with accurate measurements of core level XPS and CCV and CVV Auger spectra to determine the energy distribution of simple metal CV and VV hole states. We now consider the analysis of these profiles.

3 Auger Profiles

For the transitions we are concerned with the Auger matrix elements (4) contain the wavefunction of at least one core electron which means that the Auger profile is a probe of the local electronic structure around the site of the initial core hole. Two important characteristics of the local electronic structure will influence the shape of the Auger profile; the shapes of the local valence band densities of states, DOS, around the atomic site, and the angular momentum character of the local valence electron configuration.

In the ground state we may denote the local valence electron configuration around an atomic site in a simple metal as $ns^a np^b$ and the local DOS as $D_s^o(E)$ and $D_p^o(E)$. Since the initial core-hole state which precedes an Auger transition is usually long lived we expect that the screening response of the valence electrons will change both the local electron configuration and the local DOS around the core-hole site. We denote these features of the core-ionised site by $[C]ns^x np^y$ and $D_s^1(E)$ and $D_p^1(E)$. A crucial step in the development of our understanding of the Auger spectra of simple metals was the formulation of the final state rule which makes it clear how these various quantities influence CCV and CVV Auger profiles.

3.1 The Final State Rule

The expression for the Auger transition rate for a free atom given above, (3), is derived using the frozen orbital approximation which assumes that the wavefunctions of electrons that are not directly involved in the transition do not change in the transition. This assumption is strictly incorrect due to the difference in the potential in the initial and final states. However because of the difficulty of allowing for the effects of the non-orthogonality of the wavefunctions of spectator electrons in the initial and final states the majority of Auger transition rate calculations for free atoms have been performed within this frozen orbital framework. In a metal the use of the frozen orbital approximation is particularly inappropriate due to the efficient screening response of the conduction electrons to any changes in the local potential. Fortunately however it is possible to avoid the difficulties of a full many body treatment of Auger processes since it is now generally agreed [5-8] that Auger spectroscopy, like x-ray spectroscopy, can be analysed using a single electron description provided one works within the framework of the final state rule. The final state rule may be stated as follows:

(a) The intensity of the contribution of valence holes of different angular momentum character to the Auger profile is determined by the local atomic configuration of the initial state.

(b) The shapes of the valence s and p contributions to the Auger profile are determined by the shapes of the local DOS's in the final state.

Thus for a CCV process where both the initial and final states contain a core-hole the local electronic structure around a core-ionised site determines both the intensity and the shape of the valence s and p contributions to the Auger profile. The intensities and shapes are given by $[C]ns^x np^y$ and $D_s^1(E)$ and $D_p^1(E)$ respectively.

For CVV processes the intensities of the valence s and p contributions to the Auger profile is also given by the local electronic structure of a core-ionised site but the shapes of the valence contributions are now determined by the local DOS in the final state which has two valence holes. Since $U \simeq 0$ the final state holes are delocalised in the metal and the screening response to delocalised valence holes will be small and should not generate any significant distortion of the local DOS. It should then be a good approximation to equate the ground state DOS, $D_s^o(E)$ and $D_p^o(E)$ with the final state DOS. CVV processes produce two uncorrelated final state valence holes each of which will probe the local DOS so the CVV profile will contain all possible convolutions of the $D_s^o(E)$ and $D_p^o(E)$ DOS: $D_{ss}^o(E)$, $D_{sp}^o(E)$ and $D_{pp}^o(E)$.

Although the final state rule provides a single particle framework in which to analyse CCV and CVV Auger profiles the calculation of such profiles from first principles is quite complicated. Ideally it requires a self-consistent treatment of the local electronic structure in the ground state and in the core-ionised state and in each case the electronic structure of the valence band must be represented in an atomic orbital type basis set which lends itself to the evaluation of atomic like matrix elements coupling initial state core-holes with final state valence holes. The expressions for the Auger transition rates are themselves complicated because it is not formally possible to separate the expressions into a product of a number of separate factors. Thus the influence on the transition rates of; the summation over the possible angular momentum quantum numbers of the ejected electron, the Clebsch-Gordon algebra created by the integrals over the angular dependences of the single electron wavefunctions, the local DOS of s and p character and the radial integrals must all be evaluated together. The evaluation of the radial integrals is further complicated by the energy dependence of the valence electron wavefunctions.

In view of this complexity very few calculations of the Auger profiles of simple metals have been carried out from first principles. Indeed our present understanding of this field depends on only a handful of detailed theoretical treatments of which VON BARTH and GROSSMAN's treatment [6] of the KLV spectra of Na and the calculations of the profiles of the KVV transitions of Li and Be by JENNISON and co-workers [8,9] are particularly significant. These studies demonstrate the applicability of the final state rule to Auger processes and, apart from providing a good understanding of the particular transitions concerned, they reveal a number of general points which are useful in interpreting the Auger spectra of simple metals in general.

3.2 Simplified Expressions

VON BARTH and GROSSMAN's work [6] showed that the screening of a core-ionised site distorts the shape of the local s DOS in a simple metal, $D_s^1(E)$, by inducing a peak at the bottom of the band. However the p contribution to the screening charge is more evenly distributed across the band and the shape of $D_p^1(E)$ is similar to that of the local p DOS in the ground state, $D_p^o(E)$.

For CVV processes the work of JENNISON and co-workers on Be [8] and Li [9] established that valence hole screening effects are small and that the ground state DOS is a good approximation to the final state DOS for simple metals.

These detailed studies reveal a number of factors which make it possible to justify less rigorous treatments of Auger transition rate calculations for simple metals. To begin with the radial matrix elements were found to vary only slowly with the energy of the valence levels across the valence band. VON BARTH and GROSSMAN also found [6] that the local DOS contribution showed only a small dependence on the angular momentum of the ejected electron and on the valence band energy. If these conclusions are generally true then it is reasonable to write Auger matrix elements as a product of:

(i) a coefficient determined by the Clebsch-Gordon algebra,
(ii) the local DOS of appropriate angular momentum character,
(iii) a radial matrix element, the latter being assumed to be independent of the energy in the valence band.

The Auger profile of a CVV transition in a simple metal with local ground state and core-ionised configurations of $ns^a np^b$ and $[C]ns^x np^y$ respectively is then, apart from an arbitrary normalisation, given by;

$$A_{CVV}(E) = (ax)^2 M^2_{CSS} D_{ss}(E) + (abxy)\, M^2_{CSP} D_{sp}(E) + (by)^2 M^2_{CPP} D_{PP}(E) \qquad (5)$$

where M_{CSS}, M_{CSP} and M_{CPP} are Auger matrix elements, including both radial and angular contributions, to double hole states of ss, sp and pp character respectively. In this expression the convolutions of the DOS of valence states of s and p character are obtained from $D^o_s(E)$ and $D^o_p(E)$ after these quantities have each been normalised to 1 so only the shapes of the local DOS are required from band structure calculations. The intensity of valence s and p contributions is then determined by the transition rate matrix elements and the local atomic configurations.

Recently DAVIES and co-workers [10] used (5) in a simultaneous analysis of the profiles of the $L_{2,3}VV$ and KVV Auger transitions of Mg. Since these transitions create the same final states the contributions that determine the shape of the Auger profiles, $D_{ss}(E)$, $D_{sp}(E)$ and $D_{pp}(E)$ should be the same. These quantities were obtained by convoluting the results for $D^o_s(E)$ and $D^o_p(E)$ obtained from band structure calculations. The matrix elements were calculated using the procedure described by JENNISON [9] and the results of the calculation showed that transitions to ss valence holes make a significant contribution to the KVV profile but almost no contribution to the $L_{2,3}VV$ profile. In spite of this difference in the relative importance of matrix elements involving final state holes of different orbital character it was possible to reproduce the Auger profiles of both sets of transitions with a single value of the parameter ax/by of 1.9 ± 0.1 confirming the analysis on which (5) is based. The determination of ax/by from the profiles of these transitions gives insight into the local ground state and core-ionised configurations in Mg. Band structure calculations suggest that the local valence configuration of Mg in the ground state contains equal amounts of s and p character. One concludes that in Mg core-hole screening involves valence electrons of predominantly s character. A similar conclusion is reached from studies of CVV transitions in Li [9], Be [8] and Cu [11] and from the analysis of the CCV transitions of Na [6].

Making the same assumptions that gave (5) for CVV processes we obtain the following expression for CCV processes:

$$A_{CCV}(E) = M^2_{CCS}\, D^1_s(E) + M^2_{CCP}\, D^1_p(E) \qquad (6)$$

where the amount of local s and p valence charge is implicitly included in the unnormalised local DOS around the core-ionised site $D_s^1(E)$ and $D_p^1(E)$. This procedure is convenient since these DOS's cannot be obtained from ground state band structure calculations but require a calculation for a core-ionised state.

The expectation from VON BARTH and GROSSMAN's studies [6] that in simple metals $D_s^1(E)$ will be distorted from the ground state DOS by the appearance of a peak at the bottom of the band while the shape of $D_p^1(E)$ is essentially unchanged explains why the KLV spectra of Na, Mg and Al [12-15] are composed of two peaks with varying intensities determined by the values of the matrix elements.

Equations (5) and (6) are valuable expressions in the analysis of Auger profiles since they can be used to test calculations of local DOS and to obtain information on local screening configurations. Although their usefulness is limited by the need to determine the matrix elements approximate values of these quantities can often be determined from atomic structure calculations.

We now consider a more empirical approach to the analysis of simple metal Auger spectra which is particularly useful for understanding the profiles of CCV transitions and which offers a simple way in which they can be used to probe the local DOS of impurities.

3.3 Equivalent Cores Approximation and Impurity Screening

It is now clear that although detailed calculations have only been performed for Na we have a good understanding of the KL_1V and $KL_{2,3}V$ Auger profiles of Na, Mg and Al and that these profiles contain information on the energy distribution of the screening charge around core-ionised sites in these metals. By the equivalent cores approximation these transitions should also give information about the screening charge around impurities of atomic number Z + 1 in free electron hosts of atomic number Z. In particular, because of the difference in the matrix element contribution to the KL_1V and $KL_{2,3}V$ Auger transitions, a superposition of the profiles of these two sets of spectra gives a good indication of the distortion of $D_s^1(E)$ by the screening charge [14]. In the top part of Fig 1 we show comparisons of the spectral shape of these transitions in pure Mg and pure Al [14]. The shaded area is a rough indication of the distortion induced at the bottom of the band by the s valence charge screening the initial state core-holes in the pure elements. These comparisons give an idea of the free electron like charge distribution screening Mg^{3+} ions in an environment of Mg^{2+} ions and of Al^{4+} ions in an environment of Al^{3+} ions. By the equivalent cores approximation these results relate to the local DOS of screening charge around impurities which have a difference in local core charge, ΔZ, of +1 from that of the host; Al impurities in Mg and Si

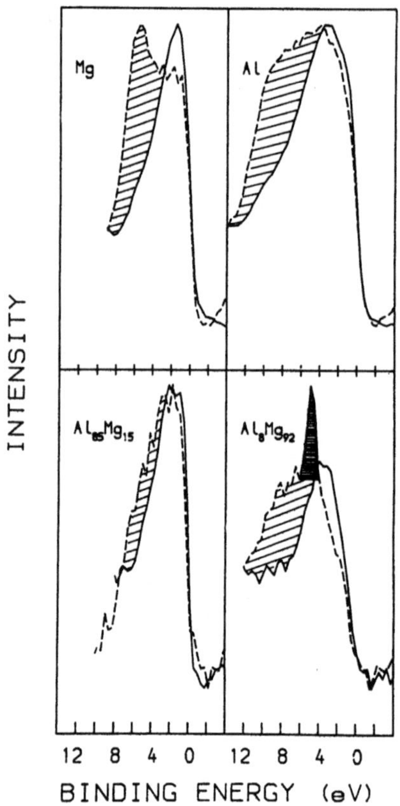

Figure 1

The full lines show $KL_{2,3}V$ spectra and the dashed lines KL_1V spectra. The diagonal shading shows the difference between these profiles. The top panels show a comparison of the KLV spectra of the pure metals. The bottom left panel compares the Mg KLV spectra of $AL_{85}Mg_{15}$. The bottom right panel is a comparison of the Al KLV spectra of Al_8Mg_{92}. The horizontal shading in this latter figure is an internal photoemission line. Figure from [14].

impurities in Al respectively. The lower parts of Fig 1 show comparisons of the spectral shapes of the same transitions but observed from alloys of Mg dilute in Al and Al dilute in Mg. In the former case the screening peak in the local s DOS has almost disappeared as we should expect since, by the equivalent cores approximation, this comparison reveals the local DOS around a normal Al site in elemental Al, ($\Delta Z = 0$). However the spectra of Al dilute in Mg show a significant increase in the distortion caused by the initial state core-hole, a distortion that is beginning to pull the maximum in the p DOS away from the Fermi energy. We should expect an increased distortion of the screening charge in this case since applying the equivalent cores approximation suggests that we now have a representation of the local DOS of the screening charge around a Si impurity in a Mg host, ($\Delta Z = +2$).

One interesting result of these studies of the CCV spectra of simple metals is that the screening of a core charge difference, ΔZ, of +1 between impurity and host is accomplished without distorting the shape of the local p DOS. We would expect that as ΔZ increases the local p DOS will be

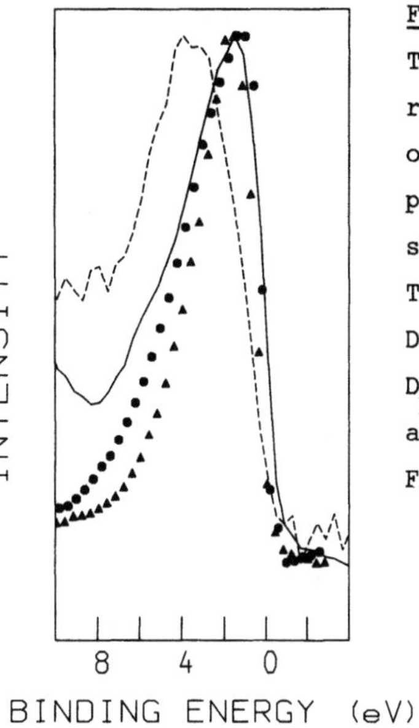

Figure 2

The triangles and the full line are respectively the KV and $KL_{2,3}V$ profiles of pure Mg. The circles show the Al KV profile of Al_5Mg_{95} and the dashed line shows the AL $KL_{2,3}V$ profile of Al_8Mg_{92}. These correspond respectively to the p DOS, $D^o_p(E)$ ($\Delta Z = 0$), $D^1_p(E)$ ($\Delta Z = +1$), $D^1_p(E)$ ($\Delta Z = +1$) and $D^2_p(E)$ ($\Delta Z = +2$) around Mg sites in a Mg environment. Figure from [14]

distorted by the screening charge. By taking advantage of the final state rule and assuming that matrix elements for both x-ray and Auger processes vary smoothly across the valence band and do not introduce any structure which is not present in the final state local DOS we can use the equivalent cores approximation to investigate the sensitivity of the shape of the local p DOS to ΔZ. In Fig 2 the triangles and the full line show respectively the Mg KV x-ray and $KL_{2,3}V$ Auger profiles of pure Mg while the circles and the dashed line show respectively the Al KV x-ray and $KL_{2,3}V$ Auger profiles of AlMg alloys dilute in Al. Both the x-ray and Auger profiles are dominated by the local p DOS in the final state. By the final state rule the Mg KV profile shows the ground state p DOS, $D^o_p(E)$ ($\Delta Z = 0$). By the equivalent cores approximation and the final state rule the pure Mg $KL_{2,3}V$ profile and the Al KV profile of Al impurities in a Mg environment should reflect the local p DOS around $\Delta Z = +1$ sites in Mg, $D^1_p(E)$. These two sets of spectra give almost identical profiles for $D^1_p(E)$ indicating that the influence of differences between Auger and x-ray matrix elements and their variation across the band is small. As expected the profile of $D^1_p(E)$ is very similar to that of $D^o_p(E)$. By the same arguments the Al $KL_{2,3}V$ profile of Al impurities in Mg should reveal the local p DOS around $\Delta Z = +2$ sites in Mg, $D^2_p(E)$. Fig 2 shows that the screening of a second core hole does distort the local p DOS, $D^2_p(E)$, the peak in the distribution being pulled back from the Fermi energy.

We now use this empirical approach to investigate the local DOS in alloys between Mg and d band elements. The KL_1V and $KL_{2,3}V$ of a large number of these alloys have been published recently [16,17]. We expect these spectra to reflect the profiles of the local final state s and p DOS in the vicinity of core-ionised Mg sites. The spectra of both transitions are very sensitive to changes in local environment arising from different alloy compositions. They show additional peaks not present in the spectra of pure Mg which can be correlated with the hybridisation between the Mg s and p bands and the d band of the other alloy constituent.

The $KL_{2,3}V$ spectra are easiest to understand since their spectral profiles should be dominated by the local p DOS. The Mg $KL_{2,3}V$ Auger and KV x-ray spectra of $MgCu_2$ and Mg_2Cu are shown in Fig 3a and 3b together with related experimental results. In pure Mg the x-ray and Auger profiles show a single peak arising from the local p DOS (Fig 2). The spectra of the two MgCu alloys show an additional peak to high binding energy the intensity of which increases with the Cu concentration. The dip between the two peaks in the x-ray and Auger profiles corresponds to the binding energy of the Cu d band as measured by XPS (Fig 3a). This observation, together with the conclusion from theoretical work [18] that the intensity of the local p DOS on a Mg site should fall in the vicinity of the d band resonance, supports the view that, as in the simple metal alloys, the Mg KV x-ray and $KL_{2,3}V$ Auger spectra show the profile of the local p DOS. A comparison of the x-ray and Auger profiles (Fig 3) shows that in both alloys the more bound peak is more intense in the x-ray profile than in the Auger profile. We will return to this difference later.

It is difficult to interpret differences between x-ray and Auger profiles due to uncertainty as to the influence of the difference between x-ray and Auger matrix elements on a comparison of spectral profiles. However there is some evidence that, apart from the Clebsch-Gordon algebra which determines the angular momentum character of the local DOS contributions to spectral profiles, this influence is small. Firstly we note the good agreement between the pure Mg $KL_{2,3}V$ profile and the KV x-ray profile of Al impurities in Mg (Fig 2). By the equivalent cores approximation and the final state rule both profiles reflect the Mg $D_p^1(E)$ and the agreement between the two profiles suggests that in each case the matrix elements have very little effect on the relationship of the spectral profile to the local DOS. Even more convincing is the agreement shown in Fig 3b between the shape of the Mg $KL_{2,3}V$ Auger profile of Mg_2Cu and the Al KV x-ray profile of $Al_{50}Cu_{50}$. In these two alloys the local environment around Mg and Al sites should be very similar. Combining the equivalent cores approximation with the final state rule and ignoring the influence of

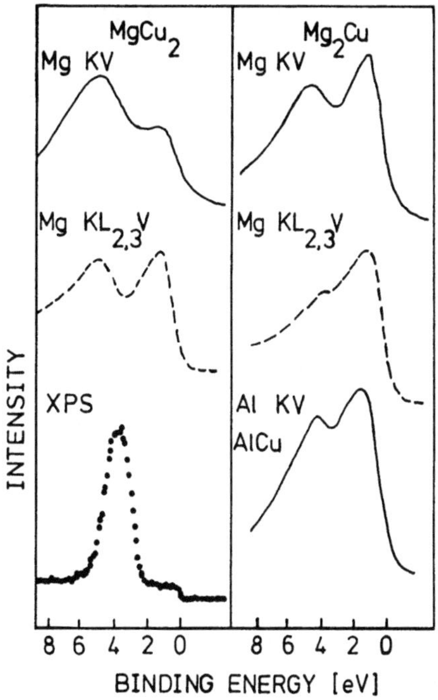

Figure 3

(a) Shows the Mg KV, Mg $KL_{2,3}V$, and valence band XPS spectra of Mg_2Cu.

(b) Show the Mg KV and Mg $KL_{2,3}V$ spectra of $MgCu_2$ and the Al KV spectrum of $Al_{50}Cu_{50}$.

Data from [17].

the difference in matrix elements and their variation across the bands leads us to expect very similar profiles for these transitions since they show respectively the local p DOS around a core-ionised Mg site, $D_p^1(E)$, or a neutral Al site in comparable MgCu and AlCu alloys. The two profiles are in good agreement and this indicates that differences in Auger and x-ray matrix elements do not have a large effect on the relationship between the spectral profiles and the shape of the local DOS.

We now consider the difference in the intensity of the two peaks on the local p DOS as revealed in the x-ray and Auger spectra of the MgCu alloys (Fig 3). Ignoring matrix element effects the Mg KV x-ray and $KL_{2,3}V$ Auger profiles show respectively the local $D_p^0(E)$ and $D_p^1(E)$ DOS around Mg sites in MgCu alloys. We should remember that this comparison of profiles does not give any information about absolute intensities, each profile shows the relative importance of the two peaks. A comparison of the x-ray and Auger spectra suggests that, unlike the situation in simple metal alloys, there is a difference between the local p DOS around ground state and core-ionised Mg sites in MgCu alloys. This difference is in the relative importance of the different regions of the p DOS in screening a Mg core hole. It would seem that in MgCu alloys the screening of a Mg core hole is achieved mainly by enhancing the local p DOS between the d band resonance and the Fermi energy.

The KL_1V Auger spectra of alloys between Mg and d band metals are more difficult to understand than the $KL_{2,3}V$ spectra since they involve roughly equal contributions from the local s and p DOS and both will hybridise with the d band. In addition experience of simple metals suggests that the screening of core holes may induce large changes in the local s DOS. The spectra show [16] that upon alloying with Ag, Pd, Cu and to a lesser extent with Ni, the most bound peak in the Mg KL_1V spectrum, which in the simple metal is associated with the s DOS, becomes more bound and more pronounced. The peak associated with the p DOS in the spectrum of pure Mg remains near the Fermi level in the alloys. There are also changes in the relative intensity and shapes of the two peaks which are associated with changes in alloy concentration. Although we do not understand these spectra in detail we can demonstrate that if we identify the most bound peak in the Mg KL_1V spectral profiles of these alloys with the peak in the $D_s^1(E)$ DOS then its position corresponds with what, by the equivalent cores approximation and the final state rule, we regard as the equivalent peak in the Al $L_{2,3}V$ x-ray profiles of the appropriate Al alloys. The following Table taken from [16] lists this comparison of the two sets of spectra.

When it is remembered that the peak positions compared in the Table correspond to broad spectral features which overlap with other components the agreement is remarkably good and is further support for the empirical approach which combines the equivalent cores model with the final state

Table 1 Binding energy of peaks in the $D_s^1(E)$ DOS around Mg site.

Mg KL_1V spectra		Al $L_{2,3}V$ spectra	
Alloy	Peak position*	Alloy	Peak position*
Mg_2Cu	5.4	$Al_{70}Cu_{30}$	5.9
MgCu	5.9	$Al_{32}Cu_{68}$	6.8
Mg_2Ni	5.5	$Al_{74}Ni_{26}$	5.9
$MgNi_2$	5.5	$Al_{25}Ni_{75}$	8.1
Mg_3Pd	5.8	Al_3Pd	6.4
MgPd	6.4	AlPd	6.9
MgAg	7.2	$Al_{35}Ag_{65}$	7.8

* Peak positions are given relative to the Fermi edge. Data from [16].

rule and neglects all but the angular momentum dependence of the x-ray and Auger matrix elements.

4 Future Developments

Clearly the empirical approach described above can be applied with every expectation of success to the interpretation of the CCV and CVV Auger and CV x-ray spectra of a large number of simple metals and their alloys. This will improve our understanding of electron screening and the hybridisation between d and sp bands in these materials. However we should remember that the development of the empirical approach depended on the results of calculations from first principles and there will be an increasing need for such calculations as experimental developments make it possible to measure the spectral profiles of more complicated and more technologically important materials. It is also important to calculate Auger profiles from first principles in order to establish the significance of many-electron contributions [19,20]. Such processes contribute a final state lifetime broadening to CCV and CVV profiles but the details of this and of other many-electron effects are not well understood. There are reasons [21] for expecting that many electron contributions will have a small effect on the Auger profiles of simple metals and the success [10] in explaining the LVV and KVV spectra of Mg in terms of matrix elements, DOS and screening effects supports this view. However so far there have not been enough theoretical treatments to enable us to establish when many-electron effects must be taken into account and when they can be ignored.

Experimentally the development of electron spectrometers with improved combinations of resolution and sensitivity will make it possible to measure the CVV and CCV profiles of more dilute systems and the spectra of weaker Auger processes in more concentrated systems. The profiles of $KL_{2,3}$-$L^2_{2,3}V$ Auger vacancy satellites arising from the decay of double core-ionised states have already been measured for Mg [22] and Al [23]. These spectra should give insight into the screening configuration and local DOS around double core-ionised sites, $\Delta Z = +2$ impurities in terms of the equivalent cores approach. Although these spectral profiles are complicated by the presence of multiplet structure arising from the term splitting of the double core holes in the initial and final states studies on free atoms give confidence that the energies and intensities of the multiplet structure components can be obtained from atomic structure calculations.

In work reported later in this conference [24] we shall see that Auger profiles serve as a useful test of the accuracy of self consistent calculations of the local DOS around core-ionised sites in simple metals and we shall also hear [25] how the orbital character of the screening charge may be revealed by calculations of its influence on the intensities of various components of an Auger profile.

Finally we suggest that it is important that the advances that have been made in our understanding of the Auger spectra of simple metals and their alloys should now be exploited in the study of the local electronic structures of interfaces and of semiconductors. The Auger spectra of these systems will contain technologically important information on band mixing, on the local DOS and on the local electron configuration and screening of impurities.

Acknowledgements

This work was supported by a twinning contract between the Universities of Liverpool, Groningen and Messina funded by the Committee for the European Development of Science and Technology (CODEST) of the Commission of the European Economic Community.

References

1 P. Weightman: Ref. Prog. Physics. 45 753 1982
2 E.J. McGuire: In Atomic Inner Shell Processes ed. by B. Crasemann, (Academic, New York 1975)
3 M. Cini: Solid State Comm. 20 605 (1976)
M. Cini: Solid State Comm. 24 681 (1977)
4 G.A. Sawatzky: Phys. Rev. Lett. 39 504 (1977)
5 D.R. Jennison: Phys. Rev. Lett. 40 807 (1978)
6 U. Von Barth, G. Grossman: Physica Scripta 28 107 (1983)
U. Von Barth, G. Grossman: Phys. Rev. B 25 5150 (1982)
7 D.E. Ramaker: Phys. Rev. B 25 7341 (1982)
8 D.R. Jennison, H.H. Madden, D.M. Zehner: Phys. Rev. B 21 430 (1980)
9 D.R. Jennison: Phys. Rev. B 18 6865 (1978)
10 M. Davies, D.R. Jennison, P. Weightman: Phys. Rev. B 29 5313 (1984)
11 D.R. Jennison: Phys. Rev. B 18 6996 (1978)
12 R. Lasser, J.C. Fuggle: Phys. Rev. B 22 2637 (1980)
13 P. Weightman: Physica Scripta (In Press)
14 P.H. Hannah, P. Weightman, P.T. Andrews: Phys. Rev. B 31 6238 (1985)
15 P. Weightman, M. Davies, J.E. Inglesfield: Phys. Rev. 34 6843 (1986)
16 M. Davies, P. Weightman: Phys. Rev. B 30 4183 (1984)
17 M. Davies, P. Weightman, D.R. Jennison: Phys. Rev B 29 5318 (1984)
18 J. Kudrnovsky, L. Smrcka, B. Velicky: In Proceedings of the International Symposium on X-ray Spectra and Electronic Structure of Matter, Munchen, (Unpublished 1972)
19 C.O. Almbladh: This Conference
20 J.N. Schulman, J.D. Dow: Int. J. Quantum Chem. 15 437 (1981)
J.N. Schulman, J.D. Dow: Phys. Rev. Lett. 47 371 (1981)
21 D.R. Jennison, F.U. Hillebrecht, J.C. Fuggle: J. Vac. Sci. Technol. A 2 1049 (1983)
22 M. Davies, D.R. Jennison, P. Weightman: J. Phys. C 17 L107 (1984)
23 P.H. Hannah, P. Weightman: J. Phys. C 18 L239 (1985)
24 P.S. Fowels, J.E. Inglesfield, P. Weightman: This Conference
25 A. Laine, G. Cubiotti, P. Weightman: This Conference

Spin Polarized Auger Spectroscopy from Magnetically Ordered Solids

M. Landolt

Laboratorium für Festkörperphysik, ETH, CH-8093 Zürich, Switzerland

1. Introduction

There are two very different aspects under which Auger spectroscopy nowadays is popular and successful: First, as a spectroscopy [1] of 2-hole final states, it reveals effective correlations and exchange potentials in solids or yields information on core-hole screening by band-electrons. Second, as a fingerprint-technique, it is very powerful and widely used for chemical analysis of surfaces and thin films. It is the purpose of this contribution to show that measuring the spin-polarization of Auger electrons from magnetized solids [2] is a quite attractive extension of traditional Auger spectroscopy, and again, I would like to isolate two different aspects: spectroscopy and surface magnetism.

In spectroscopy, the role of the magnetization of the sample primarily is to provide an array of polarized target atoms. The spin-polarization P of the emitted electrons [3], defined as $P = (N\uparrow - N\downarrow)/(N\uparrow + N\downarrow)$ with $N\uparrow(\downarrow)$ being the number of electrons in the beam with magnetic moment parallel (antiparallel) to the quantization axis of the spin analyzer, then is an extra label of the observed 2-hole state, next to energy and momentum. It should allow to better identify various final-state configurations and to further quantify the effects of exchange, correlation, L-S-coupling, or hybridization. This looks very promising; however, one is faced with some inherent problems. First, the technique is limited to magnetically ordered samples, which is, in fact, not too serious a drawback since magnetics usually are very good candidates to study many-electron effects as they, by definition, are highly correlated electron systems where the one-electron picture performs rather poorly. Second, the obtained information is very rich and thus might be too complicated. The understanding of the spectra presented in this contribution is only qualitative, at the best, and the advancement of many-electron theory is urgently awaited. Positively, one could say that a wealth of experimental information now is available where new theories can be tested, and there is good hope that a next step forward can be made, now that open-shell systems are being treated.

For surface magnetism, on the other hand, the spin polarization of Auger electrons carries element specific magnetic information, particularly if valence-band or 4f electrons are involved. The fact that the elemental origin of the Auger electrons can be unambiguously identified is a rather unique feature of a magnetometer. The spin-polarization is used as a fingerprint of the local magnetization like the intensity in traditional Auger spectroscopy for chemical analysis. An important difference, however, is that the spin-polarization, after appropriate background subtraction, does not depend on the intensity of the Auger emission and thus reveals the element specific local magnetization independently of atomic concentration. Indeed, this is very useful for surface magnetism, particularly for the study of composite systems like alloys, heterostructures or adsorbates on magnetic substrates, where most other techniques can not resolve the various magnetic sublattices. The major problem, however, with this magnetometer is the "magnetic calibration" or, with other words, the rather complex relation between the spin-polarization of Auger electrons and the magnetization at the particular atomic site. Generally, the spin-polarization is defined for an ensemble of electrons coming from a macroscopic area and thus is proportional to the long-range magnetic order, i.e. the expectation value of the z-component of the magnetic moment $\langle\mu_z\rangle$. The proportionality constant depends on the nature of the particular 2-hole final state and is, in first approximation, independent of temperature. The present difficulty is that this constant for a given Auger transition might vary with the chemical environment or coordination number of the elemental species. It is at this point that the importance of the spectroscopic understanding becomes evident. At present, only comparison of P-values with calibration values from magnetic standards is used for magnetometry, but a detailed determination of magnetic moments, e.g. on various crystal faces, would require a computation in terms of the electronic structure. I hope that the present contribution stimulates computational physics in Auger spectroscopy, where great progress was made in the recent past.

In the following, after a short description of the experiment, I will give an overview and brief discussion of the spectra of the elemental ferromagnets Fe [4], Ni [4], and Gd [5], and then present two examples of surface magnetic studies, namely Gd on Fe(100) [6] and ultrathin Fe-films on Au(100) [7].

2. Experimental

A description of the experimental arrangement can be found in Ref. 8. The concept of the experiment is quite straightforward: The sample is magnetized to a single-domain state, usually in remanence, with the magnetization lying in the surface plane. The surface then is irradiated with unpolarized electrons (20 - 3000 eV) at grazing incidence. The emitted electrons are collected at normal emission (acceptance angle $\pm$ 2.5^o), passed through an energy analyzer (cylindrical mirror), and fed into a Mott detector for spin analysis. The major difficulty lies in uniformly magnetizing the sample with magnetic fringe-fields smaller than approximately 10-100 mOe. This is achieved with a small horseshoe-electromagnet of which the sample closes the magnetic circuit. Second, the spin-detector must be of extremely high efficiency and stability since some of the higher-energy Auger lines are quite weak. In the present experiment a specially designed Mott-detector with spherical acceleration-field and large-area surface-barrier detectors is used.

Magnetic stray-fields can be elegantly circumvented by going to a finely focused beam of primary electrons, so that the spot of illumination is smaller than the magnetic-domain size. Magnetically ordered but demagnetized samples can then be studied, and the lateral distribution of element specific magnetization is readily available. An instrument of this kind with a spot size of 50 nm, a cylindrical-mirror energy-analyzer, and a copy of the present high-efficiency Mott detector is now being assembled at Stanford University.

A certain amount of ambiguity in the determination of the Auger-electron spin-polarization arises from background subtraction. The true secondary-electrons, which predominantly form the background, are strongly spin-polarized, particularly at low energies. We note that they are well used for surface magnetometry [9] without elemental resolution. The spin-polarization P_{eff} of Auger electrons alone, after background subtraction, then is given by $P_{eff} = (P - P_0)I/(I-I_0) + P_0$, with background-intensity I_0 and background-polarization P_0, respectively. If we remember that the spin polarization is a normalized quantity, $P = (N\uparrow - N\downarrow)/I$, we understand that P_{eff} does not depend on the Auger intensity $I - I_0$ or on the specific atomic concentration.

3. Spectroscopy

3.1. Transition Metals: Fe and Ni

Figures 1 and 2 present spectra [4] of the Fe MMM and LMM Auger lines, respectively. We observe a strong spin-polarization at the various Auger emissions with values in the order of $P_{eff} \approx 40\%$ (not shown for LMM). The Fe $M_{23}M_{45}M_{45}$ valence-band line might be analyzed in terms of a self-convolution of occupied d-bands. The two gain satellites, however, identified as autoionization [10] and shake-up, respectively, clearly demonstrate that a one-electron-description is not quite appropriate.

Among the Fe LMM lines the core-hole only transition $L_3M_{23}M_{23}$ is of particular interest, as it manifests the exchange coupling of the partly-filled inner shells to the net spin-moment of the 3d-electrons [11]. In the frame of a parametrized-exchange calculation, KOTANI and MIZUTA [12] have quantified the exchange-integrals; due to the great simplicity of the model, however, they obtained only qualitative agreement with the observed spectra. The treatment of open-shells systems nowadays is a great step forward and should allow for quantitative interpretation of the data.

In Ni, as presented in Fig. 3 for the LMM lines [4], the situation is markedly different. The screening of a core hole (2p or 3p) is of decisive

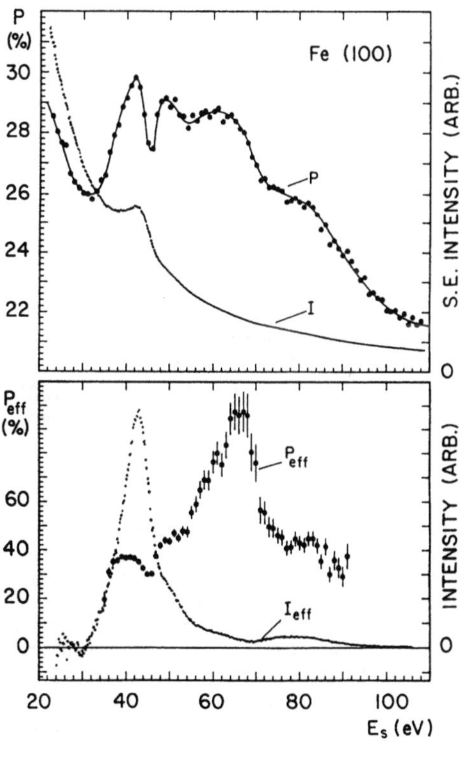

Fig.1: Spin polarization P and intensity I vs kinetic energy of secondary electrons from Fe(100), excited with electrons of 2500 eV. Upper panel: raw data; lower panel: Auger-electron signals after background subtraction.

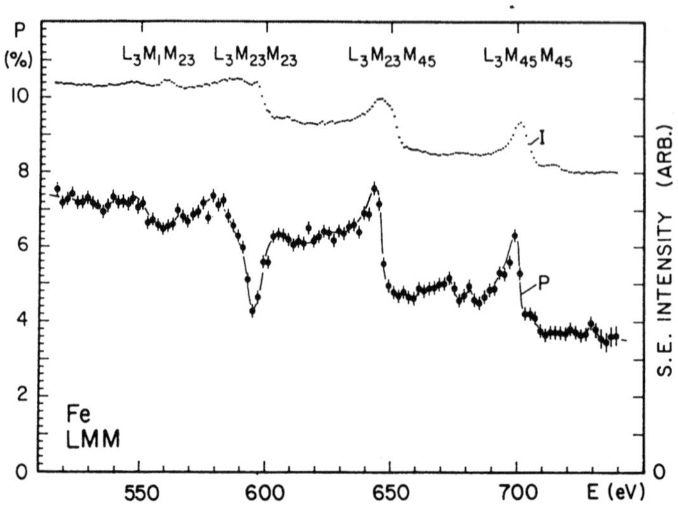

Fig.2: Spin polarization P and intensity I vs kinetic energy of secondary electrons from Fe(100), excited with electrons of 2500 eV.

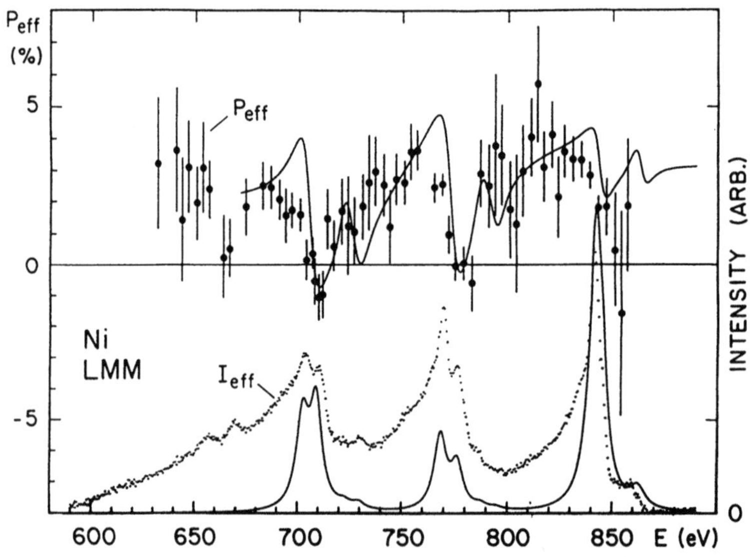

Fig.3: Auger-electron spin-polarization P_{eff} and intensity I_{eff} of Ni(100), after background subtraction. The solid lines through the data points represent the calculation for the resonant processes as described in text.

importance, as it leads to a $3d^{10}$ configuration which is nonmagnetic. Therefore, the observed spin polarization is very weak, $P_{eff} \leq 3\%$, as shown in Fig. 3. It is not completely zero, however, because part of the 2p (or 3p) holes are generated by resonant excitation to the empty 3d-states, giving fully down-spin polarized 2p (or 3p) holes. Using spin-conservation and a decay to an atomic $3d^8$ multiplet, FELDKAMP and DAVIS [13] calculated the spin polarization of all final states: P = +1 for singlets and P = -1/3 for triplets. The solid lines in Fig. 3 are the calculated spectra, with transition probabilities of Cu [14] and energetic positions from 2p, 3p and 3d binding energies of Ni [15]. To fit the data the overall polarization strength was reduced by 1/20, which then directly corresponds to the ratio of resonant to nonresonant core-hole excitation. We note that the same procedure is also successful for the $M_{23}M_{45}M_{45}$ line, where the spin-polarization and hence the resonant part of the excitation is found to be approximately three times larger, well in line with relative ELS-intensities.

In transition metals, the spin-orbit coupling is fairly weak and spin can be considered a good quantum number. Orbital momenta furthermore are quenched in the cristalline electric-field, and a treatment in the frame of pure spin-states is adequate. When going from Fe to Ni we proceed from a band-like to a more atomic-like system, and a theoretical treatment should start from both ends. To go one step further we have considered Gd, with its strongly localized magnetic 4f-shell and relaxed spin conservation because of stronger spin-orbit coupling.

3.2. Rare Earth: Gd

Gd is ferromagnetic below room temperature and has a half-filled, highly localized 4f-shell with $S = 7/2$ in the ground state. Figs. 4 and 5 present spectra of all prominent Auger lines of polycrystalline Gd [5]. As expected from the high 4f spin, we see large polarization effects, particularly with core-hole only transitions involving 4f electrons, and we can identify the spin polarization of resonant-excitation processes because of their large energy separation from the normal decay. Eventually, the comparison between corresponding 3d and 4d initial-hole Auger decays, e.g. $M_5N_{67}N_{67}$ and $N_{45}N_{67}N_{67}$, will indicate the breakdown of the L-S coupling scheme for high-Z materials.

Experimentally we used plycristalline Gd-films, evaporated onto an Fe(100) surface to achieve uniform magnetic remanence via the strong antiferromagnetic Fe-Gd coupling [16,6]. The measurements were carried out at 150 K, i.e. approximately Tc(Gd)/2. The thickness- and temperature-depen-

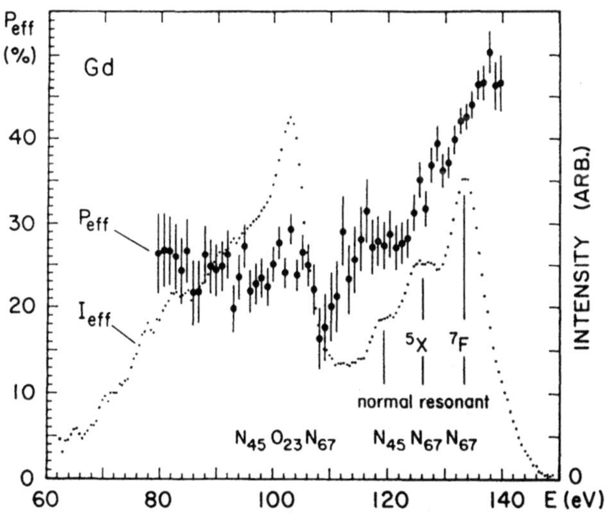

Fig.4: Auger-electron spin-polarization P_{eff} and intensity I_{eff} of polycrystalline Gd, after background subtraction. The Auger electrons are excited with electrons of 2500 eV; T = 150 K.

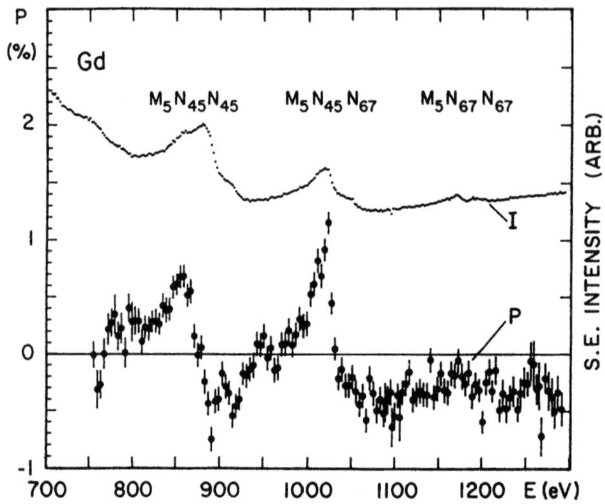

Fig.5: Spin polarization P and intensity I vs kinetic energy of secondary electrons from polycrystalline Gd, excited with electrons of 2900 eV; T = 150 K. The slightly negative background polarization is caused by secondary electrons from the Fe substrate.

dences of P [6] lead to the estimate that the observed polarization is thermally reduced to ≈ 3/5 of the ground-state value, a fact which should be kept in mind if a quantitative analysis is considered. Furthermore we note that the baseline of P in Fig. 5 is at -0.5 % because of secondary electrons coming from the Fe substrate.

The 4d core-hole transitions are presented in Fig.4. Both structures at the high-energy side are attributed [17] to resonant 4d-4f excitation followed by a super Coster-Kronig decay $N_{45}N_{67}N_{67}$. They should, within spin conservation, be fully polarized. The decay to the quintet ^5X is a spin-flip process which qualitatively manifests itself by a reduced polarization. The very intense emission at 103 eV is ascribed to $N_{45}O_{23}N_{67}$ decays. These core-hole-only transitions are well suited to be treated theoretically with molecular orbital or parameterized exchange calculations, similar to the $L_3M_{23}M_{23}$ decays of transition-metals. A puzzle is the weak polarization at 120 eV where the normal $N_{45}N_{67}N_{67}$ emission is expected to lie, with a polarization as high as on the resonant line at 133 eV.

The 3d core-hole transitions are shown in Fig.5. None of the three prominent lines involves valence electrons, and thus the entire spectrum could be analyzed in terms of an atomic model with only one single set of fit parameters. The situation, however, with the 3d core-holes is even more complicated than in the case of the 4d, since the Russel-Saunders approximation is not a good description of $3d^94f^n$ states. The spin-orbit interaction has considerably increased relative to the exchange splitting, and jj-coupling now is more appropriate. This best is illustrated with the $M_5N_{67}N_{67}$ decay, which leads to the same final state as the $N_{45}N_{67}N_{67}$ emission in Fig.4, but has a much weaker spin polarization of $P_{eff}(M_5N_{67}N_{67})$ = 18 %.

The present understanding of the Auger spin polarization in Gd is qualitative, at the best. A more detailed account of the experimental situation can be found in Ref.17. and there is hope that further efforts in theory might be stimulated by that work. For magnetometry, on the other hand, the availability of local magnetization is of great value, and to some degree the method can rely on qualitative information. In some cases we do not need a detailed knowledge of the involved Auger transition to draw direct conclusions about magnetic systems, and this will be illustrated in the next section.

4. Surface Magnetism

4.1. Magnetic Coupling of Adlayers: Gd on Fe(100)

The magnetic coupling between rare-earth (RE) and transition-metals (TM) [16] is of technological importance since some of the next-generation magnetic-recording media rely on RE-TM ferrimagnets. The study of a magnetic adlayer on a magnetic substrate, furthermore, relates to "atomic engineering" with heterostructures. Starting from these motivations the model system Gd on Fe(100) seemed appropriate to apply SPAES on for magnetic characterization.

Figure 6 shows the spin-polarization spectrum of the secondary electrons below 200 eV of approximately one monolayer of Gd on Fe(100). The Fe $M_{23}M_{45}M_{45}$ and Gd $N_{45}O_{23}N_{67}$ and $N_{45}N_{67}N_{67}$ Auger lines, respectively, can easily be identified. From our knowledge of the Auger spin polarization (cf. Sect.3) we readily conclude that the Fe and Gd magnetic moments are aligned in opposite directions. Proper background subtraction yields P_{eff} = -77 % for the resonant Gd $N_{45}N_{67}N_{67}$ transition at a temperature of 150 K.

The temperature dependence of P_{eff} of the resonant Gd $N_{45}N_{67}N_{67}(^7F)$ and Fe $M_{23}M_{45}M_{45}$ lines, respectively, is shown in Fig.7. The data represent Gd and Fe "sublattice" magnetizations at the interface, i.e. the magnetization of the outermost layer, which consists of Gd, and of basically the first Fe layer of the substrate. Fig.7 illustrates how both sublattices can be observed simultaneously by choosing the appropriate Auger kinetic energy.

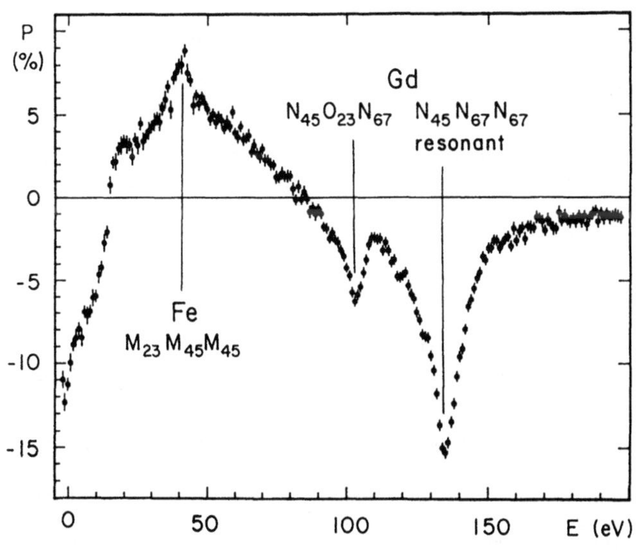

Fig.6: Spin polarization vs kinetic energy of secondary electrons from a Gd film on Fe(100), excited with electrons of 2500 eV; T = 150 K; the film thickness is ≈ 1 monolayer (2.4 Å).

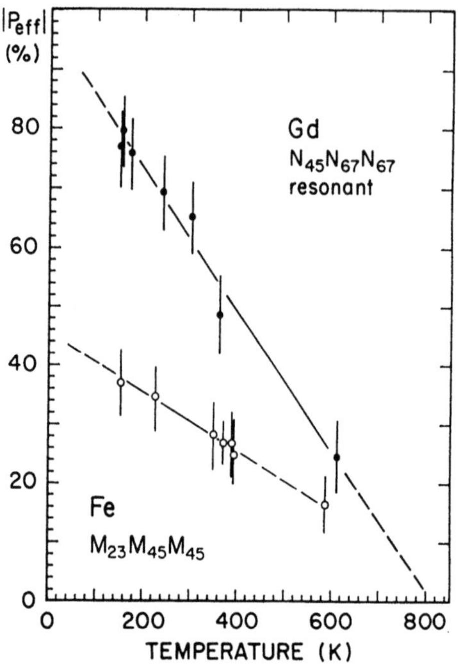

Fig.7: Absolute value of the effective spin polarization of the resonant $N_{45}N_{67}N_{67}$ Gd Auger line at 133 eV (full circles) and of the Fe

From extrapolations to low temperatures we reach the conclusion that the Gd 4f moments in the ground state are *fully* aligned opposite to the Fe magnetization, and, from comparison to clean Fe, that the Fe interface-layer at zero temperature is magnetized to saturation with approximately the same magnetic moment. The temperature dependences reveal that the presence of Gd greatly enhances the spectrum of thermal magnetic excitations at the Fe surface. Qualitatively this is ascribed to the fact that the Gd-Gd exchange interaction is much weaker than both the Gd-Fe and Fe-Fe exchange couplings. The theoretical treatment of finite-temperature magnetizations is a field of current activity, and models beyond mean-field approximation are in the stage of development.

However, it is not the objective of this paragraph to discuss magnetic properties, but rather to illustrate that the spin polarization of Auger electrons, without detailed understanding of the particular Auger process, can provide valuable information for surface magnetism.

4.2. Magnetism in Two Dimensions: Fe on Au(100).

Well ordered, i.e. epitaxial, mono- or multilayers of magnetic atoms on nonmagnetic substrates are playgrounds for computational physics to study magnetic ground-states [19,20]. Furthermore, they are candidates to experimentally investigate phase transitions in reduced dimensionality, and, moreover, the spin-wave spectrum at low temperatures in two dimensions is of considerable interest. Well suited samples for these studies are epitaxially grown Fe films on Au(100), since they exhibit long-range magnetic-order with remanence along [001] in the film plane [21] and grow in a layer-by-layer mode for the first few atomic layers [22]. We have studied the magnetic phase transition [23] and the spin-wave regime [24] by means of Spin Polarized Low Energy Electron Diffraction (SPLEED) and Spin Polarized Secondary Electron Emission (SPSEE). To discuss the magnetic results is beyond the scope of this article; here I wish to emphasize the role of SPAES in magnetic characterization of the samples. The following questions will be addressed: (i) What is the influence of the average coordination number on the saturation magnetization and ordering temperature, and (ii) what is the role of the substrate; does the hybridization provoke an induced magnetization in the substrate ?

Figure 8 presents spin polarization (raw data) and intensities (after backgrond subtraction) for ≈ 1 ML of Fe on Au(100) and clean Au. The polarization data are from one single film alone, recorded in 30 min, and represent the lower limit of useful signal-to-noise ratio. We note that averaging over a few (n) identical samples, which certainly is feasible, would reduce the noise by $\sqrt{n}$. The major ambiguity, however, again arises from the choice of the backgrounds. I therefore would like to describe the routine of background subtraction in some detail as a worst-case example. Fig 8 shows that the Fe $M_{23}M_{45}M_{45}$ and Au $O_{23}O_{45}O_{45}$ Auger lines are almost degenerate. The background intensity is obtained by fitting $I_0(E) = x \cdot E^{\gamma} + z$ [25] to the data away from the line, and P_0 is a straight line between the points where $I-I_0$ vanishes. The polarization of the Au emission in the $N_{67}O_{45}O_{45}$ line is then determined with the standard formula (see Sect.2) and found to be $P_{eff}(\text{Au } N_{67}O_{45}O_{45}) = (0\pm1)\%$. Assuming that the Au $O_{23}O_{45}O_{45}$ emission, having the same final state, is also unpolarized allows then to subtract the fairly large Au part at 40 eV from the Fe $M_{23}M_{45}M_{45}$ line by applying again the standard procedure with $P_{Au} = 0$. With the present statistics, which could still considerably be improved, we reach the error bars depicted in Fig. 9, where P_{eff} values are compiled, together with the corresponding Curie temperatures, for films between 1 and 10 ML thickness.

The thickness dependences shown in Fig.9 are ascribed to the change of coordination number when going to ultrathin films. The reduction of the

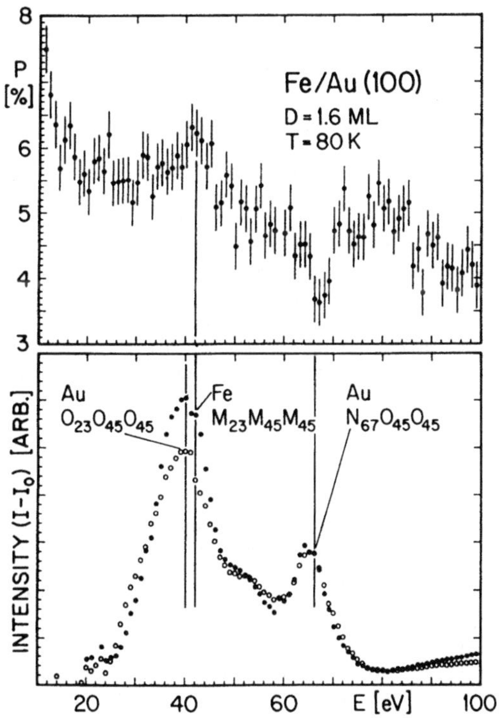

Fig.8: Spin polarization P and intensity ($I-I_0$) vs kinetic energy of secondary electrons from an epitaxial Fe film on Au(100), excited with electrons of 2500 eV. The intensities (lower panel) are from Fe/Au (full circles) and from clean Au (open circles), scaled to match at the Au $N_{67}O_{45}O_{45}$ line.

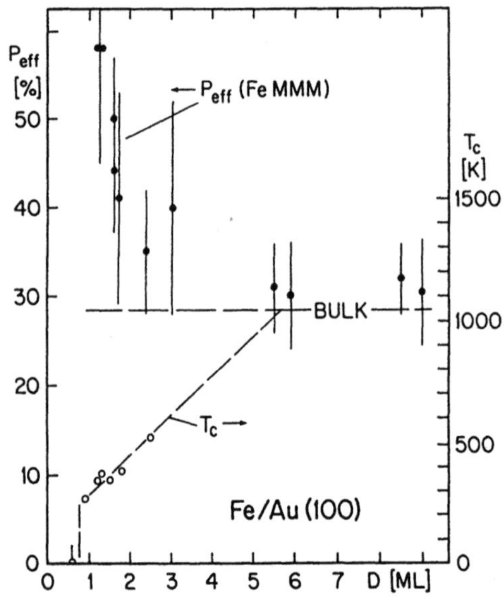

Fig.9: Spin polarization P_{eff} of the Fe $M_{23}M_{45}M_{45}$ Auger emission and magnetic ordering temperature T_C vs film thickness of epitaxial Fe films on Au(100). The P_{eff} values are reduced by a geometry factor of $\sqrt{2}$.

Curie temperature is a known phenomenon. It is ascribed to the decrease of total exchange-energy when reducing the average number of nearest neighbors. A new and striking observation, on the other hand, is the pronounced increase of remanent magnetization $\langle\mu_z\rangle$ when going below 2 ML. This is the first experimental evidence of a strong enhancement of the magnetic moment per atom of a monolayer of Fe deposited on a noble metal. This property has been predicted by densitiy-functional-computations of the magnetic ground-state of a transition-metal monolayer on a noble-metal substrate [19,20]. It is interpreted as a consequence of band-narrowing upon reducing the atomic coordination. We note that the present evidence, though very exciting indeed, is somewhat weakened by the fact that a "magnetic calibration" is missing. From simple fingerprinting we would conclude that the magnetic moment is $\mu = (4.7 \pm 1)\ \mu_B$ at 1 ML, under the assumption that the Fe surface does not exhibit an enhanced moment. A quantitative interpretation ($\mu_{th} \approx 3\ \mu_B$ for comparison [19,20]) clearly requires a calculation of the Auger emission from this particular system, but already a qualitative evidence of a strongly enhanced magnetic moment is encouraging for magnetic ground-state calculations. Finally we note that the presence of the Au substrate might not be too disturbing since the emission from the Au 5d states is found to be unpolarized.

5. Summary

Spin-polarization analysis is an interesting attribute to Auger spectroscopy from magnetically ordered samples with a potential for element-specific magnetic characterization. Two examples, namely Gd/Fe(100) and Fe/Au(100), of magnetic overlayers illustrate what can be expected without detailed understanding of the involved Auger processes. We strongly emphasize, however, that for a full blooming of the magnetic applications a quantitative, i.e. computational, analysis of the particular model-system is a prerequisite. As a starting point of the understanding we therefore have reviewed the spectra of the elemental ferromagnets Fe, Ni, and Gd, and we hope that the actual progress in Auger theory will now be directed towards the field of spin-polarized electrons.

Acknowledgements

It is a pleasure to thank H.-C. Siegmann for many stimulating conversations and permanent support. The work presented in this review was carried out together with D. Mauri, R. Allenspach, M. Taborelli, O. Paul, and K. Brunner, whom I wish to thank for a very fruitful, always exciting, and extremely pleasant collaboration. I would also like to thank the staff of the present 1st Intl. Workshop on Auger Spectroscopy and Electronic Structure for their warm hospitality in Taormina which generated a stimulating scientific athmosphere.

References

1. For a review see: J.C. Fuggle in *Electron Spectroscopy*, Vol.4, ed. C.R. Brundle and A. Baker (Academic, London 1981), p.85.
2. M. Landolt and D. Mauri, Phys. Rev. Lett. **49**, 1783 (1982).
3. For a review see: J. Kessler, *Polarized Electrons*, 2nd ed., Springer Ser. Atoms and Plasmas, Vol.1 (Springer, Berlin 1985).
4. R. Allenspach, D. Mauri, M. Taborelli, and M. Landolt, Phys. Rev **B35**, 4801 (1987).
5. M. Taborelli, R. Allenspach and M. Landolt, Phys. Rev. **B34**,6112 (1986).
6. M. Taborelli, R. Allenspach, G. Boffa, and M. Landolt, Phys. Rev. Lett. **56**, 2869 (1986).
7. O. Paul, M. Taborelli, and M. Landolt, Surf. Sci. in press (1989).
8. D. Mauri, R. Allenspach, and M. Landolt, J. Appl. Phys. **58**, 906 (1985).
9. For a review see: M. Landolt in *Polarized Electrons in Surface Physics*, ed. R. Feder (World Scientific, Singapore 1985) p.385. R. Allenspach, M. Taborelli, M. Landolt, and H.C. Siegmann, Phys. Rev. Lett. **56**, 953 (1985).
10. S.D. Bader, G. Zajak, J. Zak, Phys. Rev. Lett. **50**, 1211 (1983); and *ibid* **50**, 1713 (1983). G. Zajak, S.D. Bader, A.J. Arko, and J. Zak, Phys. Rev **B29**, 5491 (1984).
11. C.S. Fadley and D.A. Shirley, Phys. Rev. **B2**, 1109 (1970).
12. A. Kotani and H. Mizuta, Solid State Commun. **51**, 727 (1984); H. Mizuta and A. Kotani, J.Phys. Soc. Jpn **54**, 4452 (1985).
13. L.A. Feldkamp and L.C. Davis, Phys. Rev. Lett. **43**, 151 (1979).
14. E. Antonides, E.J. Janse, and G.A. Sawatzky, Phys. Rev. **B15**, 1669 (1977).
15 J.C. Fuggle and N. Mårtensson, J Electron Spectrosc. Relat. Phenom. 21, 275 (1980).
16. For a review see: K.H.J. Buschow, Rep. Prog. Phys. **40**, 1179 (1977).
17. The identification of all lines is due to: C. Rivière, F.P. Netzer, G. Rosina, G. Strasser, and J.A.D. Matthew, J. Electron Spectrosc. Relat. Phenom. **36**, 331 (1985).
18. For a review see: H. Capellmann in *Polarized Electrons in Surface Physics*, ed. R. Feder (World Scientific, Singapore 1985) p.67.
19. C.L. Fu, A.J. Freeman, and T. Oguchi, Phys. Rev. Lett. **54**, 2700 (1985).
20. R. Richter, J.G. Gay, and J.R. Smith, Phys. Rev. Lett. **54**, 2704 (1985).
21. S.D. Bader and E.R. Moog, J. Appl. Phys. **61**, 3729 (1987).
22. R. Germar, W. Dürr, J.W. Krewer, D. Pescia, and W. Gudat, Appl.Phys. **A**, in press (1988).
23. W. Dürr, M. Taborelli, O. Paul, R. Germar, D. Pescia, and M. Landolt, Phys. Rev. Lett. in press (1988).
24. M. Taborelli, O. Paul, O. Züger, and M. Landolt, J.de Physique (Paris), Colloques, in press (1988).
25. E.N. Sickafus, Phys. Rev. **B16**, 1436 (1977).

Auger Study of the Electronic Correlations in $YBa_2Cu_3O_{7-\delta}$

A. Balzarotti, M. De Crescenzi, N. Motta, F. Patella, and A. Sgarlata

Dipartimento di Fisica, Università di Roma II,
Via O. Raimondo, I-00173 Rome, Italy

Cu L_3VV and O KLL Auger spectra of the superconducting compound $YBa_2Cu_3O_{7-\delta}$, measured above and below T_c=92K, are reported. The analysis provides the electron correlation energies at both Cu and O sites.

1. INTRODUCTION

Among electronic spectroscopies, the Auger technique is one of most powerful to obtain information on the correlation energy between electrons. In this work we present the Cu L_3VV and O KLL Auger spectra of $YBa_2Cu_3O_{7-\delta}$ (YBCO) superconductor. This compound is of particular interest because its transition temperature T_c occurs around 90K, well above the liquid nitrogen temperature. Increasing experimental evidence both from photoemission [1-6] and Auger [3,7,8] spectroscopies has outlined the relevance of strong electronic correlation effects in this material. In photoemission these effects cause the appearance of final-state satellites in the valence band and shift the portion of the Cu-O band states close to the Fermi level (E_F) toward higher binding energies.

The Cu $L_3M_{4,5}M_{4,5}$ Auger spectrum presented here is dominated by the ^1G term of the $3d^8\underline{L}$ configuration ($\underline{L}$ denotes a hole on the ligand) at about 918 eV kinetic energy (KE). The other multiplet contributions are not resolved as in Cu metal. A satellite structure at 915 eV increases in intensity and shifts slightly with decreasing temperature. A concomitant increase of the satellite-to-main peak ratio with decreasing temperature has been observed in the Cu 2p photoemission [9] but this effect is difficult to be quantified [10]. A similar enhancement of the Cu-derived satellites has been found in the valence band (VB) photoemission spectra [2]. The 915 eV Auger structure is recognized as due to the $3d^8\underline{L}^2$ final-state Auger configuration of Cu^{3+} with two holes on the ligand. This assignment provides an estimation for the correlation energy of the oxygen holes. The O $KL_{2,3}L_{2,3}$ Auger spectrum consists of two components: the largest one, at 514 eV KE, is associated to the $^1D_2$ term of the $2s^22p^4$ configuration of the O^{--}.

2. EXPERIMENTAL

The samples of $YBa_2Cu_3O_{7-\delta}$ were prepared by mixing fine powders of 99.99% pure Y_2O_3, Ba CO_3 and CuO in stoichiometric ratio Y:Ba:Cu=1:2:3 and subsequently pressing them into

Springer Series in Surface Sciences, Vol. 18 Editors: G. Cubiotti · G. Mondio · K. Wandelt
Auger Spectroscopy and Electronic Structure

pellets. They were then sintered at 950 °C for 10 h in flowing pure O_2 atmosphere and then slowly furnace cooled down (~2 °C/min) to prevent loss of oxygen. The four-point method gave a sharp drop of the resistance at 92±2 K. X-ray analysis showed the presence of a single phase with orthorhombic structure. Thermal gravimetric analysis (TGA) yielded an oxygen vacancy concentration δ~0.2.

The measurements were performed in ultrahigh vacuum (UHV) at pressure in the 10^{-10} Torr range and electrons were monitored by a VG CLAM 100 hemispherical analyzer. During the chamber bakeout at 120 °C the samples were kept at room temperature to avoid oxygen losses due to heating. Before each set of measurements the samples were scraped *in situ* to expose fresh superconducting grains of correct stoichiometry.

Auger spectra were taken using a twin anode (Al and Mg) x-ray source with an energy resolution of 1.1 eV. Spectra taken after different scrapes were quite reproducible with negligible degradation of the surface, as confirmed by XPS scans. Unscraped or air-exposed samples have an oxygen concentration different from that of bulk due, probably, to surface contamination. The measurements were made at 300 K and 80 K and at a few intermediate temperatures.

Special care was adopted to minimize Cu reduction which is well-known to occur in copper oxides upon X-ray and/or uv irradiation under vacuum [11]. The presence of Cu in different oxidation states can be unambiguously inferred by an analysis of the Cu-2p photoemission and Cu L_3VV Auger lineshape [11]. To this purpose we have measured CuO pellets in the same experimental conditions as YBCO and verified that substantial reduction effects are absent.

3. RESULTS AND DISCUSSION.

3.1 Cu L_3VV spectrum.

Auger spectra of YBCO taken at T=85, 185 and 245 K are reported in Fig.1 and compared with the room temperature spectrum (solid line). For each temperature the difference spectrum is shown as well. The main structure of the L_3VV Auger spectrum peaks at 918.0±0.2 eV and does not shift with temperature. Its energy position coincides, within the experimental uncertainty, with that of CuO, indicating that copper is predominantly in the Cu^{2+} formal valence state. The spectrum is broadened compared to that of metallic copper and, in this respect, it resembles more the Cu L_3VV spectrum of CuO. In most of our samples a small subsidiary peak at ~ 915 eV KE exists around the peak maximum, in the region where the Koster-Kronig $L_2L_3M_{4,5}$ transition occurs [12]. This structure is much less evident in superconductors than in other Cu compounds, as found by many authors [9,13,14]. Lowering the temperature, this peak grows and slightly shifts to lower KE, as better evidenced in Fig.2. The temperature changes are reproducible, fully reversible and not observed in the CuO spectrum; hence we infer that they are peculiar of the superconducting oxide.

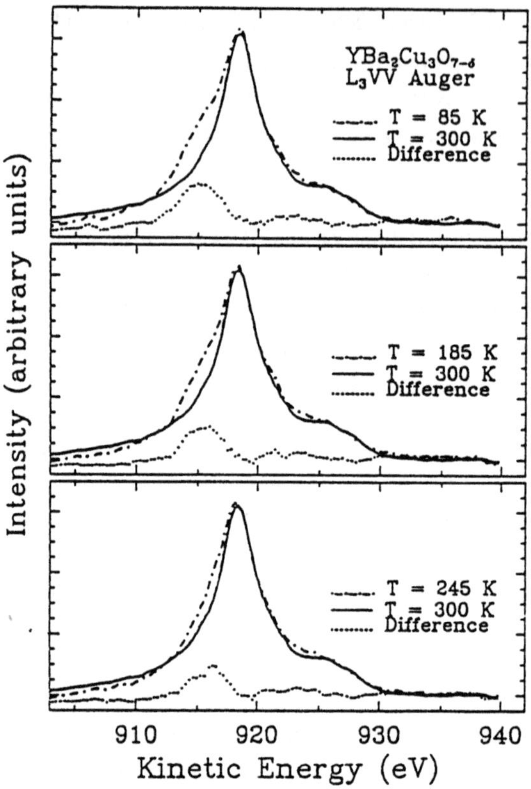

Fig.1: *L_3VV Auger spectrum of Cu in $YBa_2Cu_3O_{7-\delta}$ measured at temperatures 85K, 185K and 245K (dash-dotted curves) compared with the one measured at room temperature (solid curves). For each temperature the difference spectrum is also shown (dotted curve).*

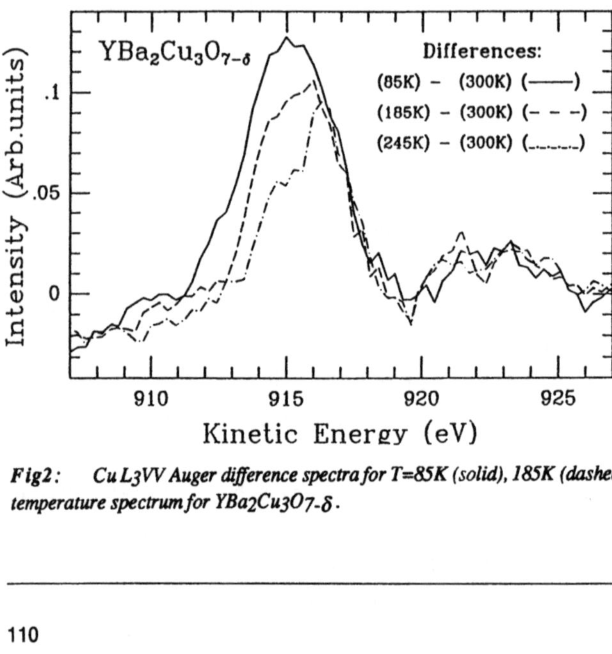

Fig2: *Cu L_3VV Auger difference spectra for T=85K (solid), 185K (dashed) and 245K (dash-dotted) and room temperature spectrum for $YBa_2Cu_3O_{7-\delta}$.*

In a solid, the kinetic energy E_k of an Auger CVV transition is given by:

$$E_k = E(C) - 2E(V) - U, \tag{1}$$

where E(C) and E(V) are the binding energies of the core level and of the centroid of the VB, respectively. The Coulomb term U describes many-body interactions such as hole-hole interaction, relaxation effects, etc. For filled, nondegenerate bands, and neglecting the non local contribution, the two-hole Auger profile is given by [15] :

$$D(E) = \frac{D^o(E)}{(1-F(E)U)^2+\pi^2U^2D^{o2}(E)}\ , \tag{2}$$

where $D^o(E)$ is the selfconvolution of the one-particle density of states and F(E) is the Hilbert transform of $D^o(E)$. By increasing U/W, where W is valence band width, one fulfills the condition leading to quasi-atomic spectra as in the case of filled bands.

The problem of evaluating experimentally the correct (Hubbard) Coulomb interaction U is serious. The major source of uncertainty in Eq.(1) is the centroid energy of the valence band. To make a quantitative comparison, we have modeled the L_3VV Auger profile of YBCO (Fig.3) using two different VB density-of-states (DOS) calculations, namely the projected DOS calculated by Mattheiss and Hamann [16] and the more recent one by Temmermann et al. [17]. The U_{dd} values which fit the position of the main Auger peak of Fig.1 are ~2 eV and ~7 eV, respectively, in the two cases. The large discrepancy between these values is attributable to

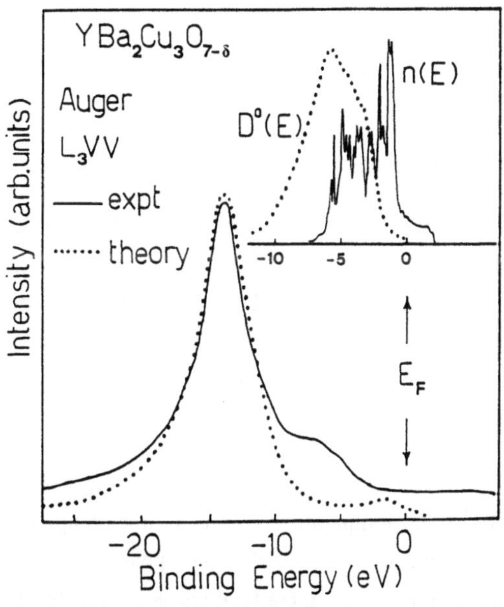

Fig3: Measured and calculated Auger L_3VV profile after background subtraction with U_{dd}=7 eV. The inset shows the theoretical total density of states (after Ref. 17) and its selfconvolution $D^0(E)$. The zero of the binding energy scale is the Fermi energy.

the different position of the centroid of the VB with respect to E_F provided by DOS calculations. In our opinion the largest correlation energy, U_{dd} ~7eV, is more reliable because it is closer to that of other estimates [18] and to the values measured for copper oxides [19].

The high correlation energy suggests that an atomic or molecular picture is adequate to describe the Auger final states. For a Cu-O cluster in YBCO, the ground state, as provided by core-level photoemission studies [10], is a superposition of $3d^9$ and $3d^{10}\underline{L}$ states for Cu^{2+} and $3d^8$, $3d^9\underline{L}$ and $3d^{10}\underline{L}^2$ states for Cu^{3+}. CuO photoemission data [19] and configuration-interaction calculations on the CuO_6^{10-} [1] cluster indicate that the divalent $3d^9\underline{L}$ multiplet has the lowest energy whereas the $3d^8$ state lies about 10 eV above E_F and the $3d^{10}\underline{L}^2$ state with two oxygen ligand holes has negligeable weight. Therefore the Cu^{3+} ground state $3d^9\underline{L}$ is mainly divalent and nearly degenerate with the $3d^9$ and the $3d^{10}\underline{L}$ states. The relevant two-hole final states are therefore $3d^7$, $3d^8\underline{L}$ and $3d^9\underline{L}^2$ for divalent Cu and $3d^8\underline{L}^2$ for trivalent Cu, respectively. The $3d^8\underline{L}$ configuration contributes with the multiplet terms ^1G, ^1D, ^1S, ^3P, and ^3F, having different weights and different U values. The ^1G term is dominant and therefore only $U_{dd}(^1G)$ has been used in the fit. The ^1S singlet component adds intensity at the low KE side of ^1G, but its weight is generally small and cannot explain the feature at 915 eV. The enhanced Auger cross section between 925-930 eV occurs in the region of the $3^9\underline{L}^2$ final state which differs by approximately U_{dd}-Δ (Δ is the energy distance between the unhybridized $3d^9$ and $3d^{10}\underline{L}$ states) from the $3d^8\underline{L}$ state.

The low KE side of ^1G contains the contributions of the $3d^7$ and $3d^8\underline{L}^2$ final states which arise from the $3d^9$ (Cu^{2+}) and $3d^{10}\underline{L}^2$(Cu^{3+}) initial states [10]. In the first case the kinetic energy splitting ΔE_k^{ms} between the main ($3d^8\underline{L}$) and the satellite ($3d^7$) lines is about $2U_{dd}$-Q, where Q is the p-d Coulomb interaction between holes at the Cu site. In the second case ΔE_k^{ms} is given by:

$$\Delta E_k^{ms}=E(\underline{c}3d^{10}\underline{L})- E(\underline{c}3d^{10}\underline{L}^2)+E(3d^8\underline{L}^2)-E(3d^8\underline{L}). \qquad (3)$$

Taking U_{dd}= 6-8 eV [10,18] and Q=8-9 eV [1], we estimate ΔE_k^{ms}>3 eV for the $3d^7$ configuration. On the other hand, from the measured splitting $E(\underline{c}3d^{10}\underline{L})$- $E(\underline{c}3d^{10}\underline{L}^2)$= 2.5-2.8 eV at the $\underline{c}$=Cu site [9], we calculate an energy separation $E(3d^8\underline{L}^2)$-$E(3d^8\underline{L})$ of the order of 4.5-5 eV to obtain the experimental value ΔE_k^{ms}=2.5-3 eV. This energy separation, caused by the interaction of p holes, provides an estimate of the hole correlation energy at the oxygen sites close to that reported previously [3,20,21].

Temperature effects in YBCO photoemission [2,8] and energy loss [22] experiments have been reported so far but the origin of the temperature effects is not completely understood. In the present case, a plausible speculation [7] is that the apparently higher correlation energy reflects a smaller relaxation or screening in the final two-hole state for the Cu atoms in the Cu-O linear chains with respect to those having square planar coordination. In this case one should find a double peak for the L_3VV Auger lineshape whose low-KE component has the larger U_{dd} value. Besides this possibilty, changes of oxygen stoichiometry with temperature could vary the relative intensity of the Auger transitions by modifying the hybridization of the Cu-O initial states.

3.2 O KVV spectrum.

The O $KL_{2,3}L_{2,3}$ spectrum of YBCO (panel (b) in Fig.4) clearly shows the two components forming the broad Auger spectrum. The highest peak at 514 eV KE corresponds to the quasi-atomic $^1D_2$ component of the final state configuration $2s^22p^4$ (two equivalent hole in the L shell) of the O^{2-} ion. The other multiplet $KL_\alpha L_\beta$ components, namely KL_1L_1 ($^1S_0$) ($2s^02p^6$), $KL_1L_{2,3}$ ($^1P_1$,^3P) ($2s^12p^5$), and $KL_{2,3}L_{2,3}$ ($^1S_0$, ^3P) ($2s^22p^4$), lie outside the spectral range examined [23]. The Auger structure at 507 eV KE [Fig. 4 (b)] has a dependence from surface conditions [10] similar to that of the 531 eV transition of the O 1s XPS spectrum [Fig.4 (a)], therefore is likely to be associated to a different oxygen coordination. From the measured binding energy of the O 1s peak and from the energy position of the oxygen VB centroid (~4eV) [17], we estimate a hole-hole correlation energy of ~4-5 eV and ~14 eV for the 514 eV and 507 eV Auger structures, respectively.

The second value suggests intra-atomic screening similar to that found in other metal oxides [24], while the first correlation energy is the effective U_{pp} of the oxygen 2p holes in

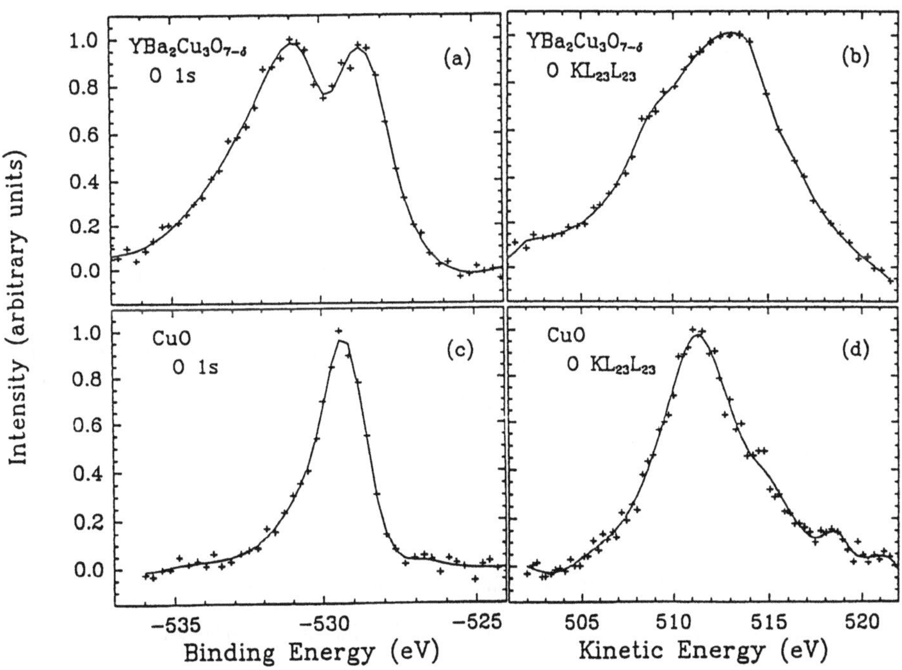

Fig4: *(a) O 1s x-ray photoemission spectrum and (b) O KLL Auger spectrum in $YBa_2Cu_3O_{7-\delta}$; (c) O 1s x-ray photoemission spectrum and (c) O KLL Auger spectrum of CuO. Background has been removed.*

the Cu-O cluster. Similar results have been reported in another study of the O KLL spectrum [3]. The present results are at variance with those reported before [10,25], in which the intensity of the two peaks was inverted. To settle the question, we measured the O 1s spectrum of pure CuO and the results are shown in Fig.4 (panels [b] and [c]). Clearly, the highest Auger peak at 514 eV in YBCO has to be associated to the two-hole states of the Cu^{2+} ion which occurs at 512 eV in CuO. It is therefore likely that the second component of the Auger profile at 507 eV KE originates in part from surface oxides due the partial disproportionation of the YBCO bulk phase, as also evidenced by the strong reduction of the 528.5 eV peak with respect to the 531 eV peak of the O 1s spectrum. Possible candidates are BaO and Y_2O_3 oxides. Clearly more experimental investigations on single crystals are desirable.

The present results emphasize the importance of the electronic interactions in the high-T_c superconducting oxides. The relationship with the mechanism of the superconductivity remains, however, to be clarified.

This work has been supported in part by the Italian CISM-GNSM.

REFERENCES

[1] A.Fujimori,E.Takayama-Muromachi,Y.Uchida, and B.Okai, Phys.Rev.B **35**, 8814 (1987).

[2] P.Thiry,G.Rossi, Y.Petroff,A.Revcolevschi, and J.Jegoudez, Europhys. Lett. **5**, 55 (1988).

[3] D. van der Marel, J. van Elp, G.A.Sawatzky and D. Heitmann, Phys. Rev.B **37**, 5136 (1988).

[4] T.Gourieux, G.Krill, M. Maurer, M.F.Ravet, A.Menny, H.Tolentino, and A.Fontaine, Phys.Rev.B **37**, 7516 (1988).

[5] R.L.Kurtz,R.Stockbauer,D.Mueller,A.Shih,L.Toth,M.Osofsky, and S.A.Wolf Phys.Rev.B **35**,8818 (1987).

[6] P.Steiner, S.Hufner, V.Kinsinger, I.Sander, B.Siegwart, H.Schmitt, R.Schulz, S.Junk, G.Schwitzgebel, A.Gold, C.Politis, H.P.Muller, R.Hoppe, S.Kemmler-Sack, and C.Kunz, Z.Physik B **69**, 449 (1988).

[7] A.Balzarotti, M. De Crescenzi, C.Giovannella, R.Messi, N.Motta, F.Patella and A.Sgarlata, Phys.Rev.B **36**, 8285 (1987).

[8] D.D.Sarma, K.Sreedhar, P.Ganguly, and C.N.R. Rao, Phys.Rev.B **36**, 2371 (1987).

[9] D.D.Sarma and C.N.R. Rao, Solid State Commun. **65**, 47 (1988).

[10] A.Balzarotti, M. De Crescenzi, N.Motta, F.Patella and A.Sgarlata,Phys.Rev. **B38**, xxx, (1988).

[11] T.H.Fleish and G.J.Mains, Appl.Surf.Science **10**, 51 (1982).

[12] G. van der Laan, C.Westra, C.Haas, and G.A.Sawatzky, Phys.Rev.B **23**, 4369 (1981).

[13] B.Heinrich, K.Myrtle, N.Alberding, W.B.Xing, O.Rajora, K.B.Urquhart, and S.Gygax, Proc. 34thAVS-AIP Conf. (1987), to be published.

[14] D.D.Sarma and C.N.R.Rao, J.Phys.C-Solid State Phys. **20** 9 (1987).

[15] M.Cini, Solid State Commun **24**, 681 (1977); G.A.Sawatzky, Phys.Rev.Lett. **39**, 504 (1977).

[16] L.F.Mattheiss and D.R.Hamann Solid State Comm. **63**, 395 (1987).

[17] W.Temmerman, Z.Szotek, P.J.Durham, J.M.Stocks, P.A.Sterne, J. Phys. F: Metal. Phys. **17**,L319 (1987).

[18] J.C.Fuggle, J.Fink and N.Nucker, Proc. of the Int. Adriatico Res. Conf., Trieste July 1988, ed. S.Lundqvist, E.Tosatti, M.P.Tosi, and Yu Lu (World Sci. Singapore), to be published and Int. J. Mod. Phys.B, to be published.

[19] M.R.Thuler, R.L.Benbow, and Z.Hurych, Phys.Rev. B **26**, 669 (1982).

[20] G.Wendin, J. de Physique Colloq. **C9**, xxx, (1987).

[21] H.Chen, J.Callaway and P.K.Misra, Phys.Rev.B **38**,195 (1988).

[22] A.Balzarotti, M. De Crescenzi, N.Motta, F.Patella and A.Sgarlata, Solid State Commun. (to be published).

[23] P.Legare, G.Maire, B.Carriere, J.P.Deville, Surface Sci. **68**, 348 (1977).

[24] P.Humbert and J.P.Deville, J. Electron Spectrosc. Relat. Phenom. **36**, 131 (1985).

[25] A.Balzarotti, M. De Crescenzi, N.Motta, F.Patella and A.Sgarlata, Proc. of the Int. Adriatico Res. Conf., Trieste July 1988, ed. S.Lundqvist, E.Tosatti, M.P.Tosi, and Yu Lu (World Sci. Singapore), to be published and Int. J. Mod. Phys.B (to be published).

Valence Electron States in Gd-Silicides via Si-$L_{2;3}VV$ Auger Lineshape Analysis

M. Sancrotti[1], *E. Vescovo*[1], *L. Calliari*[2], *and F. Marchetti*[2]

[1]Istituto di Fisica, Politecnico di Milano,
Piazza Leonardo da Vinci 32, I-20133 Milan, Italy
[2]Istituto per la Ricerca Scientifica e Tecnologica, I-38050 Povo, Italy

We report the Si-$L_{2,3}VV$ lineshapes from the whole series of Gd-silicides (Gd_3Si_5; GdSi; Gd_5Si_3) and compare them with the corresponding spectra of Ca-silicides ($CaSi_2$; CaSi; Ca_2Si). The evolution of the Si-$L_{2,3}VV$ lineshape upon increasing the metal content is strictly similar for both the series of silicides. Thus analogies in the distribution of valence electron-states of the two families of silicides are put forward. On the basis of the already assessed experimental and theoretical knowledge of valence electron-states in Ca-silicides we suggest the formation of Si-3p/Gd-5d hybrids as responsible of the chemical bond in Gd-silicides along with the dehybridization of the Si-3s derived states which tend toward an atomic-like configuration in the Gd-rich compound.

1. Introduction

In the last decade there have been many experimental and theoretical contributions addressing the issues of the valence electron-states and chemical bond in Transition-Metal (TM) silicides [1-3]. The relevance of these problems is clearly related to the possibility of modelling out properties of the reactive Si/TM interfaces via the well characterized bulk TM-silicides. Semiconductor/metal interfaces are in turn extremely important for scientific and technological reasons (e.g.ohmic contacts, Schottky barriers). In this context Si/Rare-Earth (RE) interfaces have more recently attracted increasing attention because of their peculiar properties such as high thermal stability and low Schottky barrier values. Up to now spectroscopic and structural investigations have focused onto the growth mechanisms of some Si/RE interfaces whereas the issue of the distribution of the valence electron-states at the Si/RE intermixed interfaces and in RE-silicides can be considered almost completely uncovered [2-3]. The main reason of why the state of the art is so limited is related to the presence of the highly localized and atomic-like RE-4f states in the energy range where the valence electron-states partecipating to the chemical bond fall. Actually the high values exhibited by the 4f-states photoionization cross-sections make the 4f-derived features to dominate the valence states in photoemission spectra over a very wide photon energy range. This limitation is obviously more and more serious while the 4f-occupation number is increased. One way for overcoming this difficulty can be looked for in Si-$L_{2,3}VV$ Auger Lineshape Spectroscopy. Actually the Si-$L_{2,3}VV$ transition is well known to be band-like and intraatomic ,i.e. it reflects the distribution of valence electron states projected onto the Si-sites

Springer Series in Surface Sciences, Vol. 18 Auger Spectroscopy and Electronic Structure
Editors: G. Cubiotti · G. Mondio · K. Wandelt

[4-5]. Thus no disturbing 4d-derived contributions appear in the measured Si-$L_{2,3}VV$ Auger spectra. This possibility has been recently fruitfully exploited for the Si/Yb intermixed interface [6]. Actually a strong coupling between the Yb-5d derived states and Si-3p ones has been suggested as responsible of the chemical bond. These conclusions were drawn via a joint analysis of the Si-$L_{2,3}VV$ lineshape from the Si/Yb interface and the series of Ca-silicides, combined with well assessed theoretical (LMTO calculations) and experimental (synchrotron radiation photoemission) findings on the valence electron-states in Ca-silicides.

In this communication we report original Si-$L_{2,3}VV$ Auger lineshape results from the whole series of Gd-silicides. The comparison with the corresponding spectra of Ca-silicides reveals an extremely close similarity in the evolution of the Si-$L_{2,3}VV$ lineshape upon increasing the metal content for both the series of silicides. Thus hybrids between the Si-3p-derived states and the Gd-5d states are suggested to be responsible of the chemical bond in Gd-silicides.

2. Experimental

Polycrystalline Ca-silicides and Gd-silicides were prepared by melting stoichiometric amounts of the high purity atomic species in inert gas atmosphere. The structure was checked via X-ray diffraction. Extra phases resulted to be confined below 3%. Clean surfaces were prepared by scraping the samples with a diamond file (base pressure $7*10^{-11}$torr). For the Si(111)/Yb interface experiments, Si(111) surfaces were prepared by cleavage and high purity Yb was evaporated from a tungsten helix. The Yb coverage was checked with a quartz microbalance. The Auger spectra were excited with an electron-gun (primary energy=1.5KeV) and measured via a Cylindrical Mirror Analyzer operated in the derivative mode (V_{pp}=2V). All measurements were done in Ultra High Vacuum ($p=1*10^{-10}$torr). Contaminants were estimated to be below 2% at the end of the Auger spectra acquisition.

3. Results and discussion

In Fig 1 the raw Si-$L_{2,3}VV$ lineshapes for the whole series of Ca- and Gd-silicides are collected. An horizontal background line has been assumed for each spectrum and referred to as zero of the vertical scale.

Fig 2 shows the spectra derived upon integrating those of Fig 1. Also indicated are the spectroscopic assignements (the character of the two holes created via Auger emission) relative to the main features of the Si-$L_{2,3}VV$ lineshape as expected from the self-folded weighted Density Of States (DOS) which will be discussed further below.

Very strong similarities are clearly found out when comparing the evolution of the Si-$L_{2,3}VV$ lineshape of Fig 1 and Fig 2 for increasing metal content for both the silicide series. More precisely the detailed lineshapes appears extremely close each other for each couple of silicides of corresponding stoichiometries ($CaSi_2$ vs Gd_3Si_5; CaSi vs GdSi; Ca_2Si vs Gd_5Si_3). The general trend can be described in the following way.

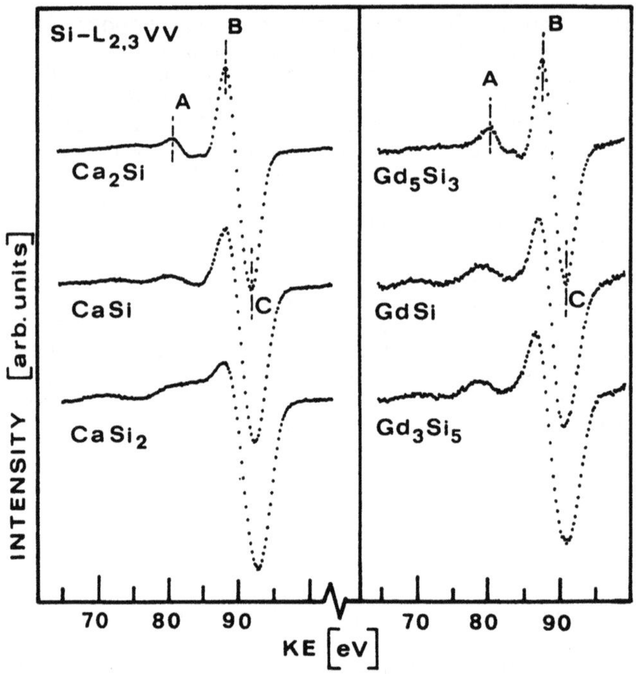

Fig. 1 - Si-$L_{2,3}VV$ lineshape acquired in the derivative mode from the wohle series of Ca-silicides (left panel) and Gd-silicides (right panel)

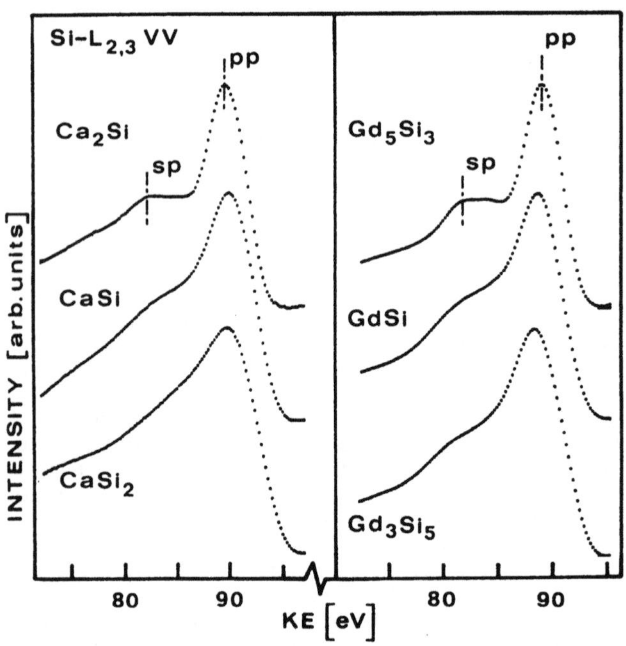

Fig. 2 - Si-$L_{2,3}VV$ lineshape (integrated form) from the whole series of Ca-silicides (left panel) and Gd-silicides (right panel)

i) The positive over negative amplitude ratio of the pp main feature of Fig 1 tends toward 1 upon increasing the metal content. This fact is reflected in an evolution of the pp structure of Fig 2 toward a more symmetric lineshape.

ii) The energy range defined by the markers B and C in Fig 1 progressively decreases while going toward the metal rich silicides. The corresponding counterpart of Fig 2 is a decreasing width of the pp feature.

iii) An increasingly better pronounced feature (A) emerges in the lower energy range corresponding to the sp final state emission.

All these properties have been already discussed for the set of Ca-silicides and successfully interpreted in terms of their valence electron-states calculated within the LMTO method in the Atomic Scheme Approximation (ASA) [7].

The starting point is to recognize that the Si-$L_{2,3}VV$ lineshape exhibits band-like nature in elemental silicon and in a wide set of metal silicides, i.e.it is approximately related to the self-folded DOS projected onto the Si-sites and weighted for the different partial DOS's in order to mimic matrix-element effects, these latter being strongly dependent on the symmetry of the final-state hole [4-5; 8-14].

This property has been fruitfully exploited in order to get an insight into the valence electron-states and chemical bond of many TM-silicides. The corresponding results did consistently agree with experimental findings derived via other experimental approaches, such as Cooper minimum photoemission and resonant photoemission, and theoretical calculations [15-16]. The chemical bond was recognised to be mainly related to the formation of bonding and antibonding hybrids between the Si-3p and the TM-nd derived states straddling the so called nobonding TM-nd states. In parallel the Si-3s states were found to progressively decouple from the other states moving toward higher binding energy values upon incresing the metal content [1-3].

This electron-states scheme cannot be given a priori for granted in the case of Alkaline-Earth (AE)- and RE-silicides because the occupation of the nd-symmetry states in the free-AE- and RE-atoms is nearly zero and because the electronegativity differences between silicon and metal atoms are remarkably higher than for TM-silicides. Nevertheless the above mentioned systematic investigation on Ca-silicides has revealed that Si-3p and Ca-3d coupling is at least partly responsible of the chemical bond whereas the Si-3s states tend to increasingly dehybridize up to the extreme case of Ca_2Si where they can be envisioned as atomic-like [6; 17-20].

On the basis of the strong analogies recognisable upon comparing the Si-$L_{2,3}VV$ lineshapes of Ca-silicides and Gd-silicides, we induce that the general trend in the chemical bond and valence electron-states distribution must be substantially similar for the two sets of silicides. This means that the evolution of the pp-feature in the Si-$L_{2,3}VV$ lineshapes of Fig's 1 and 2 while increasing the Gd-content basically reflects the tendency of the occupied Si-3p states to move linked to the Gd-5d states and to disperse on an energy range whose width continously decreases moving toward the Gd-rich compound. The complementary description is that the Gd-5d states result to be progressively pulled down below the Fermi level along with the increasing mixing with the Si-3p states while one goes toward the Si-rich silicide.

It should be noticed that similar conclusions were drawn in the case of Si/Yb intermixed systems [6].

In strict analogy to what assessed for the Ca-silicides, the evolution of the sp-feature of the Gd-silicides is interpreted in terms of the progressive decoupling of the Si-3s states with respect to the states of different character while moving toward the Gd-rich compound.

In this connection the extreme limit of metal-rich compounds deserve a particular discussion. Actually it was found that the dehybridization tendency leads the Si-3s states toward an atomic-like configuration in Ca_2Si. This atomic-like character determines in turn a strong hole-hole repulsive interaction in the final state of the sp-Auger emission, i.e. it is responsible of the breakdown of the band-like picture at the Si-$L_{2,3}VV$ lineshape of Ca_2Si. A stimulating question is whether such a correlation effect occurs in other systems or it is peculiar of Ca_2Si. Actually we found essentially the same Si-$L_{2,3}VV$ lineshape in Gd_5Si_3 (Fig's 1 and 2) and from the Si(111)+12A Yb interface [6] which was estimated to exhibit a mean Yb_2Si stoichiometry via photoemission. These results would suggest that many-body effects at the sp-emission are a property common to different metal-rich RE-silicides. Actually an independent measurement of the energy location of the Si-3s states via photoemission (as done in the case of Ca_2Si) would be desirable in order to draw unambigous conclusions with regard to this point.

4. Conclusions

We have reported the Si-$L_{2,3}VV$ lineshapes from the whole series of Gd-silicides. The evolution of the Si-$L_{2,3}VV$ lineshape for increasing Gd-content is found to be strongly similar to that corresponding to the set of Ca-silicides. Thus, based on the well assessed information on the valence electron-states of Ca-silicides, we interpret the chemical bond of Gd-silicides as basically due to the formation of Si-3p/Gd-5d hybrids. Complementarily, the Si-3s-derived states are suggested do decouple from the states of different symmetry and to be pushed down at higher binding energy values upon increasing the Gd-atomic concentration.

Acknowledgements - We are greatly indebted with O.Bisi for very stimulating discussions. This work was partially supported by CISM and GNSM.

References

1. C.Calandra, O.Bisi and G.Ottaviani: Surf. Sci. Rep. 4, 271 (1985)
2. Giorgio Rossi: Surf. Sci. Rep. 7, 1 (1987)
3. L.Braicovich: In Surface Properties of Electronic Materials ed. by D.A.King and D.P.Woodruff, The Chemical Physics of Solid Surfaces and Heterogeneous Catalysis, Vol.5 (Elsevier, Amsterdam 1988)
4. D.R.Jennison: Phys. Rev. Lett. 40, 807 (1978)
5. P.J.Feibelman and E.J.Mc Guire: Phys.Rev. B17, 690 (1978)
6. M.Sancrotti, I.Abbati, A.Rizzi, L.Calliari, F.Marchetti and O.Bisi: Surf. Sci. 189/190, 300 (1987); M.Sancrotti, A.Rizzi and F.Marchetti: Phys. Rev. B37, 3120 (1988)

7. M.Sancrotti, I.Abbati, L.Calliari, F.Marchetti, O.Bisi, A. Iandelli, G.L.Olcese and A.Palenzona: Phys. Rev. 37, 4805 (1988)
8. J.A.Roth and C.R.Crowell: J. Vac. Sci. Technol. 15, 1317 (1978)
9. P.S.Ho, G.W.Rubloff,J.E.Lewis,V.L.Moruzzi and A.R.Wil liams: Phys. Rev. B22, 4784 (1980)
10. S.D.Bader, L.Richter, M.B.Brodsky, W.E.Brower and G.V.Smith: Solid State Commun. 37, 729 (1981)
11. U.del Pennino, P.Sassaroli, S.Valeri, C.M.Bertoni, O.Bisi and C.Calandra: J.Phys. C16, 6309 (1983)
12. J.Zak and S.D.Bader: Phys. Rev. B27, 6649 (1983)
13. G.Rossi and I.Lindau: Phys. Rev. B28, 3597 (1983)
14. L.Calliari, M.Sancrotti and L.Braicovich: Phys. Rev. B30, 4885 (1984)
15. G.Rossi, I.Abbati, L.Braicovich, I.Lindau and W.E.Spicer: Solid State Commun. 39, 195 (1981)
16. J.H.weaver, A.Franciosi and V.L.Moruzzi: Phys.Rev. B29, 3293 (1984)
17. D.D.Sarma, W.Speier, L.Kumar, C.Carbone, A.Spinsanti, O.Bisi, A.Iandelli, G.L.Olcese and A.Palenzona: Z.Phys. 71 69 (1988)
18. O.Bisi, L.Braicovich, I.Lindau, A.Iandelli, G.L.Olcese and A.Palenzona: to be published
19. C.Chemelli, L.Braicovich, F.Ciccacci and M.Sancrotti: to be published
20. O.Bisi: In these Proceedings

Correlation Effects in Photoemission and Auger Spectra of Palladium

M. Cini [1] *and C. Verdozzi* [2]

[1] Dipartimento di Fisica, Università di Roma II, Via O. Raimondo, I-00173 Rome, Italy
[2] Istituto Metodologie Avanzate Inorganiche, C.N.R., I-00016 Monterotondo Scalo, Rome, Italy

Abstract. *We extend the graph-theoretical expansions of the Self-Energy in the second -order Approximation and of the two-holes Green's function in the Bare Ladder Approximation to degenerate bands in the intermediate coupling scheme, including the Spin-Orbit Interaction. We apply the formalism to Pd, and obtain good agreement with experiments .*

1 Introduction

In order to understand the spectra of Pd we need to extend the previous theoretical treatments in several ways, because the Pd d band is degenerate and partially filled, and further it displays important spin-orbit effects. Orbital degeneracy has already been dealt with by one of us [1], but the treatment was restricted to filled bands and included no relativistic corrections. The open-bands problem has been discussed by several Authors [2-7] with special emphasis on the case when the hole density n_h is small. It is obvious that the closed-band limit, which is solved exactly [1,8,9], should be a good starting point for the general problem. For small n_h, the photoemission spectrum is well represented [2,10] by GALITZKII's Low Density Approximation [11](LDA). Although this point has been questioned, we believe that the results of our cluster calculations [6], where exact solutions are obtained, confirm the validity of LDA in a physically interesting range of parameters. The cruder second-order approximation to the self-energy, which has been widely used by TREGLIA and coworkers [3] was found to be much more limited in scope, but still useful when interaction effects are weak. Therefore, we use a second-order theory for Pd in the present work. However, previous extensions to degenerate bands were based on approximations or model simplifications that we want to avoid here. Concerning the Auger line shape, we found [6] that it can be well described by the ladder approximation provided we use bare one-hole propagators (Bare Ladder Approximation or BLA). Casting the BLA in the matrix form of [1] enables one to describe multiplet effects properly. However, we still need to include the Spin-Orbit Interaction, that was neglected in all previous treatments. In Pd, n_h is fairly small and interaction effects, while important, are not too severe. Therefore we consider it a good testing

Springer Series in Surface Sciences, Vol. 18 **Auger Spectroscopy and Electronic Structure**
Editors: G. Cubiotti · G. Mondio · K. Wandelt

ground for these new developments, and a prototype of a variety of systems, like transition metals and alloys, where similar problems arise.

2 Generalized Theory including the Spin-Orbit Coupling

We extend the graph-theoretical expansions of the LDA and BLA theories to degenerate orbitals and bands. Working in the intermediate coupling scheme, we are able to include the effects of spin-orbit coupling without having to resort to further simplifying assumptions. For small n_h, the Auger line shape theory can be formulated with an Anderson model Hamiltonian, where interactions are restricted to a single site in the solid (say, site 0), where the Auger decay occurs. For the present purposes, we write the line shape in the form

$$A(\omega) = \frac{1}{\pi} \; \mathrm{Re} \sum_J \sum_{\lambda_J} I(\lambda_J)\, \Phi(\lambda_J,\lambda_J,\omega) \quad . \tag{1}$$

Here, the first summation is over total angular momenta J of two holes, while the second is over intermediate coupling eigenvectors λ_J; the spectral intensities $I(\lambda_J)$ are those obtained by atomic calculations, and the ϕ matrix is the two-holes Green's function. The bare one-hole propagator S^0_j depends on the hole angular momentum j and its convolutions $\phi^0_{ij} = S^0_i \otimes S^0_j$ yield the noninteracting ϕ^0 matrix in the jj scheme. The Coulomb interaction is then included in ϕ using essentially the same formalism as in [1], although the details will be given elsewhere [12]; we transform to the intermediate coupling scheme by a unitary transformation. The self-energy Σ is calculated with the same degenerate Anderson model as the Auger line shape, and is expressed in jj coupling. So, Σ is labelled by the hole angular momentum J_0 and is independent of J_{0z} by symmetry.The first-order contribution must be omitted, since we have to compensate for the average level shift of site 0 due to the interaction (see [2]). In the present work, we gain a first insight on the combined effects of correlations and the Spin-Orbit coupling by studying the second order contributions. There are a direct and an exchange term and the self-energy turns out to be [12]

$$\Sigma^{(2)}_{J_0} = \sum_{J,J_1} \frac{(2J+1)(1+\delta_{J_0,J_1})}{(2J_0+1)} \sum_{(J_2J_3)} X(J_1J_2J_3) \Big|\sum_{LS} U(LS) \langle J_1 J_0 J | LSJ\rangle \langle LSJ | J_2 J_3 J\rangle\Big|^2 \quad (2),$$

where X also depends on ω and is given by

$$X(J_1J_2J_3) = \int_{-\infty}^{\infty} S^0_{J_1}(\omega')\, \phi^0_{J_2J_3}(\omega+\omega') \frac{d\omega'}{2\pi} \quad . \tag{3}$$

In (2), the sum over J_2 and J_3 is constrained in such a way that the two indices occur in the same arbitrary, but fixed order as used to define the two-holes states as Slater determinants. The index J_1 is coupled in a similar way to the angular momentum index J_0 of the hole. Since S^0 depends on its J index because of the Spin-Orbit interaction, Σ also contains such effects. Each S^0_J is obtained from a "bare" one-hole density of states $N^0_J(\omega)$, which is normalized such that

$$\int_B^T |N^0_J(\omega)| d\omega = 1, \quad J = 3/2, 5/2, \tag{4}$$

where B and T are the bottom and top of the band, respectively; integrating between the Fermi level and T yields n_h. The "bare" N^0 actually comes from selfconsistent calculations and contains interaction effects in so far as they are contained e.g. in calculations at the Hartree-Fock or even Local-Density level. The self-energy correction Σ we are considering here is due to the local hole-hole scattering correlations of the ladder type which are not included in the standard treatments. Thus, there is no risk of counting the interactions twice.

3 Application to Pd Metal

We apply the above formalism to $M_{4,5}N_{4,5}N_{4,5}$ Auger transitions and $N_{4,5}$ photoemission from Pd metal. N^0 is taken from the work of CHRISTENSEN [13], who used a selfconsistent relativistic generalization of the linear muffin-tin orbital method. His results for N^0 imply n_h = 0.0475 for each (J_0, J_{0z}) state, and show an evident dependence of N^0_J on J. The Slater integrals , the lifetime broadening (Lorentzian with FWHM=1.0 eV) and the instrumental broadening (Gaussian with FWHM =0.5eV) were taken from WEIGHTMAN ET AL. [14]. The atomic value of the Spin-Orbit parameter ζ_{4d} =0.18 eV was also taken from [14] and is in good agreement with the mean-across-the-band value found by CHRISTENSEN [13]. For the intensity ratio we assumed $I(M_5N_{45}N_{45})/I(M_4N_{45}N_{45})$ = 1.0:0.3, which leads to a good agreement with experiments [15] in the atomic limit. The theoretical Auger spectrum was obtained by composing the M_4 and M_5 contributions separated by the 5.3 eV splitting of the primary hole states. Finally, we added a theoretical background obtained by convolving the theoretical spectrum with the inelastic part of the experimental energy-loss spectrum [16]. The Slater integral F^0 was our only adjustable parameter, and the same value was used for Auger and photoemission spectra. In Fig.1 we present the results obtained with F^0=2.63 eV. The lifetime broadening was reduced to FWHM=0.9 eV, which is still well justified by the arguments of Ref.[14].It may be seen that the "fingerprint" is well reproduced. Some disagreement on the low kinetic energy side is probably due to the difficulty of properly including all loss processes and to the neglect of the asymmetric shape of the core-hole (see the discussion of this point in [2]). A similar agreement for Pd was obtained more simply [14] by regarding the contributions of the various Spin-Orbit split components as independent and just superimposing them. This simplified treatment of multiplets is justified for highly symmetric environments [8,4] in the absence of Spin-Orbit effects, and the present results validate it for Pd; however, it might not be adequate for metals with higher ζ values,in which cases the present detailed theory should be necessary.

In the photoemission case we used strictly the same parameters as above, although we did not attempt to add any loss contributions. Theory and experiment are compared in Fig.2; the "bare" results are also shown.We remark that interaction effects are important, although not so severe as to seriously impair the use of the second-order self-energy. The interacting theory

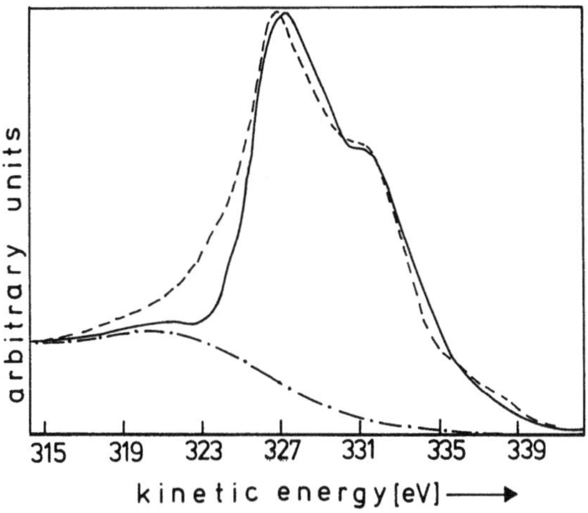

Fig.1 Comparison of the calculated Auger line shape (solid line) with the experimental spectrum of Ref.[14] The theoretical background is also shown as the chain line

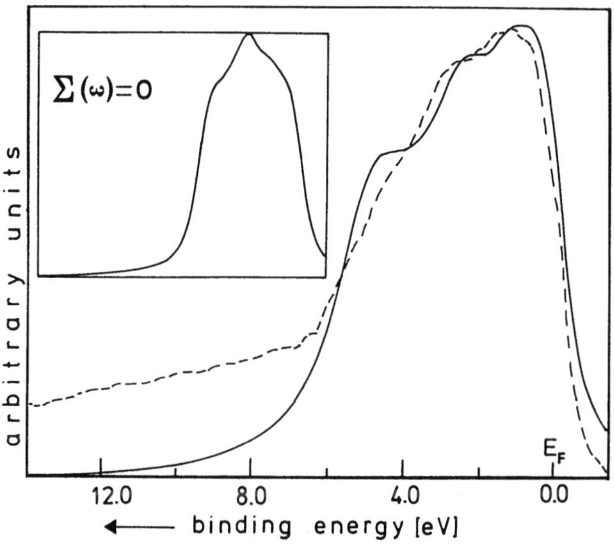

Fig.2. The experimental photoemission spectrum of Ref. [17] (dotted) compared to present theory (solid). The non-interacting line shape is shown in the inset

reproduces quite well the skewness of the line and all the main experimental features, which instead are missed completely by the "bare" one. The intensity of the shoulders also compares well, except that the -5 eV feature is somewhat exaggerated. We regard these preliminary results as quite encouraging, while the inclusion of the LDA corrections to infinite order and of an appropriate background are deferred to a forthcoming paper [12].

4 Conclusions

By the single adjustable parameter F^0 we have succeeded in giving a good interpretation of both photoemission and Auger line shapes of Pd. This confirms the usefulness of the generalized theory presented here, and also supports the ability of the BLA to describe the spectra from partially filled bands. We expect that the present theoretical scheme should be equally successful with a variety of transition metals and alloys, which may constitute more stringent tests for the theory than Pd because of stronger interaction effects and/or higher ζ.

REFERENCES.

1 M.Cini, Phys.Rev. B17,2788 (1978)

2 M. Cini, Surface Sci. 87, 483 (1979)

3 G.Treglia, M.C.Desionqueres F.Ducastelle and D.Spanjard, J.Phys.C14, 4347 (1981)

4 A.Liebsch, Phys. Rev. Lett. 43, 1431 (1979); Phys. Rev. B23, 5203 (1981)

5 V.Drchal and J.Kudrnowsky, J. Phys. F14, 2443 (1984)

6 C.Verdozzi, Thesis, Universita' degli studi di Roma 'La Sapienza', A.A. 1984-85; M.Cini and C.Verdozzi, Solid State Commun.57 , 657 (1986) and Nuovo Cimento 9D, 1 (1987)

7 C.Presilla and F.Sacchetti, J. Phys. F17 , 779 (1987)

8 M. Cini, Solid State Commun. 20, 605 (1976) and Solid State Commun. 24, 681 (1977)

9 G.A.Sawatzky, Phys. Rev. Lett. 39, 504 (1977)

10 D.R.Penn, Phys. Rev. Lett. 42, 921 (1979)

11 V.Galitzkii, Sov. Phys. JETP.7, 104 (1958)

12 M.Cini and C.Verdozzi, to be published

13 N.E.Christensen, J. Phys. F8, L51 (1978)

14 P.Weightman and P.T.Andrews, J. Phys. C 13, L821 (1980); P.Weightman, H.Wright,S.D.Waddington , D.Van der Marel, G.A.Sawatzky, G.P.Diakuu, and D.Norman, Phys. Rev. B36, 9098 (1987)

15 S.Aksela, M.Harkoma and H.Aksela, Phys. Rev. A29, 2915 (1984)

16 F.U.Hillebrecht, J.C.Fuggle,P.A.Bennett,Z.Zolnjerek, Phys. Rev. B27, 2179 (1983)

17 J.C.Fuggle, F.U.Hillebrecht, R.Zeller,Z.Zolnjerek,P.A.Bennett, Phys. Rev. B27, 2145 (1982)

Recent Successes in Understanding the Electronic Structure of CuPd Alloys

S.D. Waddington [1], *H. Wright* [1], *J. Evans* [1], *P. Weightman* [1], *W. Folkerts* [2], *D. van der Marel* [2], *C.F. J. Flipse* [2], *and G.A. Sawatzky* [2]

[1] Physics Department and Surface Science Research Centre, University of Liverpool, Oxford Street, P.O. Box 147, Liverpool L69 3BX, UK

[2] Physical Chemistry Department of the Material Science Centre, University of Groningen, NL-9747 AG Groningen, The Netherland

1 Introduction

The Cu-Pd and Ag-Pd alloy systems have been an important testing ground for developments in the area of electronic structure calculations. These systems might be expected to have similar electronic structures since they both form simple substitutional alloys, with the fcc lattice structure, over the whole range of alloy compositions. However, there has been considerable controversy concerning the electronic structures of the two alloy systems. The theoretical results of WINTER and co-workers [1] indicate that the two systems have very different electronic structures, the component d bands being 'split' in Ag-Pd and 'mixed' in Cu-Pd. The theoretical results for the electronic structures of Ag-Pd alloys have been shown to be in agreement with the results of a number of experimental techniques. This work is concerned with the resolution of the controversy surrounding the electronic structure of the Cu-Pd system.

We begin with the observation that the Pd DOS given by the self-consistent-field Korringa-Kohn-Rostoker coherent potential (SCF-KKR-CPA) calculations [1] do not agree in detail with the results of a number of experimental probes of the Pd DOS including; Auger spectroscopy [2], UV excited photoemission [3] and synchrotron excited photoemission near the Pd Cooper minimum [4]. An obvious difference between the Ag-Pd and Cu-Pd systems is the difference in radii of the atoms. Although Ag and Pd are similar in size, Cu is significantly smaller as can be concluded from the interatomic distances in Pd and Cu metal (2.72Å and 2.55Å respectively). However, in order to preserve translational symmetry, the SCF-KKR-CPA calculations assume that an alloy has an average lattice spacing weighted by the relative concentrations of the respective components. The results of these calculations could thus be in error in systems where there are significant local distortions of the lattice spacing. Local distortions should be most easily observed in dilute systems and should be revealed by the EXAFS technique.

2 EXAFS studies of $Cu_{99}Pd_1$ and $Ag_{99}Pd_1$

The lattice constants of Cu, Pd, Ag, $Cu_{99}Pd_1$ and $Ag_{99}Pd_1$ were measured by x-ray diffraction using the Debye-Scherrer method. For Ag the results show that the addition of 1 at. % Pd causes a reduction in the average nearest-neighbour distances of $\simeq 0.0024 \pm 0.0004$ Å while the addition of 1 at. % Pd to Cu results in an expansion of the average nearest-neighbour distance of $\simeq 0.0009 \pm 0.0005$ Å. However, it is possible that these small changes in average nearest-neighbour distances produced by the addition of Pd may mask larger changes in local nearest-neighbour distances around Pd sites. Consequently we determined the local near-neighbour distances directly by performing EXAFS experiments on these materials [5], a technique ideal for determining near-neighbour distances. In $Cu_{99}Pd_1$ the EXAFS data yield an increase in the nearest-neighbour distance around a Pd site of 0.05 ± 0.01 Å, a 2% expansion which is confined to the first shell of neighbours around a Pd site. In $Ag_{99}Pd_1$ the EXAFS data is consistent with the view that there is no change in the host lattice around Pd sites. We now examine the effect of such a local expansion on the electronic structure of a dilute alloy system.

3 Influence of a local lattice expansion on the impurity DOS of a $Cu_{95}Pd_5$ alloy

There is general agreement that the Pd impurity DOS in Cu consists of a narrow state about 1.5eV below the Fermi level and a considerable Pd 4d DOS distributed throughout the Cu 3d band. There is, however, disagreement about the shape of the distribution of the Pd impurity DOS within the Cu 3d band. The Cu and Pd DOS are thoroughly mixed in CuPd so that it is difficult to deduce the Pd DOS experimentally from photoelectron results which reflect the total DOS. When analysing photoelectron spectra it is also difficult to allow for the contribution of electron energy loss processes which make a strong contribution to the spectrum at the bottom of the combined d band which is the region of most interest. Therefore it is virtually impossible to deduce from photoemission results exactly what the experimental Pd DOS is like within the band.

We use as our experimental probe of the Pd DOS, the profile of the Pd $M_{4,5}N_{4,5}N_{4,5}$ Auger transitions shown in Fig. 1 (full curve) as measured for $Cu_{95}Pd_5$ by WEIGHTMAN et al [6]. The procedure for calculating the Auger profile from the DOS uses the Cini-Sawatzky [7,8] formalism, described in detail by WEIGHTMAN et al [5]. The only free parameter in the calculation of the Auger profile is $U(^1G_4)$, the correlation energy of two holes localised in a $^1G_4$ term. Work on a number of Pd alloys shows that this parameter has a value in the range 2.8 eV to 3.1 eV and that this parameter is independent of alloy composition. In other alloy systems, notably $Ag_{95}Pd_5$ [9] and $Cd_{97}Ag_3$ [10], the Auger profile of the impurity has been shown to be a good test of models of the local impurity DOS. For CuPd alloys it has been shown [3] that by arbitrarily reducing the Pd DOS at the bottom of the band from that given by a SCF-KKR-CPA calculation, agreement between the theoretical Auger profile and experiment is improved. We now consider the effect of a local expansion of the Cu lattice around a Pd site on the Pd DOS, an effect that is not allowed for in either previous impurity calculations [11] or the

SCF-KKR-CPA calculations [1]. The results of our analysis of the EXAFS data show that there is a local lattice expansion of about 2% around Pd sites in Cu. In order to estimate the effect of this expansion on the local DOS we need to relate the local electronic structure to the nearest-neighbour distance. We adopt HARRISON's [12] scheme relating the hopping matrix element T between d states on atoms A and B to the d-state radii r^a_d and r^b_d and the interatomic distance R^{ab} by

$$T^{ab}_{dd} \propto \frac{(r^a_d r^b_d)^{3/2}}{(R^{ab})^5} \tag{1}$$

Having determined how the local lattice expansion changes the hopping matrix element between the impurity and the host we can use VAN DER MAREL et al's [3] modification to the Clogston-Wolff model to determine the local impurity DOS. In this model Hamiltonian the difference between the hopping matrix element between host-impurity and host-host is represented by a parameter t where

$$t = \frac{T_{Pd-Cu} - T_{Cu-Cu}}{T_{Cu-Cu}} \tag{2}$$

Assuming no lattice relaxation gives t = 0.6 and a DOS generated by the impurity model shown in the inset of Fig. 1a. The Auger profile calculated from this DOS is shown by the dashed line in the main part of figure 1a. From the EXAFS results and (2) we expect a value of t = 0.4 giving the Pd DOS shown in the inset to Fig. 1b. As can be seen from a comparison of the insets to Figs. 1a and 1b the local lattice expansion causes a reduction in the local Pd DOS at the bottom of the band. The Auger profile calculated from the new DOS is shown by the dashed line in Fig. 1b. The reduction in the Pd DOS at the bottom of the band caused by the expansion greatly improves agreement between experiment and the calculated Auger profile.

We have shown that the disagreement between theoretical and experimental results for the electronic structure of dilute CuPd alloys arises from the influence of a local lattice expansion. We now consider concentrated alloys.

4 Concentrated Cu-Pd alloys

Given the understanding achieved for dilute alloys it is clearly pertinent to establish the Pd DOS in concentrated alloys. In recent work [4] synchrotron radiation was used to measure the photoelectron spectrum of

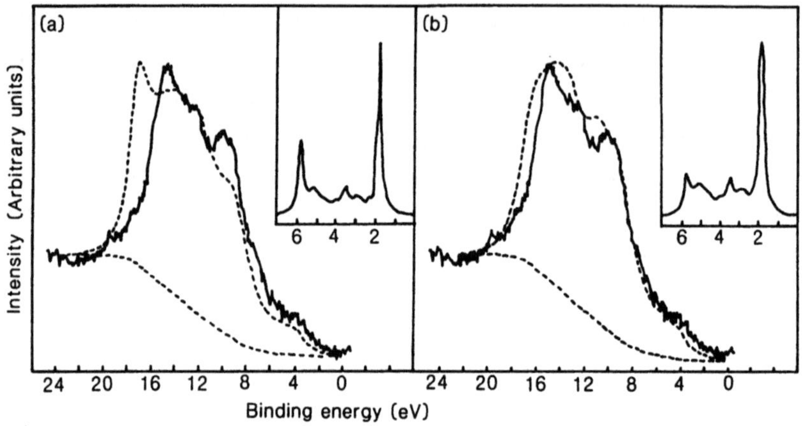

Figure 1

The insets of (a) and (b) show the theoretical Pd DOS calculated with t = 0.6 and 0.4 respectively. In the main parts of (a) and (b) the measured Pd $M_{4,5}N_{4,5}N_{4,5}$ Auger transition (full lines) for $Cu_{95}Pd_5$ is compared to the theoretical profiles (dashed lines) calculated using the theoretical DOS with $U(^1G_4)$ = 3.0eV. Figures from [5].

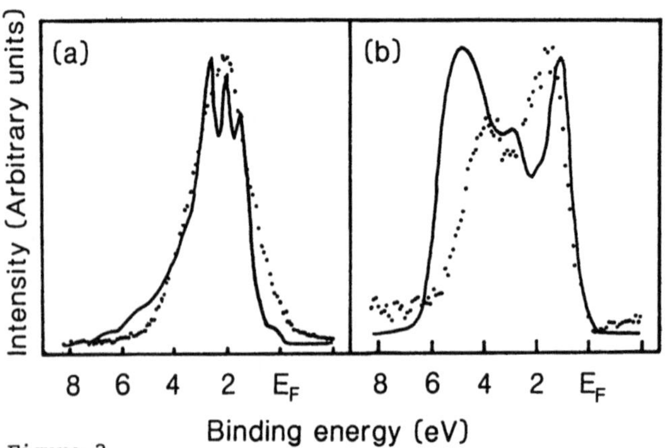

Figure 2

The empirically deduced Cu and Pd DOS are shown by the dots in (a) and (b) respectively. The theoretical results of [1] are shown by the full lines and have been broadened by a Gaussian of 0.5 eV (FWHM). Figures from [4].

$Cu_{75}Pd_{25}$. By taking advantage of the Cooper minimum in the Pd photoelectron cross-section, it was possible to deduce the local DOS of Cu and Pd in this concentrated alloy. These empirically deduced DOS are then compared with the results of the SCF-KKR-CPA calculations in Fig. 2. As can be seen, good agreement is obtained between theory and experiment for

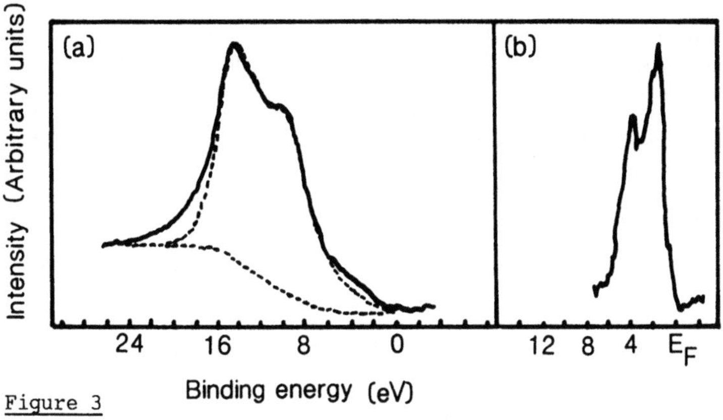

Figure 3

(a) The solid line shows the measured Pd $M_{4,5}N_{4,5}N_{4,5}$ Auger transition of $Cu_{75}Pd_{25}$. The dotted line shows the theoretical lineshape derived using the DOS shown in (b).

(b) The empirically deduced Pd DOS in CuPd [4].

the Cu DOS (Fig. 2a). However the theoretical results are shown to overestimate the Pd DOS (Fig. 2b) at the bottom of the band in a similar way to the results of impurity calculations. In this work we show that these empirical results are a good representation of the Pd DOS by using them to calculate the Pd $M_{4,5}N_{4,5}N_{4,5}$ Auger profile in $Cu_{75}Pd_{25}$. This is done according to the formalism described in [5] for dilute alloys and is shown by the dashed line of Fig. 3a. As can be seen the calculated profile is in good agreement with the experimental result shown by the solid line of Fig. 3a. An excellent reproduction of the experimental spectrum is obtained showing that the empirical Pd DOS deduced from the synchrotron work has the correct general distribution of intensity with energy.

5 Conclusions

EXAFS measurements of Cu, Ag, Pd, $Ag_{99}Pd_1$ and $Cu_{99}Pd_1$ have revealed that there is a 2% local expansion of the Cu lattice around Pd impurity sites, whereas there is no such relaxation around Pd sites in Ag. We have used a modified Clogston-Wolff model to show that by considering the influence of such a local expansion on the electronic structure of a dilute alloy system gives a Pd DOS consistent with the results of Auger spectroscopy. This shows that the discrepancy between the results of the calculated impurity DOS and experiment arises from the neglect of the local lattice expansion.

For concentrated alloys the Pd DOS deduced empirically from photoemission measurements of the valence band in the vicinity of the Cooper minimum shows a similar discrepancy with the SCF-KKR-CPA results to that found for dilute alloys. We have confirmed the accuracy of the empirical DOS in concentrated alloys by showing it is consistent with the Auger profile. We suggest that the reason for the failure of first-principles calculations to predict the Pd DOS of CuPd is similarly due to the neglect of lattice relaxation in this system. It is possible that the local lattice relaxation could be included in the first principles type of calculation within a self-consistent framework by using a cluster calculation. Having demonstrated the importance of this local relaxation we suggest that in concentrated systems it may be the driving force behind the well known ordering behaviour observed in such systems as $Cu_{75}Pd_{25}$.

References

1 H. Winter, P.J. Durham, W.M. Temmerman, G.M. Stocks: Phys. Rev. B33, 2370 (1986)
2 M. Davies, P. Weightman: J. Phys C17, L1015 (1984)
3 D. van der Marel, J.A. Jullianus, G.A. Sawatzky: Phys. Rev. B32, 6331 (1985)
4 H. Wright, P. Weightman, P.T. Andrews, W. Folkerts, C.F.J. Flipse, G.A. Sawatzky, D. Norman, H. Padmore: Phys. Rev. B35, 519 (1987)
5 P. Weightman, H. Wright, S.D. Waddington, D. van der Marel, G.A. Sawatzky, G.D. Diakun, D. Norman: Phys. Rev. B36 9098 (1987)
6 P. Weightman, P.T. Andrews, G.M. Stocks, H. Winter: J. Phys. C16, L81 (1983)
7 M. Cini: Solid State Commun. 24, 681 (1977)
8 G.A. Sawatzky: Phys. Rev. B39, 504 (1977)
9 M. Vos, G.A. Sawatzky, M. Davies, P. Weightman, P.T. Andrews: Solid State Commun. 52, 159 (1984)
10 P.H. Hannah, P. Weightman: J. Phys. F16, 1015 (1986)
11 P.J. Braspenning, R. Zeller, A. Lodder, P.H. Dederichs: Phys. Rev. B29, 703 (1984)
12 W.A. Harrison: In Electronic Structure and the Properties of Solids (Freeman, San Francisco, 1980)

Resonant Auger Emission at the La 3d-Excitation Threshold

H. Hinkers and H. Merz

Physikalisches Institut der Universität, Wilhelm-Klemm-Str. 10,
D-4400 Münster, Fed. Rep. of Germany

Abstract The La $M_{45}N_{45}N_{45}$-Auger emission of solid LaB_6 is investigated near the M_{45}-excitation thresholds. At threshold the Auger emission is resonantly enhanced due to a resonance in the M_{45}-excitation cross section. The data are compared with corresponding Appearance Potential Spectroscopy data.

Introduction

From X-ray emission spectroscopy and bremsstrahlung isochromat spectroscopy of light rare earths it is known that the X-ray emission is resonantly enhanced at excitation energies near the thresholds of the M_{45}-core levels. In the case of La and Ce the resonant X-ray emission process has been subject to detailed investigations /1-3/. But only little information exists about threshold effects in the competing Auger emission process /4,5/.

The Auger electron spectroscopy (AES) process is most commonly described in terms of a two-step-model. In the excitation step a core hole is created by electron or photon impact. The core electron is either ejected to the continuum leaving a positively charged ion behind or - in the case of rare earth compounds - may be excited to the 4f-level resulting in a highly excited but neutral atom /6/. In the decay step the core hole is filled up with an outer-shell electron and another electron is ejected to the continuum and detected by the energy analyzer.

The excitation is most commonly performed with electrons of energies well above the excitation threshold of the corresponding core level to be excited. Under such standard conditions the shapes as well as the energy positions of the Auger features do not depend significantly on the excitation energy and the variation in Auger intensity with varying excitation energy is rather small. On the contrary, near the M_{45}- exitation thresholds of La and Ce it is expected that the core level excitation cross section exhibits singularities or resonances which should then influence the Auger emission. This is investigated in the following with Disappearance Potential Spectroscopy (DAPS) which probes the core level excitation cross section by measuring the yield of the elastically scattered electrons as a function of excitation energy /7/. Furthermore a detailed investigation of the near-threshold $M_{45}N_{45}N_{45}$-Auger emission at the M_{45}-levels of La is presented. The Auger emission data have been taken as excitation curves with fixed Auger energy E_A as a function of the excitation energy E_P. A series of excitation curves has been taken at different Auger energies to get a three-dimensional presentation of the Auger intensity $I(E_A,E_P)$. The interpretation of the Auger data is supported by the DAPS data.

Experimental

A total of 65 excitation curves has been measured with an acquisition time of 2h for each curve. Although the measurements have been performed in ultrahigh vacuum of about $2 \cdot 10^{-10}$ mbar the sample had to be cleaned several times during the experimental run to ensure indentical surface conditions throughout the acquisition of data. The measurements have therefore been performed on LaB_6 which can easily be cleaned in situ by scraping with a diamond file. Cleanliness was checked by conventional AES.

The excitation curves have been taken in a derivative mode with lock-in techniques. This offers the possibility to discriminate between energy loss and Auger features by modulating the potential of either the electron source or the sample. The energy resolution of the energy analyzer (CMA) was about 0.6%.

Results and discussion

In fig.1 the derivative of the intensity $I(E_P,E_A)$ of the emitted electrons is plotted versus both the excitation energy E_P and the energy E_A of the emitted electrons near the La $M_{45}N_{45}N_{45}$-Auger emission.

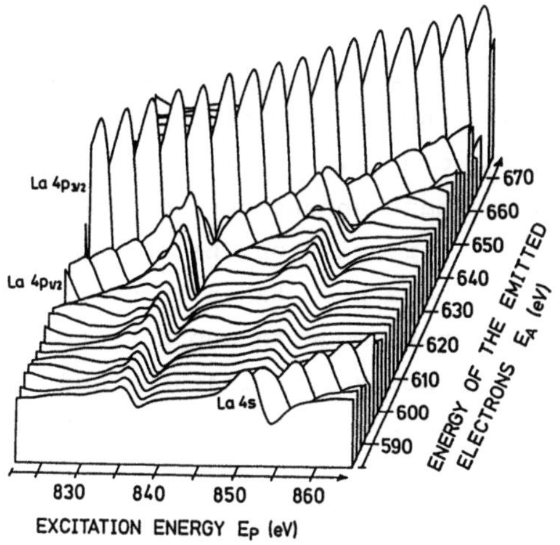

FIG.1. Intensity of the emitted electrons of LaB_6 near the La $M_{45}N_{45}N_{45}$-Auger structure at excitation energies near the M_{45}-excitation thresholds. The data are presented as excitation curves in derivative form.

The intensity distribution is dominated in this energy region by two different kinds of spectral features:

- Energy loss structures of nearly constant intensity in diagonal direction;
- Auger electron spectra parallel to the E_A-axis with strong intensity variations in the direction of E_P.

Both spectral features are superimposed in some energy regions and it is suggested in the following that the underlying processes - energy loss process and Auger process - are not affected by each other and can be treated as an incoherent superposition.

The energy loss features which are predominantly caused by La 4s- and 4p-core electron energy loss processes have been investigated in detail by NETZER et al. /8/ and by STRASSER et al. /9/. Therefore the following discussion is mainly confined to the Auger features.

In fig.1 the derivative of the intensity $I(E_P,E_A)$ was taken by applying the modulating voltage to the electron source. In a second experimental run the modulating voltage has been applied to the sample in order to remove the energy loss features resulting in the derivative of the intensity with respect to the diagonal direction. The remaining spectral features are dominated by Auger structures. A small contribution of additional structures results from the background of inelastically scattered electrons whose intensity undergoes abrupt changes at the M_{45}-excitation thresholds. This contribu-

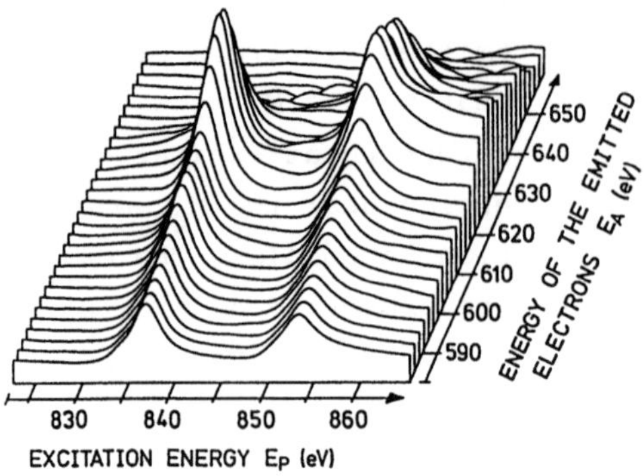

FIG.2. Intensity of the $M_{45}N_{45}N_{45}$-Auger electrons near the M_{45}-excitation threshold illustrating the resonant enhancement of the Auger emission.The curves have been taken in derivative mode to suppress the energy loss features and have been integrated numerically.

tion was found to be identical with the corresponding DAP spectrum and has been removed by subtraction together with a smoothly varying background. For further discussion the data have been integrated numerically using an interpolation method and the result is shown in fig.2.

The figure clearly demonstrates the resonant behaviour of the Auger emission near the M_{45}-excitation thresholds. In the case of the M_5-level the onset of the Auger structure is found near E_p=831.5eV and the intensity reaches its maximum at $E_p=(837.1\pm0.2)$eV. Increasing E_p by further 10eV results in a reduction of the intensity by a factor of about five. Below the excitation threshold of the M_4-level the Auger intensity solely results from M_5NN emission. The onset for M_4NN emission is found near E_p=848.5eV and reaches its maximum at $E_p=(853.8\pm0.2)$eV. The corresponding Auger spectra at these selected excitation energies are shown in fig.3. They have been taken from fig.2 as sections parallel to the E_A-axis. We suggest this resonance phenomenon to result from the resonant excitation of the La 3d-core electron. At the excitation threshold transitions with both the incident electron and the excited core electron entering the 4f-level $3d^{10}4f^0+e^- \rightarrow 3d^9 4f^2$ lead to enhanced core hole production.

The major part of the $3d^9 4f^2$ excited states decays via nonradiative transitions and, as a consequence, the total Auger intensity should be roughly proportional to the rate of core hole production. To support this idea fig.2 has been integrated with respect to E_A to obtain the total Auger intensity as a function of excitation energy and the result is compared in

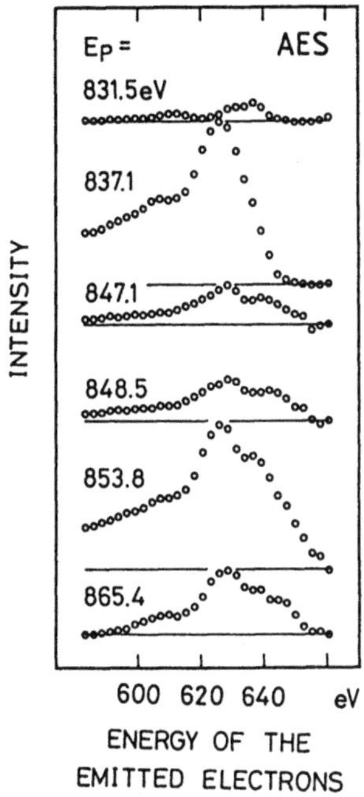

FIG.3 (left) La $M_{45}N_{45}N_{45}$ Auger spectra of LaB_6 at some selected energies taken from the data of fig.2.

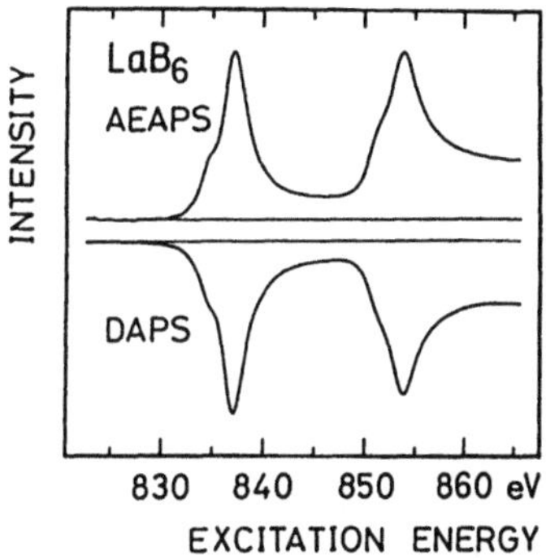

FIG.4. (above) Auger yield (AEAPS) and yield of the elastically reflected electrons (DAPS) near the La M_{45}-threshold

fig.4 with the corresponding DAP spectrum. Both spectra have been deconvoluted with the energy window of the apparatus and a smooth background has been subtracted from the DAP spectrum. The results turn out to be nearly identical. Both curves reveal the same fine structure which in the case of DAPS has been related to the multiplet structure of the excited $3d^9 4f^2$-configuration /10/. By variation of the excitation energy in the near-threshold region the $3d^9 4f^2$-multiplet terms are gradually scanned and the excitation curves in fig.4 should roughly represent the complete multiplet structure of the $3d^9 4f^2$-configuration. This is indeed not the fact: From comparison with intermediate coupling calculations it has been shown that the excitation probability is strongly term-dependent /10/ and for both spin orbit levels the low energy parts of the $3d^9 4f^2$-multiplet structures are drastically suppressed and reduced in the excitation curves of fig.4 to small shoulders on the low energy sides of the main peaks.

In the decay step transitions $3d^9 4d^{10} 4f^2 \rightarrow 3d^{10} 4d^8 4f^2 + e^-$ are expected to dominate the Auger spectra. There may be further contributions resulting from competing transitions like

$3d^9 4p^6 4f^2 \rightarrow 3d^{10} 4p^5 4f^1 + e^-$. But a detailed investigation of this problem needs a better energy resolution of the Auger spectra. In conclusion the Auger emission near the La M_{45}-excitation thresholds is dominated by resonant transitions involving the formation of quasidiscrete $3d^9 4f^2$-states in the Auger excitation step. Thus, the $3d^9 4f^2$-configuration plays an important role not only in resonant X-ray emission but also in near-threshold Auger emission.

This work has been supported by the Deutsche Forschungsgemeinschaft.

1. R.J. Liefeld, A.F. Burr, M.B. Chamberlain, Phys.Rev. A9, 316 (1974)
2. M.B. Chamberlain, A.F. Burr, R.J. Liefeld, Phys.Rev. A9, 663 (1974)
3. F. Riehle, Thesis, University of Karlsruhe 1977
4. M.B. Chamberlain, A.F. Burr, R.J. Liefeld, J.Vac.Sci. Technol. A2,1169 (1983)
5. M.B. Chamberlain, A.F. Burr, R.J. Liefeld, J.ac.Sci. Technol. A2, 973 (1984)
6. J.C.Rivière, F.P. Netzer, G.Rosina, G. Strasser, J.A.D. Matthew, J.Electron Spectrosc.Relat.Phenom.36, 331 (1985)
7. J. Kirschner, P. Staib, Appl.Phys. 6, 99 (1975)
8. F.P. Netzer, G. Strasser, G. Rosina, J.A.D. Matthew, J.Phys.F: Met.Phys. 15, 753 (1985)
9. G. Strasser, F.P. Netzer, J.A.D. Matthew, Solid State Comm. 49, 817 (1984)
10. H. Hinkers, R. Stiller, H. Merz, W. Drewes, H.-G. Purwins, Physica Scripta 36, 166 (1987)

On the Theory of the Auger CVV Lineshape of Graphite

M. Cini[1] *and A. D'Andrea*[2]

[1]Dip. di Fisica, II Università di Roma, Tor Vergata, Via Orazio Raimondo, I-00137 Rome, Italy
[2]Ist. di Metodologie Avanzate Inorganiche, C.N.R., I-00016 Monterotondo Staz., Rome, Italy

1 Introduction

Recently a quite large number of papers /1-5/ has been devoted to Auger CVV line shape of Graphite. The increasing interest in this topic is due to the following reasons: a) much of the attention is in the context of the novel Graphite intercalation compounds /2,3/ that show interesting electronic and structural properties,moreover b) the Graphite Auger CVV spectrum represents the infinite limit of the aromatic series, benzene, naphatalene etc... and finally c) the Auger spectrum of Graphite is itself of interest since it shows new dynamical effects, namely: the shakedown exciton satellite /4/ and the intrinsic plasmon satellite /6/ .

The first attempt to obtain an accurate Auger CVV line shape of Graphite was reported by SMITH and LEVENSON /1/ in their pioneering work in 1977. From the experimental Auger spectrum the background of secondary electrons was removed utilizing a method developed by SICKAFUS /7/ and the resulting line shape was deconvoluted by an electron energy loss spectrum of 263 eV elastic peak. The Auger spectrum shows a broad peak at 240 eV electronic kinetic energy that the authors attribute to the ω_p=27eV plasmon loss peak (see Fig.2 of /1/) .

MURDAY et al. /3/ in a 1981 paper search for a quantitative accord between the theory and SMITH and LEVENSON experimental results. They deduced the Graphite partial density of states (PDOS) components : σ_s , σ_p , and π_p from experimental X-ray photoelectron (XPS) and angular dependent X-ray emission (XES) spectra. The relative areas of the individual partial density of states were normalized to provide a $\sigma_s \sigma_p^2 \pi_p$ configuration. MURDAY et al. /3/ obtained a fair accord between experimental results and the theory without including Coulomb repulsion and screening effects in the two holes final state, but a mistake in the self-fold of PDOS makes the agreement between theory and experiment fortuitous as pointed out in a recent paper by HOUSTON et al./5/ .

In that paper /5/ the authors performed a very accurate experimental Auger CVV spectra of POCO-Graphite (or amorphous Graphite) samples.The experimental Auger spectrum was corrected for the effects of secondary-electron background and extrinsic losses and it was placed on an absolute energy scale.

The RAMAKER /5,8/ theoretical interpretation of the Auger spectrum proceeds using MURDAY et al. one-particle local density of states in the frame of closed shells by using CINI's theory /9/,but assuming an "intuitive" , let us say an "arbitrary" , multiplet decomposition. In fact the fold contributions $\sigma * \sigma$, $\sigma * \pi$ and $\pi * \pi$ are considered separately and the Auger matrix elements normalized per electron were scaled by the matrix elements of Neon. The accord with experimental results was obtained by using as fit parameters the effective two holes repulsion energies (see Fig.8 of /5/) . The discrepancy between theory and experiment in the low binding energy side of the spectrum was attributed to an intrinsic plasmon satellite effect.

We think that the arbitrary multiplet decompositions and the Auger matrix elements assumed by atomic Neon makes not fully justified the physical considerations on Graphite behaviour sketched in HOUSTON et al. paper/5/. Moreover since the bands are open we must extend the matrix theory according to the Bare Ladder Approximation (BLA) /10,11/. In this way we may check the applicability of BLA to widely open bands. Obviously this aim requires in principle the correct application of the matrix theory /10/.

In the present work we perform Auger CVV calculation by using the static CINI's theory /10/ with the correct multiplet analysis. The Auger matrix elements of Carbon are taken by Mc GUIRE /12/ and effective two holes Coulomb repulsions of static theory are taken as fit parameters. Finally a detailed plasmon satellite analysis will be performed.

2 Static screening in Auger CVV theory

In the case of closed bands one of us /10/ performed a tensorial analysis of Auger CVV in presence of two final holes Coulomb repulsion. The Auger current is given as function of the matrix density of states in the two final holes configuration,

$$J = \sum_{L_1L_2L_3L_4\sigma} X(L_1L_2L_3L_4) D_{L_1L_2L_3L_4\sigma}(\omega) \tag{1}$$

where L=(l,m) denotes a single-particle orbital and the quantities X are given by,

$$X(L_1L_2L_3L_4) = \sum_{L_f m_c} A_{L_fL_cL_1L_2} A_{L_fL_cL_3L_4} \tag{2}$$

while A are the Auger matrix elements,

$$A_{L_fL_cL_1L_2} = \left\langle \varphi_{L_f}(1)\, \varphi_{L_c}(2) \left| \frac{1}{r_{12}} \right| \varphi_{L_1}(1)\, \varphi_{L_2}(2) \right\rangle \tag{3}$$

It is well known that the non-interacting two hole matrix density of states $D_0(\omega)$ can be expressed in terms of the folds of single particle density of states.The Green function $\overleftrightarrow{G}(\omega)$ matrix is given in terms of its independent-particle $\overleftrightarrow{G}_0(\omega)$ counterpart and Coulomb repulsion matrix $\overleftrightarrow{U}$, namely,

$$\overleftrightarrow{G} = \overleftrightarrow{G}^0 \left[\overleftrightarrow{1} + \overleftrightarrow{U}\,\overleftrightarrow{G}^0 \right]^{-1} \tag{4}$$

In the case of Graphite symmetry reduces the Green matrix in block form and we need to invert no more than 4x4

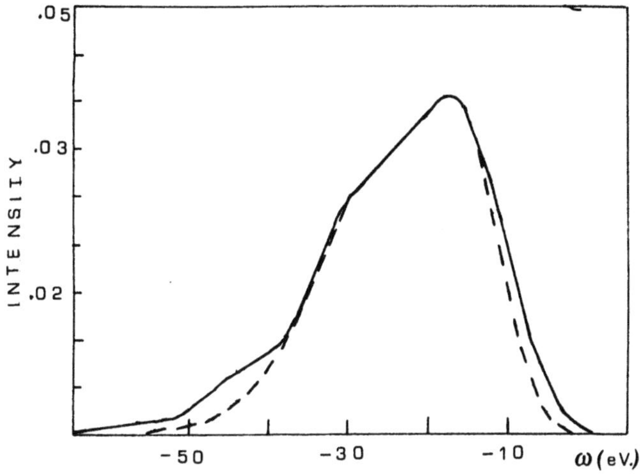

Fig.1 Comparison of calculated Auger line shape by static theory (dashed curve) and experimental spectrum (solid curve).

matrices. A similar analysis was successful in the case of Tetramethyl-Silane /13/. The static calculation proceeds by the following steps: a) the Auger matrix elements are computed by Mc GUIRE's /12/ atomic radial elements, b) the single-hole densities of states are taken diagonal respect to atomic quantum numbers and are given by MURDAY et al. /3/ and c) the screened repulsion matrix U is taken as fitting parameters.

In Fig.1 we show the results obtained by static screening theory.

The agreement with experimental curve is very good and anyway it is not worse than that obtained by HOUSTON et al./5/, but the values of the two holes screened repulsions result in strong contrast with /5/. While HOUSTON et al. found very small values of Coulomb repulsion components: pp,sp and ss (0.6, 1.5 and 2.2 eV. respectively) we obtain different values for pp (5.5 eV) and for ss,sp components (11 eV); a similar trend is found in Silicon by Bisi's /12/ calculation reported in the present Meeting.

The residual discrepancies between theory and experiment are avoided introducing near the Fermi level a long lived valence-core exciton as shown by HOUSTON et al. /4/, while the discrepancy on the tail of the curve can be cured by a plasmon satellite ω_p=27 eV far from the maximum of the peak. HOUSTON et al./5/ showed that this feature is much narrower than the Auger peak, thus it is unlikely that this feature can be identified as due to the initial state core-hole filling. Moreover a feature like that was computed by the present authors /6/ in the case of dynamical screening effects in Auger CVV line shape . That calculation was performed by assuming rectangular one-particle density of states and final two holes bare Coulomb interaction, while the plasmon screening was modelled by displaced harmonic oscillator where g_0 was the linear hole-plasmon coupling (see Fig. 3 of /6/).

3 Dynamical screening in Auger CVV theory

Let us consider two final holes created in Auger CVV decay in a crystalline solid. If g_0^2/ω_p is the screening energy due to linear hole-plasmon interaction, the two final holes can be found in two different configurations, namely : a) two holes in the same solid site i with screening energy 4 g_0^2/ω_p and b) two holes in different sites i,j with screening energy 2 g_0^2/ω_p. Therefore we take the following model for hole-plasmon interaction Hamiltonian

$$H_{h\text{-}p} = \left[2g_0\sum_i n_{i+}n_{i-} + g_0\sqrt{2}\,(1-\sum_i n_{i+}n_{i-})\right](b^+ + b) \tag{5}$$

where b^+ creates a plasmon of energy ω_p.

The exact solution of this problem for one S-valence band case was given by the present authors /6/ in 1984. Now we must generalize the exact solution to the degenerate band case, for doing this all the quantities that appear in the one band solution (Green functions ,Coulomb repulsions and excitation amplitudes) must assume matrix form. The two holes local Green function developed for the first plasmon satellite results,

$$\begin{aligned}\overleftrightarrow{G}(\omega) = &\left[\overleftrightarrow{1} + \frac{g\,\gamma}{\omega_p}\overleftrightarrow{G}_1(\omega)\right]\left[\overleftrightarrow{1} - g^2\,\overleftrightarrow{G}_0(\omega)\overleftrightarrow{G}_1(\omega)\right]^{-1}\overleftrightarrow{G}_0(\omega) +\\ &+\left[\frac{g\,\gamma}{\omega_p}\overleftrightarrow{G}_0(\omega) + \left(\frac{\gamma}{\omega_p}\right)^2\right]\left[\overleftrightarrow{1} - g^2\,\overleftrightarrow{G}_1(\omega)\overleftrightarrow{G}_0(\omega)\right]^{-1}\overleftrightarrow{G}_1(\omega)\end{aligned} \tag{6}$$

where $\overleftrightarrow{1}$ is the identity matrix , $\overleftrightarrow{G}_0(\omega)$ is the zero order two holes Green function given by (4) for the dynamical Coulomb repulsion matrix,

$$\overleftrightarrow{U}_d = \overleftrightarrow{U}_s + \overleftrightarrow{1}\left[6 - 4\sqrt{2}\right]g_0^2/\omega_p \tag{7}$$

and the quantities $g = g_0\,(2-\sqrt{2})$, $\gamma = g_0\,(\sqrt{2}-1)$ are connected to hole-plasmon linear coupling g_0 /6/ Finally the Green function $\overleftrightarrow{G}_1(\omega)$ is the zero order Green function $\overleftrightarrow{G}_0(\omega)$ computed at energy $\omega_1 = \omega - \omega_p$.

In Fig.2 we compare the statical and dynamical theoretical spectra by using as Coulomb repulsion parameters the same values of static theory and after we have rescaled them by (7) for the dynamical case , the hole-plasmon linear coupling value results g_0 =13 eV.

We would underline the good accord obtained by dynamical theory and experimental result also in the low binding energy side of the spectrum. The discrepancy in the high energy side is cured by introducing at Fermi energy a long lived valence-core exciton level /4/.

Finally in Fig.3 we compare our dynamical Auger spectrum in Grafite and HOUSTON et al. experiment results,

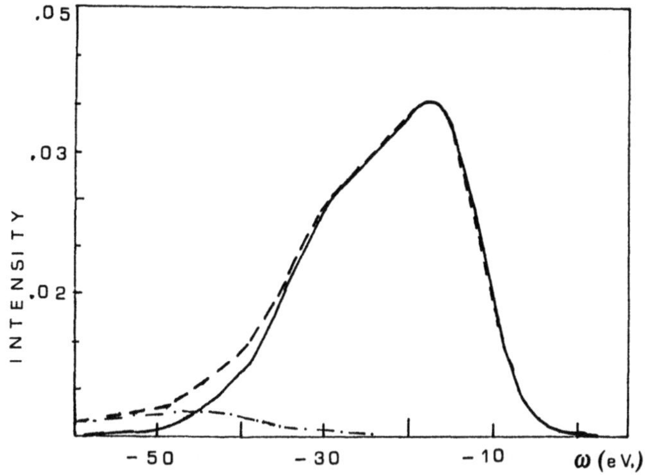

Fig.2 Computed Auger CVV spectra by static theory (solid curve) and dynamical theory (dashed curve), the dashed-dotted curve is plasmon contribution.

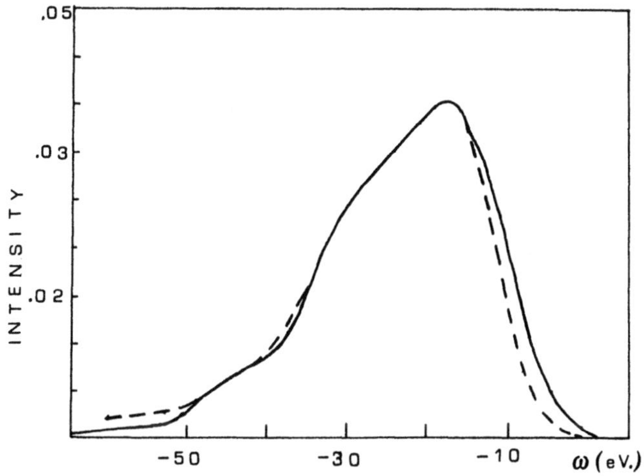

Fig.3 Comparison of calculated Auger line shape by dynamical theory (dashed curve) and experimental spectrum (solid curve).

4 Conclusion

The main finding of the present work concerns the good accord obtained between the experimental Auger CVV spectrum of Graphite and the CINI's multi-band theory /10/. Moreover the CINI-D'ANDREA Auger CVV dynamical theory /6/ has been generalized to multi-band case for taking into account of final state dynamical screening effects. This dynamical effect explains the satellite and improves the accord with the experiments.

REFERENCES.

1- M.A.Smith and L.L.Levenson, Phys.Rev.B16,2973 (1977).

2- P.Oehafen,P.Pfluger and H.J.Guntherodt,Sol.St.Comm.32, 885(1979)

3- J.S.Murday,B.I.Dunlap,F.L.Hutson and P.Oehafen, Phys.Rev.B 24, 4764 (1981)

4- J.E.Houston,D.E.Ramaker,J.W.Rogers.R.R.Rye and F.L.Hutson, Phys.Rev.Lett.56,1302 (1986).

5- J.E.Houston,J.W.Rogers,R.R.Rye,F.L.Hutson and Ramaker, Phys.Rev.B 34,1215 (1986).

6- M.Cini and A.D'Andrea,Phys.Rev.B29,6540 (1984).

7- E.N.Sickafus, Rev.Sci.Instrum.42, 933 (1971).

8- D.E.Ramaker,J.S.Murday,N.H.Turner,G.Moore,M.G.Legally and J.Huston,Phys.Rev.B19, 5375 (1979).

9- M.Cini,Sol.St.Comm. 24,681 (1977).

10- M.Cini,Phys.Rev.B 17,2788 (1978).

11- M.Cini and C.Verdozzi, Il Nuovo Cimento D9,1 (1987)

12- E.J.Mc Guire, Phys.Rev. 185,1 (1969).

13- M.Cini,F.Maracci and R.Platania,J.de Physique C9,781(1987).

14- V.Bisi, in the present Proceedings.

Surface Auger and Double Hole Shifts in Ni(100)

*N. Mårtensson, A. Nilsson, A. Stenborg, and D. Parashkevov**

Department of Physics, Uppsala University, Box 530,
S-751 21 Uppsala, Sweden

1. Introduction

Surface core level shifts (SCS) in photoelectron spectra have been studied extensively during the last decade both experimentally and theoretically /e.g. 1-4/. The experimental discrimination between the surface and the bulk signals from single core holes has been greatly aided by the use of synchrotron radiation. By tuning the excitation energy it is possible to vary the surface sensitivity whereby the surface contributions to the spectra can be separated. Using a model based on the assumption of complete screening of the core ionized site a consistent description of the SCS has been obtained /5/.

Also Auger spectroscopy is a most surface sensitive technique with an appreciable intensity from the first atomic layer. Using the same type of theoretical treatment as has been successfully used for the single hole states, surface shifts are predicted also for the Auger process and these will be of the same order of magnitude as the single core hole shifts /6/. So far, there has been no experimental verification of this prediction. The Auger electrons have constant kinetic energies which limits the possibility to vary the sampling depth. Furthermore, the Auger lines are often inherently broader than the single hole lines which makes it more difficult to resolve small shifts. However, the presence of surface shifts, even if they are not resolved, can have important consequences for the interpretation of the Auger spectra.

The final state of the Auger process is a doubly ionized state, while the initial state has a single core hole. In Auger processes involving localized final states the surface shift is determined by the difference in surface shifts of the single and double hole states /6/. Double-hole states can also be generated directly in a photoemission experiment due to electronic shake-up processes. Such doubly ionized states have been identified in spectra from several metals both in the valence and the core regions /e.g. 7,8/.

The L_3VV Auger spectrum in Ni is dominated by the atomic multiplet levels of the final state d^8 configuration. In the Ni valence band spectrum, satellites have been identified which originate from similar localized d^8 configurations. Also in the core level spectra from Ni one finds satellites which can be interpreted as due to double hole states with properties similar to the Auger final states. In the present paper we report surface shift measurements for the ($2\underline{p}$) single and ($2\underline{p}3\underline{d}$) double hole states for Ni(100). By measuring at different angles the surface sensitivity can be varied due to angle dependent scattering cross-sections. It has been shown that the peak intensities are enhanced at angles corresponding to axes connecting neighbouring atoms due to strong elastic scattering in the forward direction /9/. In the present case this leads to a change in the bulk intensity by a factor of 2.8 between angles of maximum and minimum intensity whereas the surface intensity remains constant. Using these results the surface Auger shift can then be derived for the Ni(100) surface.

In connection with the present measurements we will also discuss the theoretical treatment of surface shifts of double hole states and Auger transitions /6/.

2. Experimental

The experiments were performed in a high resolution XPS system. The X-ray source consists of a high speed rotating anode and a 25 crystal monochromator on a Rowland circle arrangement. The electron spectrometer contains a hemispherical analyzer with 36 cm mean radius and a multichannel detection system. The overall resolution in the present measurements was 0.3 eV.

The Ni(100) crystal was cleaned by a combination of Ar^+ sputtering and annealing in vacuum or oxygen. The sample was heated by radiation from a tungsten filament on the manipulator and cooled by admission of liquid nitrogen through a capillary. All measurements were performed at a temperature of 150 K.

3. Theoretical Model

In Fig. 1, processes leading to single and double hole states are schematically represented for bulk and surface ionizations. The energy of a photoionization process is given by the difference in total energy of the initial and final states. Using a Born-Haber cycle this energy difference can be decomposed into a number of terms. When a shift between different metallic situations, such as the shift between a core ionization process in the bulk and at the surface of a metal, is considered, only a few of these terms change and have to be treated explicitly. All other terms cancel in the shift expression. A considerable advantage with this decomposition is that the important terms to a good approximation can be treated within an equivalent core approximation for the final state. In this way the surface shift for a single hole state corresponds to the energy difference between Z-metals with a Z+1 substitutional impurity in the bulk and at the surface /5/. This energy difference is represented in Fig. 1. An interesting aspect of this is that the surface shift measures the surface segregation energy of a Z+1 impurity in the Z-metal /5/.

The surface shift for a double ionization process is represented at the top of Fig. 1. The shift is determined by the energy difference between a Z+2 impurity in the bulk and at the surface of the Z-metal. As seen from Fig. 1 this corresponds to the surface segregation energy of a Z+2 impurity /6/.

The leading terms in surface shift expressions will be the difference in surface energies (γ) of the metals representing the initial and final states /5/. For the single hole state the surface shift can approximately be expressed as

$$\Delta E_C^{S,B} \approx \gamma_{Z+1} - \gamma_Z \qquad (1)$$

and analogously for the double hole state

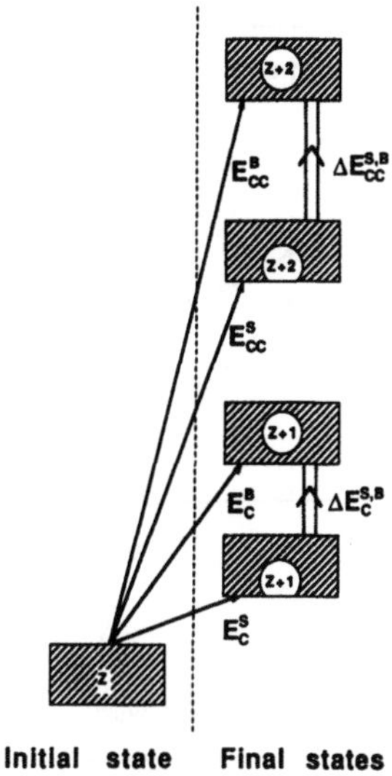

Fig. 1 Schematic representation of the energy terms determining the single and double hole surface shifts. E denotes an ionization energy and ΔE a shift. S and B denotes surface and bulk, respectively and c and cc are used for one and two core holes. The single and double hole states are represented within the equivalent core approximation.

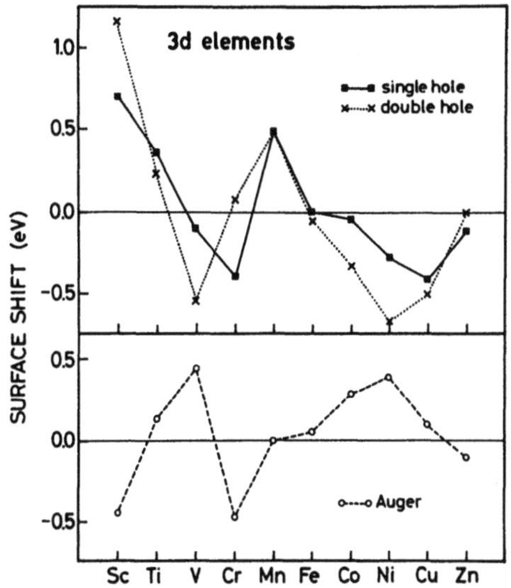

Fig. 2 Calculated single and double hole surface shifts (top) and Auger energy shifts (bottom) for the 3d transition elements. For the hole states the binding energy differences (surface - bulk) are given. The Auger shifts refer to kinetic energy shifts.

$$\Delta E^{S,B}_{cc} \approx \gamma_{Z+2}-\gamma_Z \qquad (2)$$

The shift expression for the Auger process (kinetic energy shift) is obtained by combining (1) and (2);

$$\Delta E^{S,B}_{Auger} \approx \gamma_{Z+1}-\gamma_{Z+2} \qquad (3)$$

In addition to the equivalent core approximation, (1)-(3) are approximate in the sense that they neglect the fact that the final state atoms are created as impurities in the Z-metal lattice and not in sites surrounded by equivalent final state atoms. These approximations will only give relatively small errors in the calculated surface shifts /5/.

In Fig. 2 the calculated surface shifts for the 3d transition elements are plotted for single hole states, double hole states and for Auger transitions. A problem with the calculation of surface shifts for the 3d metals using (1)-(3) is that the surface energies for some of these metals are not accurately known, neither from experiment nor from theory. The presently used surface energies are essentially derived from heats of vaporization /10/. The heat of vaporization, however, is affected by the energy position of the Hund's rules ground state for the free atom relative to the state corresponding to the metallic situation. This purely atomic effect may lead to significant errors in the so derived surface energies, especially in the middle of the transition series.

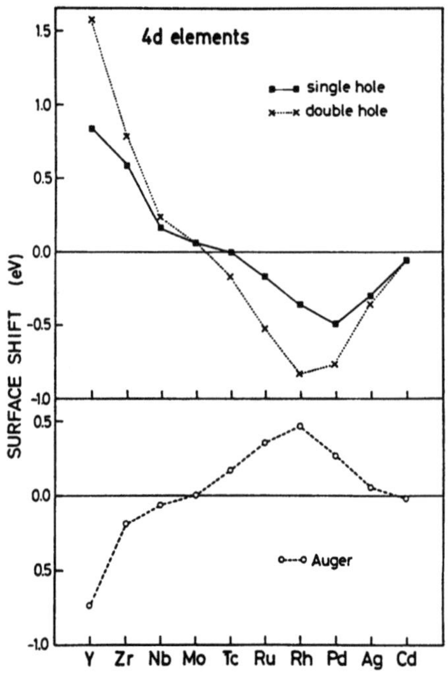

Fig.3 Calculated single and double hole surface shifts (top) and Auger energy shifts (bottom) for the 4d transition elements.

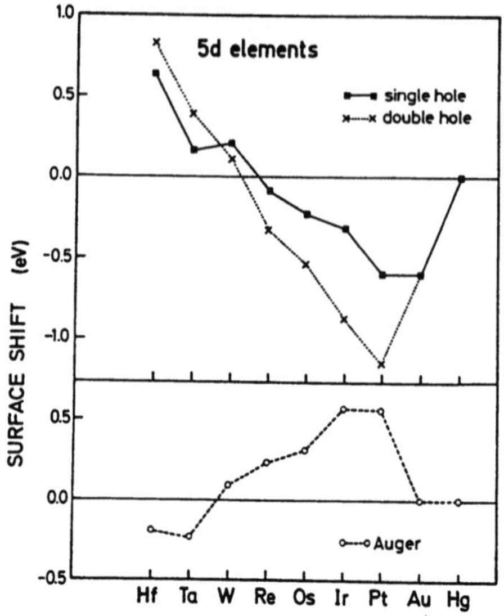

Fig. 4 Calculated single and double hole surface shifts (top) and Auger energy shifts (bottom) for the 5d transition elements.

In fact, the strong variations of the calculated surface shifts in the middle of the series in Fig. 2 are most probably due to such artifacts. It should also be remembered that many of the 3d transition metals are magnetic and that it is not well established what their electronic structure will be at the surface. If, for some reason, there would be a large change in the number of d-electrons (or in their character) between the bulk and the surface, fairly large shifts might be expected. In spite of these difficulties the shifts in the beginning and at the end of the series can be expected to be rather well described by the utilized surface energies. The general trend which is expected is that there are positive shifts for the early transition metals and negative shifts for the elements at the end of the series. For the single hole states this behaviour is well established. Note, however, that these possible errors in the calculated shifts for the 3d-metals are not due to shortcomings of the presently used theoretical model but to a lack of knowledge of the surface energies. On the other hand, (1)-(3) provide interesting possibilities to derive information on the surface energies from measured surface shifts.

Figs. 3 and 4 show the calculated single hole, double hole and Auger surface shifts for the 4d and 5d transition series. These curves show the same overall behaviour as the curves for the 3d elements. However, the rapid variations in the middle of the series are much less pronounced for the 4d and 5d elements as can be expected from the lesser importance of the multiplet effects.

4. Results and Discussion

In Fig. 5 the XPS spectra obtained for two different emission angles from a Ni(100) surface are shown. The figure displays the Ni2$p_{3/2}$ main line and the 2p3d double hole satellite. The spectra were recorded at polar angles 45° and 60° along the (010) direction. An intensity enhancement by a factor of 2.2 is observed in the spectrum recorded at 45° over the one at 60°. This can be explained by variations in the bulk signal whereas the signal from the outermost surface layer remains constant with these angular variations. At 45° the electron analyzer is viewing the crystal along (110) atomic rows and the bulk intensity is enhanced relative to other directions due to strong elastic scattering in the forward direction. We denote the 45° spectrum bulk sensitive and the 60° spectrum surface sensitive.

The spectra can be considered to consist of three different contributions, the bulk and surface spectra including their shake-up and shake-

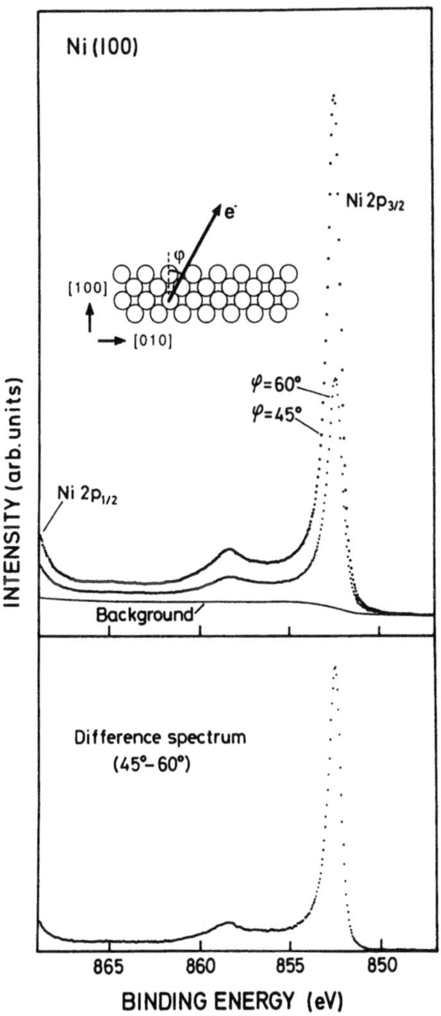

Fig. 5 Ni2p spectra measured at two different polar angles. The spectral intensities are normalized according to the background at the low binding energy side. With this normalization the difference spectrum in the lower part of the figure represents the background free bulk emission, see text. The background function is obtained by subtracting this difference spectrum from the measured spectra.

off tails and an inelastic background. The forward scattering enhancement occurs only for the bulk signal, whereas the emission from the topmost surface layer is unaffected by such effects. Also the inelastic background will be largely isotropic due to multiple scattering events. This implies that the surface and background intensities will be proportional to each other. By taking the difference between the bulk and the surface sensitive spectra a simultaneous cancellation of the surface component and the inelastic background can be obtained. The resulting difference spectrum will in this way be a pure bulk spectrum without background but with the shake-up and shake-off contributions. This difference spectrum is shown at the bottom of Fig. 5. The background function in Fig. 5 is obtained by subtracting the pure bulk spectrum from the measured spectra until the intensity under the main line disappears. This background function is smoothed in order to remove effects due to differences in the shake-up satellites in the surface and bulk partial spectra.

The spectrum from the surface layer can be obtained by subtracting the pure bulk spectrum from the measured surface sensitive spectrum. The problem, however, is to know the appropriate amount of bulk intensity in the measured spectrum. In order to estimate this amount it was investigated how the satellite region in the 60° spectrum could be reproduced by a bulk spectrum, a shifted bulk spectrum to represent the surface and a background contribution. From this analysis a surface contribution of 35±5% in the surface sensitive spectrum was estimated. Using this intensity ratio the pure surface spectrum was then obtained by subtraction.

The derived bulk and surface spectra are shown in Fig. 6. A surface shift of -0.31 (6) is obtained for the single hole state. This can be compared with the value obtained by Egelhoff /11/ of -0.43 eV using a similar procedure but with lower energy resolution in the measurement. For the doulbe hole satellite the surface shift is -0.7(1) eV.

The determined shifts agree well with the calculated shifts in Fig. 2, -0.28 eV and -0.67 eV, respectively. This supports the presently used treatment of the single and double hole states. For the derived Auger

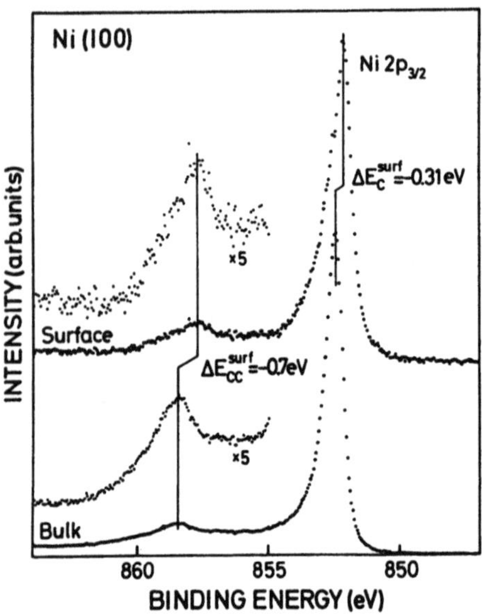

Fig. 6 Constructed bulk and surface $2p_{3/2}$ spectra for a Ni(100) surface. The peak position shifts for the main line (single hole state) and the satellite (double hole state) are marked.

shift an equally good agreement is obtained between experiment (0.4 eV) and theory (0.39 eV). From the previously established agreement between theory and experiment for the single hole states /1,5/ and the presently obtained agreement for the double hole state in Ni, we conclude that also the Auger shifts are correctly predicted with the present theory. This implies surface shifts of the order of 0.5 eV for most metals.

5. Summary and Conclusions

In the Ni2p region the surface and bulk partial spectra have been separated. For the $2p_{3/2}$ main line a surface shift of -0.31(6) eV is determined. For the 2p3d two-hole satellite the surface shift is -0.7(1) eV. The difference in shifts for the single and double hole states immediately shows that there are surface shifts also for the Auger process. The measured shifts are well reproduced by a theoretical treatment based on the complete screening of the ionized states.

6. Acknowledgements

The assistance of J.O. Forsell and H. Rydåker is gratefully acknowledged. This work was supported by the Swedish Natural Science Research Council, the Swedish Board of Technical Development and the Swedish Institute.

7. References

1. B. Johansson and N. Mårtensson, Helv. Phys. Acta 56, 405 (1983).

2. J. Lecante and C. Guillot, Helv. Phys. Acta 56, 427 (1983).

3. P.H. Citrin and G.K. Wertheim, Phys. Rev. B27, 3176 (1983).

4. D. Spanjard, C. Guillot, M.C. Desjonguères, G. Tréglia and J. Lecante, Surf. Sci. Rep. 5, 1 (1985).

5. B. Johansson and N. Mårtensson, Phys. Rev. B21, 4427 (1980).

6. N. Mårtensson, P. Hedegård and B. Johansson, Physica Scripta 29, 154 (1984).

7. S. Hüfner and G.K. Wertheim, Phys. Lett. 51 A, 299 and 301 (1975).

8. N. Mårtensson and B. Johansson, Phys. Rev. Lett. 45, 482 (1980).

9. S. Kono, S.M. Goldberg, N.F.T. Hall and C.S. Fadley, Phys. Rev. B22, 6085 (1980).

10. A.R. Miedema, P.F. de Châtel, In Theory of Alloy Phase Formation, ed. by L.H. Bennett, Proc. 108th AIME Annual meeting (New Orleans 1979) p. 344.

11. W.F. Egelhoff, Phys. Rev. B30, 1052 (1984).

* Permanent address:
Department of Physics, Higher School of Chemical Engineering
8005 BURGAS, Bulgaria

Auger Spectra Induced by Ar^+ Bombardment on Silicon and Silicides

S. Valeri and R. Tonini

Dipartimento di Fisica, Università di Modena,
Via Campi 213/A, I-41100 Modena, Italy

Impact of ions of several keV on solids produces inner-shell excitations with consequent emission of Auger electrons. A significant portion of these electrons comes from particles sputter-ejected from the surface. Therefore the lineshape of Auger peaks involving outer shell electrons is a superposition of band-like and atomic-like structures. We investigated the Ar^+ stimulated Auger electron emission from silicon and Ni-silicides of different stoichiometry. The dependence of such emission on ion energy, target stoichiometry and acceptance geometry was determined. The atomic-like structures dominate in metal-rich silicides, and are enhanced at grazing take-off angle. The parents of atomic Auger electrons are mainly neutral-excited sputtered atoms in Si and Si-rich silicides, while in metal-rich silicides there is a dominant contribution from sputtered ions. The Doppler shift of the Auger peaks indicates that a flux of metal and Si atoms is anysotropically ejected from the surface, with a kinetic energy of more than 1 keV.

1. Introduction

Under heavy ion impact in the kilo-electron-Volts (keV) range, electron emission may take place from a solid surface, due to ionizing collisions experienced by primary ions and energetic recoil atoms. Auger emission related to the decay of the inner-shell vacancies created in either the target atoms and incident ions results in structures superimposed on the continuous tail of secondary electrons in the energy distribution curve [1,2]. The inner-shell vacancy production which precedes the Auger electron emission is explained according to the Fano-Lichten model of molecular orbital promotion [3,4]. Auger electron emission by low energy ($1 \div 5$ keV) ion bombardment is observed in noble, transition and free-electron metals, and in semiconductors [2,5-14]. Emission from the previously implanted projectiles (alkali metals or rare gases) is also reported [6,11,15]. Light solids like Si, Mg and Al were preferentially investigated, due to the unexpected features in the ion-induced Auger emission (IAE), which are quite different from the corresponding electron-induced emission (EAE). However, most of these studies have dealt with pure elements [2,5-11] and very few results were reported for compound and alloys [13,16-20].

In this paper, we present the results of a study on Ar-ion induced Si Auger emission in silicides of different stoichiometry. The IAE silicon spectra consist of sharp atomic-like peaks superimposed on a broad solid-state peak. The atomic-like structures are associated with the Auger decay of sputtered atoms, and strongly depend on the incident and take-off angle: they lead to information on the sputtering mechanism, on the state charge and on the spatial and energy distribution of the sputtered particles. Deviation from a cosine-shaped polar distribution of the sputtered atoms, and the energy distribution dependence on the emission angle were recently reported [21,22], suggesting

Springer Series in Surface Sciences, Vol. 18 Editors: G. Cubiotti · G. Mondio · K. Wandelt
Auger Spectroscopy and Electronic Structure

that the random collision cascade theory [23] is inadequate for a detailed description of the sputtering process. We focused on the spatial and energy distribution of sputtered metal and Si atoms as deduced by IAE spectra, and we compare these results with the corresponding ones obtained by directly collecting the sputtering products on carbon films arranged in suitable geometries near the bombarded silicide surface. Ni silicides have been chosen for this investigation, because of the detailed knowledge of their sputtering behaviour [24].

2. Experiment

Silicides have been prepared by e^- beam heating a calibrated amount of metal and silicon powder. All the samples have been characterized in composition and purity by RBS and XRD: the deviation from the nominal stoichiometry was less than 2 % and the C and O concentration was less than the sensitivity of the techniques used. They were mechanically polished to obtain a flat, mirror-like surface, and subsequently introduced in the UHV system. Conventional Auger depth profile has been used to check the in-depth homogeneity. The exciting probe for the IAE spectra was a differentially pumped Ar^+ gun. The ion energy ranged from 1 to 5 keV, and the ion current was about $2x10^{-1}$ μA over a spot of about 0.2 mm in diameter. A commercial cylindrical mirror analyzer (CMA) operating in the first derivative mode (0.3 % resolution) was used to obtain the dN(E)/dE form of the ion-induced electron distribution. By mechanical shielding, we selected two distinct acceptance cones, corresponding to polar angles of 12° and 72° with respect to the surface normal. Conventional Auger spectra were also recorded, using as exciting probe the electron gun (E_p=3 keV, I_p=0.1 μA) coaxial to the CMA.

In a separate experiment, silicides were bombarded under the same conditions described before and the sputtering products were collected by two carbon covered stainless steel supports placed in the plane of the sample and in the plane of incident ions and centered at the point where the beam axis intercepts the target. The in-depth and lateral metal and silicon distributions in the carbon film were subsequently determined by electron excited Auger spectroscopy.

3. Results and Discussion

In fig. 1 are shown the Auger spectra of NiSi at "normal" take-off angle excited by electrons and ions. The as-recorded dN(E)/dE form is presented, together with the N(E) form obtained by the previous one by numerical integration and background subtraction with a spline method. The EAE spectra are in good agreement with previous results [24]. A comparison with the IAE spectra reveals significant differences: i) the metal peak is reduced in intensity and shifted by a few eV toward lower kinetic energy (K.E). ii) the Si peak presents several structures in the 65÷ 95 eV region. Referring to the dN(E) form in which the different structures are better resolved, we outline that only a small shoulder can be identified at about 92 eV, in correspondence to the main peak in the EAE spectrum. Consequently, we attribute all the relevant peaks in the IAE spectrum to the atomic-like components, in spite of a consolidated tradition which assigns the 94 eV peak to the band-like component [8]. The separation of these two contributions can be conveniently done in the N(E) spectra by subtracting the EAE spectrum from the IAE one, in the assumption [1,2,5,10] that the bulk contribution is satisfactory represented by the EAE spectrum. Figure 2 (left panel) shows the integral form of IAE

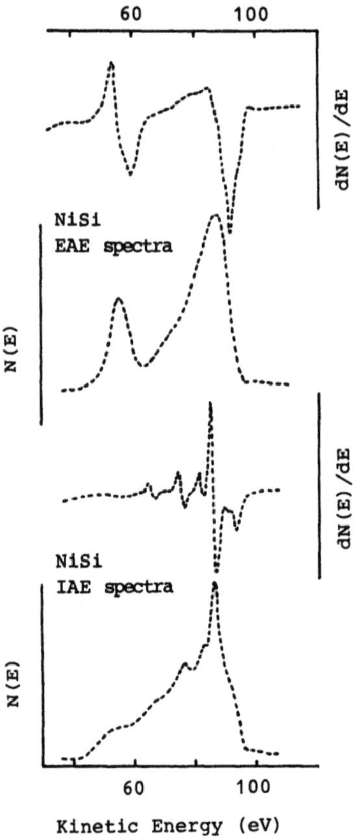

Fig. 1 - Electron-excited and ion-excited Si and Ni Auger spectra from NiSi at normal take-off angle.

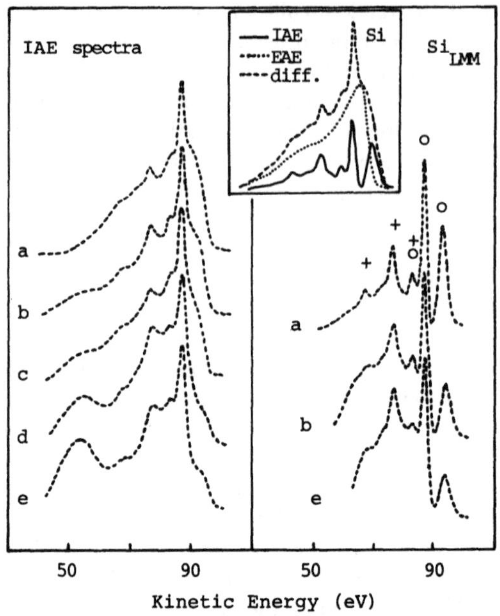

Fig. 2 - Left: IAE spectra of silicon (a) and Ni-silicides of different stoichiometry [$NiSi_2$ (b), NiSi (c), Ni_3Si_2 (d), Ni_3Si (e)]. The inset shows the procedure for the subtraction of the bulk-like contribution [the $Si(L_{23}VV)$ spectrum] from the IAE Si (L_{23}-related) spectrum in pure Si. Right: the results of such a subtraction, i.e. the atomic-like part of the IAE spectrum, are shown for Si, $NiSi_2$ and Ni_3Si. (o) and (+) indicate the peaks associated to excited neutral atoms (Si^+) or ions (Si^o) respectively.

spectra of Si and silicides. The separation procedure is illustrated in the inset of fig. 2 for the Si case. The results of such a procedure are presented in Table Ia and in fig. 2 (right panel). Table Ia shows the percent weight of the "atomic" component of the Si (L_{23}-related) IAE spectra for Si and silicides. It is about 35% in silicon, and increases on going from Si to silicides and also from Si-rich to metal-rich silicides. The Si 2p core hole lifetime is assumed not to vary appreciably with Si content, therefore the results of table Ia are interpreted in terms of the larger energy transfer to Si atoms in a shallow region (= a reduced time spent by Si atoms in the solid) in metal-rich silicides with respect to Si-rich silicides and pure Si [20].

In fig. 2 (left panel) are shown the atomic-like components of the Si (L_{23}-related) IAE spectra for Si and two Ni-silicides of different stoichiometry ($NiSi_2$ and Ni_3Si). In the generally accepted interpretation [1,2,9,18,19] the atomic peaks are assigned to transitions either in Si neutrals with a 2p vacancy (Si^o) and in Si ions with a single (Si^+) or a double (Si^{2+}) 2p vacancy. The Si^{2+} related structures do not appear in fig. 3; it is located between 100 and 120 eV and is very weak with respect to Si^o and Si^+

Table I: (a) percent weight of the "atomic" component of the Si(L_{23}-related) IAE spectra for Si and silicides. The estimate accuracy is about 10 % of the reported values .
(b) percent weight of Si^{2+} yield with respect to the total yield in Si and silicides.

	Si	$NiSi_2$	NiSi	Ni_3Si_2	Ni_3Si
a)	35	51	53	62	65
b)	0.8	1.0	2.5	3.7	6.0

peaks. The energy position of the main Si peaks is insensitive to the stoichiometry . However on passing from Si-rich to metal-rich silicides the intensity of the Si^+ related spectral region (65÷ 85 eV) increases with respect to the Si^o related region (85-100 eV). A neutral excited atom can be produced by capture of a valence- band (VB) electron by an escaping ion with a single 2p vacancy, or in the collision event in a sort of autoionization process [1]. The efficiency of the first process depends on the free electron density and therefore should be higher in metal-rich silicides. As far as the second process is concerned, it is related to the density of Si 2p empty states above the Fermi level, which is reduced in silicides at increasing metal content. Therefore our results suggest that the neutral excited atoms are mainly created in the collision process.

As already mentioned, the Si^{2+} peak is rather weak. It is hardly distinguishable in the N(E) spectral form and can be easier detected and studied in the derivative spctra, as shown in fig. 3a (upper panel)for the $NiSi_2$ target. In spite of its low intensity, it can provide important informations on the collision events. In fact the neutral excited and single ionized atoms can originate from both symmetric (Si-Si within the cascade) or asymmetric (Ar-Si at the surface) collisions. However, the doubly ionized atoms only originate in asymmetric collisions [18,19]. Therefore, the Si^{2+} peak is a fingherprint of the occurence of such direct projectile-target interaction. The peak shape does not change appreciabily with ion energy, therefore the peak-to-peak height in the dN(E)/dE spectrum allows an accurate determination of the Auger yield. Both the total and the Si^{2+} related Auger yields in $NiSi_2$ are reported in fig. 3 vs primary Ar^+ energy (E_p) in the 1 to 5 keV range at a constant ion density of about $3x10^2$ $\mu A\ cm^{-2}$. All the values are normalized to the corresponding ones at 5 keV.

A threshold energy of nearly 1 keV is observed for the Auger emission. The total yield dependence on E_p shows two different regimes, separated by a transition region between 2 and 3 keV. The yield increasing rate is strongly enhanced in the second region ($E_p \geq$ 3 keV) with respect to the first one ($E_p \leq$ 2 keV). We obtained similar results in silicides of different stoichiometry and in pure Si in agreement with those already reported for Ar ions on Al [5], silicon [10] and silicides [20]. The enhanced efficiency observed at high E_p has been often attributed to the onset of asymmetric collision contribution to Si 2p excitation. Althought this interpretation is still controversial, it is strongly supported by the Si^{2+} yield dependence on E_p shown in fig. 3. Indeed the high energy Si peak associated with the asymmetric collisions is not observed below 3.2 keV, and at increasing energy its functional dependence on E_p strongly rensembles that of the total yield. We notice that in Cu-Al and Fe-Al alloys, an energy threshold

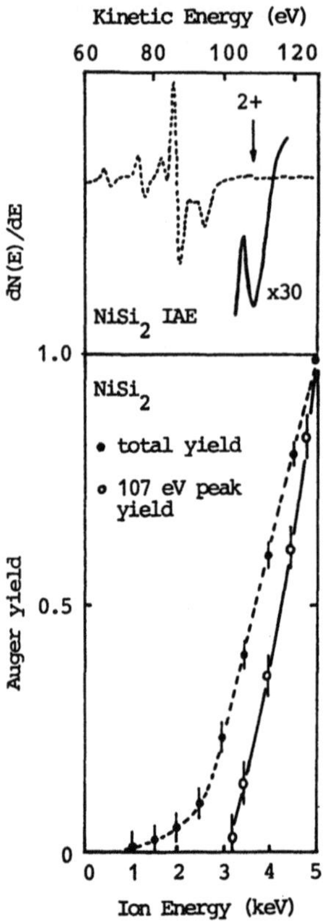

Fig. 3

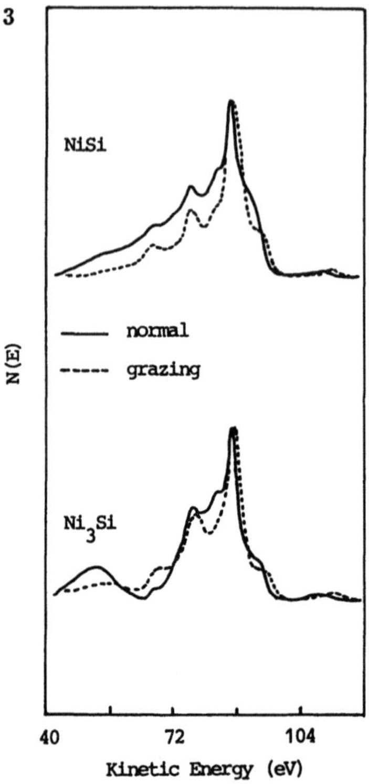

Fig. 4

Fig. 3 - Upper panel: IAE spectrum of $NiSi_2$. Arrow indicates the peak assigned to transitions in Si ions with a double 2p vacancy (Si^{2+}). Lower panel: total (•) and Si^{2+}-related (o) Auger yield vs projectile energy in $NiSi_2$. Each set of values are normalized to the corresponding one at 5 keV.

Fig. 4 - Normalized Si IAE spectra from NiSi and Ni_3Si recorded in normal (—) and grazing (- - -) geometry.

of nearly 4 keV has been determined for the corresponding Al^{2+} related Auger peak [18,19]. However, no detailed Al^{2+} yield dependence on E_p above the threshold has been reported.

In table Ib the percent weight of Si^{2+} yield with respect to the total yield is shown, at E_p=5 keV, for silicon and silicides. In this case both Si^{2+} and total yield were determined from the area of the corresponding regions in the N(E) spectra, because of the different shape of Si(L_{23}-related) IAE spectrum in samples with different stoichiometry. This further reduces the accuracy of Si^{2+} yield measurement, which can now be evaluated to be 30÷ 50 % of the reported values. In spite of this bad accuracy, the Si^{2+}-to-total yield ratio shows a dependence on Si content at a fixed E_p. We can conclude that the lower the Si content in the target, the higher the contribution from asymmetric collisions with respect to the symmetric collision in the cascade.

We will focus now on the angular dependence of the Auger lineshape. Fig. 4 shows the IAE spectra of $NiSi_2$ and Ni_3Si recorded either in "normal" or "grazing" geometry respectively. The Doppler shift, due to the distribution of velocity components of the decaying atoms in the direction of observation, is responsible for the spectral broadening observed in "grazing" with respect to "normal" geometry. A Doppler broadening of the IAE spectra at grazing take-off angles has been reported by several authors [6,7,10,16] and interpreted as evidence of a flux of high energy excited atoms anysotropically ejected from the surface , superimposed to the sputter-induced isotropic emission of low energy atoms. The broadening was found to depend on the K.E. of the parent atoms and on the acceptance geometry [6,7]. This dependence is expressed by $\Delta E = 0.07\sqrt{E(eV)}\cos\theta$, where ΔE is the Doppler shift and θ is the angle between the direction of motion of the parent atom and of the electron. From the measured broadening (fig. 4) and the experimental geometry we conclude that a flux of Si atoms with K.E. of about 1.4 keV is anisotropically emitted in a direction close to the projection of the ion beam on the sample surface. The anisotropy and high energy of this flux, and the strong enhancement of the S^{2+} peak at grazing take-off angle, suggest that this sputtered Si flux mainly originates in asymmetric Ar-Si collisions.

A further and independent check of this spatial anisotropy of the sputtered particles in silicides comes from the Auger analysis of the metal and Si atoms collected in the carbon films as described in the Experiment section. Fig. 5 shows the Ni and Si angular distribution in the incident plane (polar plot at azymutal angle=0°) and in the plane perpendicular to it (azimutal plot at polar angle= 50°), as deduced by the IAE Ni_{LVV} and Si_{KLL} peak-to-peak height (H_{pp}) measurements along the carbon strip. The H_{pp} of both Si and Ni atoms have been normalized to the higher value. These results refer to atoms emitted from $NiSi_2$ sample bombarded at oblique incidence (55° with respect to the surface normal) with 5 keV Ar^+ beam. Other investigated silicides show a very similar behaviour.

The angular distribution shows a marked anisotropy, being peaked toward a near specular direction with respect to the incident beam, in good agreement with the conclusions drawn from the angular dependence of the IAE lineshape. The measured distributions for both Si and Ni strongly deviate from the cosine distribution expected on the basis of the random cascade theory [23]. We outline that deviations from this law

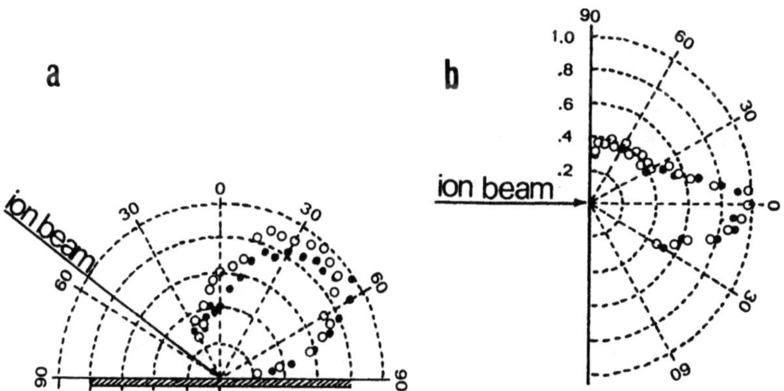

Fig. 5 - Polar (a) and azimutal (b) distribution of Ni (•) and Si (o) atoms sputter-ejected from $NiSi_2$ target, bombarded with 5 keV Ar^+ at oblique incidence (55°). The Ni and Si intensities are normalized to the higher value.

are widely reported in the literature for either normal or oblique incidence, and they are generally interpreted as a failure of the random cascade theory [21,22]. This result confirms that the anysotropic emission originate in asymmetric collision at the surface.

By sputtering with Ar ions of 1 keV we also studied the implantation profile of metal and Si in the carbon film. Comparison of these profiles with Monte-Carlo simulations of metal and Si implantations at different energies enables us to obtain indications on the angular distribution of the energy of particles sputtered from the silicides surface. For silicon atoms ejected from $NiSi_2$ we obtained a value of about 1.2 keV in correspondence of a polar angle of about 50° and azimutal angle of 0°. We believe that the agreement with the value obtained by the Doppler shift is quite good, taking into account the uncertainities associated with the Doppler-derived and profile-derived results, which we estimate to be about $\pm 10\%$ and $\pm 30\%$ respectively.

We aknowledge the guidance received from G.Ottaviani in this work. We also would to thank E.Angeli for helpful technical assistance. This work was supported by CICAIA, Universita' di Modena, and by CNR and MPI.

4. References

1. E.W.Thomas: Vacuum **34**, 1031,(1984).
2. J.Mischler and N.Benazeth: Scann.Electr.Microscopy **II**,351,(1986).
3. U.Fano and W.Lichten: Phys.Rev.Lett. **14**,627,(1965). W.Lichten, Phys. Rev. : **164**,131,(1967).
4. M.Barat and W.Lichten: Phys.Rev. **A6**, 211,(1972).
5. O.Grizzi and R.A.Baragiola:Phys.Rev.A **35**,n.1,135,(1987).
6. S.V.Pepper and P.R.Aron: Surf.Sci. **169**,14,(1986).
7. R.A.Baragiola,E.V.Alonso and H.J.L.Raiti:Phys.Rev. **A25**,1969,(1982).
8. K.Saiki and S.Tanaka:Jap.J.of Appl.Phys. **21**,L529,(1982).
9. R.Whaley and E.W.Thomas: J.Appl.Phys. **56**, 1505.(1984).
10. K.Wittmaack: Surf.Sci. **85**, 69,(1979).
11. K.Wittmaack: Phys.Lett. **74A**, 197, (1979).
12. K.O.Legg,W.A.Metz and E.W.Thomas: J.Appl.Phys. **51(8)**, 4437, (1980).
13. F.Chuizhen,Y.Zhenjiang and C.Xukang: J.Vac.Sci.Technol.**A5(4)**, 1206,(1987).
14. E.V.Alonso,M.A.Alurralde and R.A.Baragiola: Surf.Sci. **166**,L155, (1986).
15. G.Zampieri and R.Baragiola: Phys.Rev. **B33(1)**,588,(1986).
16. M.Iwami,S.C.Kim,Y.Kataoka,T.Imura,A.Hiraki and F.Fujimoto:Jap.J.Appl. Phys. **19**,1627,(1980).
17. Y.Yamauchi,I.Ogoh,R.Shimizu and H.Hashimoto : Jap.J.Appl.Phys. **24**, n.2, L157 , (1985).
18. P.Viaris de Lesegno and J.F.Hennequin: Surf.Sci. **103**,257,(1981).
19. J.F.Hennequin,R.L.Inglebert and P.Viaris de Lesegno: Surf.Sci. **140**, 197, (1984).
20. S.Valeri,R.Tonini and G.Ottaviani: Phys.Rev. **B15** in press.
21. G.Brauer,D.Hasselkamp,W.Kruger and A.Scharmann: Nucl.Instr.and Meth. **B12**, 458, (1985).
22. M.Hautala and H.J.Whitlow: Nucl.Instr. and Meth., **B6**,466,(1985).
23. P.Sigmund:In Sputtering by particle bombardment 1,ed. by R.Behrisch, Topics Appl.Phys. vol.47 (Springer,Berlin,Heidelberg,1981)p.9.
24. S.Valeri,U.del Pennino,P.Sassaroli and G.Ottaviani: Phys.Rev. **B28**, 4277, (1983).

Auger and Secondary Emission Electronic Structural Characteristics in Metals and Insulators

H.G. Maguire

Department of Electrical and Electronic Engineering,
Trent Polytechnic, Nottingham NG1 4BU, UK

Secondary electron emission measurements performed in a conventional UHV LEED/Auger chamber are used to investigate band structural characteristics of type 1A diamond and Be surfaces prepared under UHV conditions. The Auger spectra complements the techniques of X-ray photoemission and soft X-ray emission. For diamond, various regions in the secondary spectrum delineate electron band structural characteristics whilst for Be the Auger spectra are even more critical in this respect.

1. Introduction

Diamond (space group O_h^7) is the classic model of an insulating covalent crystal on which numerous band structure calculations have been performed, the most recent indicate a splitting of the conduction band into two bands separated by a well-defined energy gap PAINTER et al [1]. Provided the primary beam is below the K-shell energy ≃ 284eV any subsidary maxima in the secondary distribution is due to promotion from occupied valence band states to unoccupied final conduction band states. Also, because of the narrow 1s width (<1eV) greater structural detail should be revealed by excitation from this level.

The true Auger spectrum of clean Be has been the subject of considerable debate PATTINSON et al [2] particularly in relation to surface contaminants (oxygen) but not in terms of valence band DOS distribution.

2. Insulators - Diamond

The Auger spectrum obtained from a clean diamond surface has been discussed elsewhere in detail MAGUIRE [3], in which it is shown that Auger emission occurs from predominantly p-states near the top of the band probably because - unlike 2p->1s which is allowed - 2s->1s is a forbidden transition. However, no information is available on the energy-dependent transition matrix elements for the Auger experimental emission intensity. No Auger plasmon losses are observed unlike the lowered losses in the near elastic region, showing that unlike Si the process (two step) in diamond is slow since the collective oscillations have died away before emergence of the Auger electron.

The K-shell ionisation loss spectrum of diamond is shown in Fig 1 which reflects structurally reproducible features in DOS calculations by KEOWN [4] and PAINTER et al [1]. Features A, B, C, and D reflect JDOS between final states and the initial 1s level which because of its narrow width and providing the transition probability is not significantly energy dependent should reflect structure in the final DOS, as observed. Discontinuity of the trace in (c) separating peaks B and C coincides with the energy gap

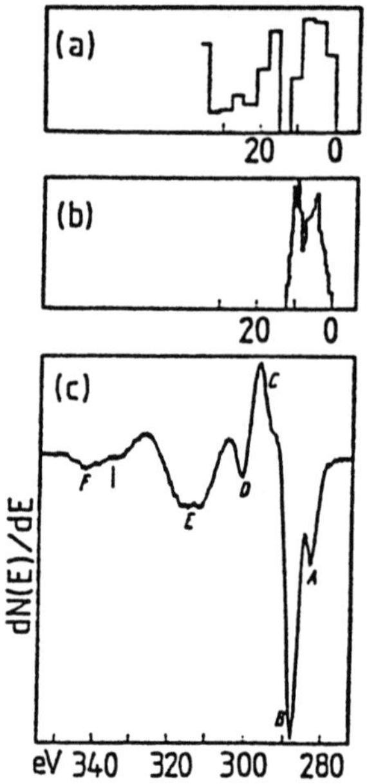

Fig 1 K-shell edge spectrum

verifying theoretical predictions. Tailing of C into the energy gap is most probably due to the effects of lifetime broadening and core-hole perturbation of final states resulting in tailing into the energy gap. Features E and F coincide with bulk plasmon effects but are not due solely to this because (1) their width and profile do not correspond exactly with bulk and surface plasmons observed in the near elastic region - also a one-step process, (2) the occurence of structure in these features. Undoubtedly the major contribution to near edge structure is due to these collective oscillations but other factors such as the JDOS and the matrix element $\rho = \langle f|e^{iqr}|i\rangle$ reflecting bound electron transitions. Coulomb attraction between the positive core hole and electrons below the appropriate bound levels may produce excitonic effects where the final state is a band state not directly included in the band structure. Also plasmon dispersion due to single electron excitations across the band gap at large scattering angles as here resulting in energy and wave vector conservation allow the plasmon to give up its energy directly in exciting electron-hole pairs across the Fermi surface resulting in Landau damping. In addition, for these large scattering angles optically forbidden transitions may occur, the sum of all these effects giving rise to divergences from plasma resonances only as observed in features E and F. Extended structure observed - not shown - does not strictly represent electronic but rather atomic structure and occurs as an intensity modulation ≃ 5%.

The total calculated width of valence and conduction bands is ≃ 60eV, and primary beam energies greater than this but less than the 1s level (284eV) would be expected to populate final states. Because of momentum transfer indirect transitions are expected BAUER [5] their only observable effect being a broadening in the structure due to predominatly direct processes. Close agreement is observed between structural maxima in Fig 2 and symmetry points and lines above the vacuum level in the calculated energy bands of Fig 3, PAINTER et al [1] indicating excitation via direct transitions involving zero or small momentum changes. Such structure will not accurately reflect a DOS function since it is convoluted by the large valence bandwidth, and also due to relaxation processes involving inelastic

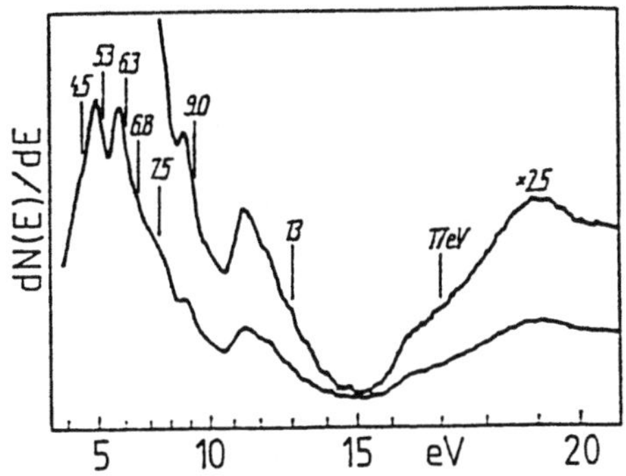

Fig 2 Secondary emission distribution showing superimposed subsidiary maxima resulting from interband transitions

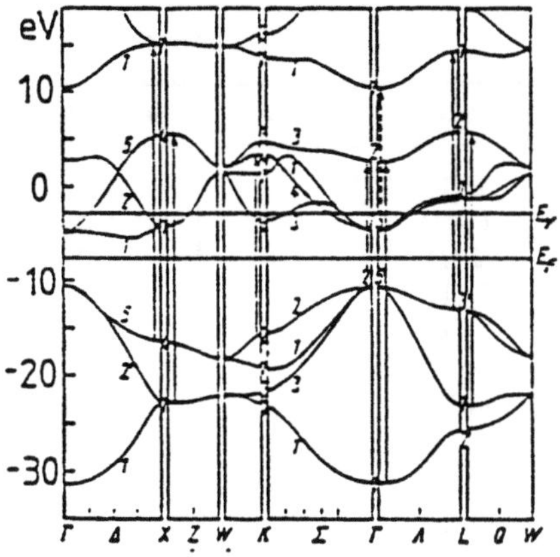

Fig 3 Energy band structure - transitions

electron-electron and electron- phonon scattering and possible Auger processes before emission. Possible inter-band transitions contributing to the structure in Fig 4 may be identified from selection rules for electric dipole transitions obtained from the crystal symmetry described by the space group isomorphic with the group of the symmetry transformations of the Hamiltonian describing the interaction between E (longitudinal) of the incident and crystal electrons. The Eigenfunctions are characterised by irreducible representations of the group in matrix form BOUCKARET et al [6] whose traces form the character tables for the appropriate point groups. The selection rules are derived from BASSANI [7]

$$\alpha_i^j = \frac{1}{h} \sum_R \chi_j(R)\chi_{xyz}(R)\chi_i(R) \quad \begin{cases} = 0 \text{ forbidden} \\ \\ \neq 0 \text{ allowed} \end{cases} \qquad (1)$$

where h is the number of transformations contained in the group of the Hamiltonian $\chi_j(R)$ and $\chi_i(R)$ the characters of the j-th or i-th irreducible representations of the transformation R, and $\chi_{xyz}(R)$ the characters of the irredicible representation to which the co-ordinates belong. Character tables at symmetry points and lines in the Brillouin zone STREITWOLF [8] allow calculation of the transition probabilities by substitution into (1). For cubic crystals the co-ordinates x, y, z transform according to Γ_{15} and at the zone centre, Γ-point

$$\langle \Gamma_{2'}, \Gamma_{15} | \Gamma_{15} | \Gamma_{25'} \rangle \neq 0, \quad \text{also} \quad \langle \Gamma_{2'}, \Gamma_{15} | \Gamma_{15} | \Gamma_1 \rangle \neq 0 \qquad (2)$$

only transitions to $\Gamma_{2'}$ are observed under present experimental conditions because it lies above the vacuum level. Possible transitions at various symmetry points and lines in an elementary region in the Brillouin zone are calculated (Table 1) providing good agreement between the energy band calculations of Fig 3 and the structural maxima observed in Fig 2 in which separation in maxima observed at about 9 and 13eV correspond with transitions to the top of the first and bottom of the second conduction bands respectively. The main contribution to the broad structural feature at 13eV will be due to allowed transitions extending over regions in k-space along the directions $\Gamma - \Delta, \Sigma, \Lambda$.

Table 1 Transition energies calculated from Fig 3 to first and second conduction bands

conduction band state	calculated	observed	possible transitions
$W(1)$; $K_2(1)$	4.3	4.5 (0.5)	$W_{1,2} - W_{1,2}(1)$; $K_{1,3} - K_1(1)$
$\Gamma_{2'}(1)$	5.4	5.3 (0.5)	$\Gamma_{25',1} - \Gamma_{2'}(1)$
$K_4(1)$	6.2	6.3 (0.5)	$K_{1,2} - K_4(1)$
$\Sigma_3(1)$	6.4	6.8 (0.5)	$\Sigma_{1,2,3} - \Sigma_3(1)$
$K_3(1)$	7.4	7.5 (0.5)	$K_{1,2,4} - K_3(1)$
$L_{2'}(1)$; $X_4(1)$	8.5	9.0 (0.5)	$L_{1,2,3} - L_{2'}$; $X_{1,2,3} - X_4(1)$
$\Gamma_1(2)$	13.0	13.0 (1.0)	$\Gamma_{15} - \Gamma_1(2)$; $\Lambda - \Lambda(2)$; $\Sigma_{1,3} - \Sigma_1(2)$; $\Delta_{1,5} - \Delta_1(2)$
$K_1(2)$; $\Sigma_1(2)$	17.0	17.0 (1.0)	$\Sigma_{1,3} - \Sigma_1(2)$; $W_{1,2} - W_{1,2}(2)$; $X_{1,4} - X_1(2)$; $K_{1,3} - K_1(2)$
$K_3(2)$	18.8	19.4 (0.5)	$K_{1,2,3} - K_3(2)$
$K_2(2)$	22.0	22.2 (0.5)	$K_{2,3} - K_2(2)$
$\Gamma_{25'}(2)$	23.5	23.5 (0.5)	$\Gamma_{15} - \Gamma_{25'}(2)$
$X_4(2)$	27.0	27.0 (0.5)	$X_{1,4} - X_4(2)$

3. Metals - Beryllium

The surface of Be foil is inevitably coated with oxide immediately on insertion into the UHV chamber prior to any cleaning procedure. Heating to 1050°C for five minutes produced the two major lines shown in Fig 4(a) which degraded with time due to system ambients. The two line spectrum was subsequently regenerated by further heating and/or evaporation of fresh Be onto the Be foil surface. Lowered losses were observed corresponding to bulk plasmon excitations associated with the individual peaks. Ar ion

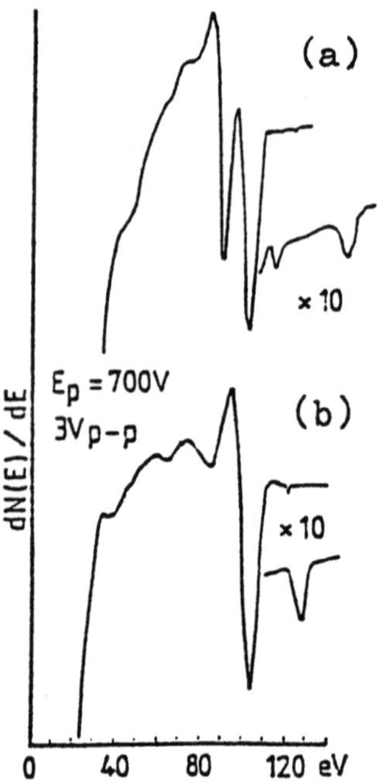

Fig 4 Auger spectra from Be (see text)

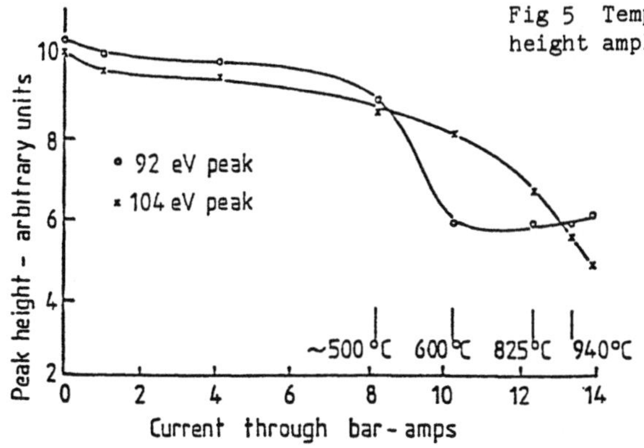

Fig 5 Temperature variation in peak height amplitudes of Fig 4(a)

bombardment of the clean Be surface produced the dramatic changes shown in Fig 4(b) resulting in one major peak of width equivalent to the two major peaks but in addition severely contaminated with Cl, Ar, C and O. The two peak spectrum could be regenerated by the above procedure. The only available temperature dependent Auger spectrum from Fig 4(a) is shown in Fig 5 and is the only temperature dependent Auger electron emission available. Early work on the appearance of extra lines in the x-ray pattern report the existence of a β-phase occuring at $\simeq$ 500°C KOSOLAPOV et al [9] whilst thermal emf and electrical resistivity measurements also indicated an allotropic transformation at about $\simeq$500°C BUNCE et al [10]. Various techniques of differing sensitivities including specific heat and dilatometric measurements also indicate such a transition and it is suggested that the presence of impurities is responsible for a large-celled hexagonal form of the α-phase. It seems likely that such a phase transition is indicated by the temperature dependent Auger spectra since the divergence in peak amplitudes occurs at about the same temperature reported by various investigations/techniques and the maximum divergence in peak amplitudes shown also occurs at approximately the same temperature as the maximum intensity of the extra lines in the x-ray spectrum $\simeq$650°C, reported above.

Controvesy exists as to the true Auger spectrum of Be. The results here unequivocately identify the two line Auger spectrum from a clean uncontaminated Be surface. Existence of only two peaks shows that the downward transition 2s->1s ejects electrons from 2s-like states in the valence band but not from 2p-like states. This is probably because 2s->1s is a forbidden transition and provides a potential for the up-electron of the form

$$\int \psi_{2s}{}^*(r_1)\psi_{1s}(r_1)\ \frac{e^2}{r_{1,2}}\ d^3x_1 \qquad (3)$$

falling off as $\exp-(r_2/a_{1s})$ where a_{1s} is the radius of the K electron and will only eject electrons which penetrate the K shell. The 2s electrons do, thus giving rise to the 1s2s2s transition at 92eV. The 104eV peak is assigned as a 1s2p2p transition which is allowed and the two peaks observed clearly indicate that the p and s electron density distributions are sufficiently well separated. No other fine structure effects were observed in the energy distribution across the band. However, evaporation onto a cold (room temperature) substrate, or Ar ion bombardment of a thermally cleaned surface, appeared to have created a disordered phase in which the 2s and 2p distributions were insufficiently well separated, thus giving one peak only, of the same width as the two peaks. Drastic changes in the valence-bandwidth is not possible (as observed) since this involves energies of the order of electron volts, while the energy of disordering is comparable to latent heats of fusion ≃0.1eV. In the disordered state (after Ar ion bombardment) a given atom sits in an unsymmetrical position with respect to its neighbours and to satisfy boundary conditions at the edge of the Wigner-Seitz cell the 2s wave function at the bottom of valence band will be admixed with 2p, then any down-transition from the valence band bottom now has additional 2p nature and is accompanied by an up-transition from the p-state density maximum near the top of the band giving rise to Auger emission at energies between the two peaks in the crystaline case as observed here.

Attendent with the Auger spectra of Fig 4 is also observed minor peaks - 2-3% - due to stimulated de-excitation of the emerging Auger electrons by energy absorbtion from collective bulk oscillations generated within the solid. Such bulk plasmon gains are not due to phonon interactions since their energies are of an order of magnitude 10^{-4} smaller than the energies of collective oscillations observed here. The differing Auger spectra above result from differing electron distributions which do not affect the volume electron density and hence the bulk plasma vibrations as observed in both cases. It is not possible to determine the mechanism of plasmon gain either by complete relaxation of the vacancy before emission - 2-step process - or whether relaxation and ionisation may be treated as an entity. The results in Fig 4 are at variance with those of WEIGHTMAN [11] and is most probably due to lack of instrumental resolution.

4. Conclusions

Secondary electron emission provides a method for evaluating the electron structural characteristics for both wide-gap (insulators) and zero-gap (conductors). In particular, estimates of the valence band width from the Auger emission spectra fall within the range of values found for both insulators and metals but no detailed conclusions may be drawn regarding its intensity variation.

For diamond the K-shell ionisation spectrum confirms predicted conduction band structure for two separate conduction bands, but it is not possible to explain the intensity variation across these bands. Also, for primary energies below the K-shell provides a means for investigating the energies of symmetry points and lines in an elementary region of the Brillouin zone and the present work endorses band structure calculations. For Be both the one- and two-peak spectrum are compatible with surface treatment and that the 92eV peak in the two-peak spectrum is due neither to surface oxidation nor absorbed and/or surface segregation of bulk impurities.

The usefulness of secondary and Auger emission from well characterised surfaces has clearly demonstrated this facility in characterising electron structural characteristics and is more sensitive than the normally better known external probe techniques of electron and x-ray diffraction which give an overall integrated response as opposed to the techniques demonstrated here which involve probes of the individual electronic structure of the atoms from which the Auger electrons emerge.

5. References

1. G.S. Painter, D.E. Ellis and A.R. Lubinsky, Phys. Rev., B4 (1971) p.3610.

2. E.B. Pattinson and M. Suleman, J. Phys F, Vol.1 (1971) L.24; Ibid., Vol.3, p.497.

3. H.G. Maguire and P.D. Augustus, Phil. Mag. Vol.30 (1974) p.95.

4. R. Keown, Phys. Rev., Vol.150, (1966) p.568.

5. E. Bauer, Z.Phys., Vol.224, (1969) p.19.

6. L.P. Bouckaret, R. Smoluchowski and E.P. Wigner, Phys.Rev., Vol.50 (1936) p.50.

7. G.F. Bassani, Rend. Scuola Internaz. Fis. 'Enrico Fermi', XXXIV. Corso (1966) p.33.

8. H.W. Streitwolf, Gruppentheorie in der Festkorperphysik, Akademische Verlagsgesellschaft Geest & Portig, Leipzig (1967).

9. G.F. Kosolapov and A.K. Trapoznikov, Z. Dristallogr. A, Vol.94 (1936) p.53.

10. E.J. Bunce, A.J. Martin and P.D. Tilbury, J.less-common Metals, Vol.4 (1962) p.191.

11. P.J. Weightman, Proc. 1st Int. Wkshp. Auger Specty. & Electnc. Structure (1988).

XPS and Auger Spectra of Some Differently Substituted Indoles

R. Girlanda[1], *G. Martino*[1], *A.M. Mezzasalma*[1], *G. Mondio*[1], *G. Saitta*[1], *and R. Giorgi*[2]

[1]Istituto di Struttura della Materia, Università di Messina, GNSM-CISM Università di Messina, Messina, Italy
[2]ENEA, Centro Ricerche Casaccia, Laboratorio Superficie e Metallurgia Fisica, Rome, Italy

1. Introduction

In this work, some preliminary results concerning XPS and Auger spectra of a number of organic compounds are presented. In fact, they are to be considered as a starting point in the study of the electronic properties of a group of charge transfer compounds built up by Tetracyanoquinodimethane (TCNQ) and a set of differently substituted Indoles. In such compounds, TCNQ acts as an acceptor molecule while Indoles represent the donor counterparts /1,2/. More specifically the Indoles' Nitrogen atom transfers part of its electronic charge towards the TCNQ molecule, so realizing a charge transfer bond. Thus, the knowledge of the valence properties of these materials, is important for a better understanding of the electronic properties of these donor-acceptor complexes. The recent interest in this type of organic compounds is mainly due to their remarkable electrical features, in particular, a sharp metal-insulator transition must be expected as in the case of the TTF-TCNQ complex.

Here we present the XPS and AES spectra of the two Indoles $C_{22}H_{19}N$ (N-methyl 2,3-diphenyl 5-methyl Indole) and $C_{20}H_{15}N$ (2,3 diphenyl Indole) whose basis structures are shown in Fig. 1 and 2, respectively.

They are characterized by a common base formed by the Indole molecule with different substituents in 1,2,3 and 5 positions. A CNDO type calculation of the electronic energy levels has allowed us to compare the experimental results with the predictions of the theory, and good agreement has been found between the XPS spectra of the compounds and the corresponding calculated valence band density of states (DOS).

The Auger profiles seem to be well described by a selfconvolution of the valence DOS without invoking matrix element effects or Coulomb interaction between the two holes in the final state.

Fig. 1

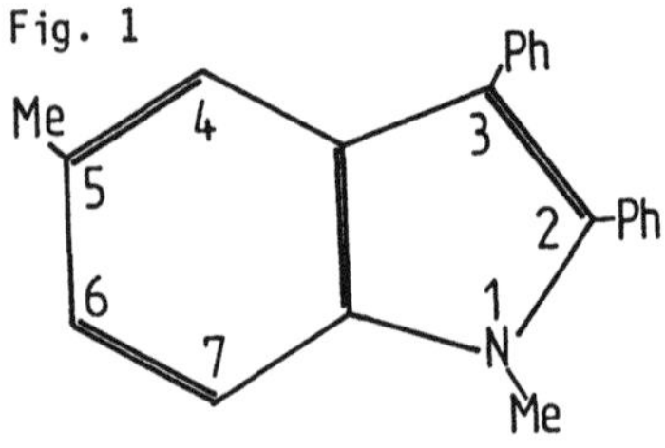

Fig. 2

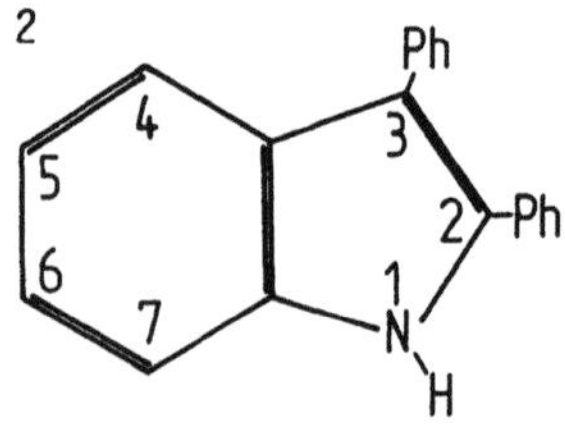

Fig. 1 The basis structure of the compound $C_{22}H_{19}N$ (N-methyl 2,3-diphenyl 5-methyl Indole)

Fig. 2 The basis structure of the compound $C_{20}H_{15}N$ (2,3 diphenyl Indole)

Springer Series in Surface Sciences, Vol. 18 Auger Spectroscopy and Electronic Structure
Editors: G. Cubiotti · G. Mondio · K. Wandelt

2. Theory

The electronic energy levels of $C_{22}H_{19}N$ and $C_{20}H_{15}N$ have been calculated by a Complete Neglect of Differential Overlap (CNDO) method. CNDO provides a useful approximation to the Hartree-Fock equations for the treatment of the electronic energy levels by explicit inclusion of electron and core interactions. When applied to a small cluster model or to a molecular unit, it shows the advantage of interpreting the parametrization on the grounds of the main interactions between atoms /3-5/.

The parametrization that we have adopted to carry out the calculations is the same as the one used for calculating the electronic energy levels in TCNQ /6,7/. For the details we refer to references /2/ and /6/. Here we only briefly summarize the main features.

We have adopted a modified Sichel and Whitehead scheme /8/ using the Mataga formula for the interatomic electron-repulsion integrals γ_{AB} /9/ and the Pople expressions for the bonding parameters β_A and β_B /10/. These CNDO parameters are those that allow a good fit of the energy gap of the material as deduced by optical absorption and reflectivity measurements. The resonance integrals ($\beta_{\mu\nu}$) have been taken as:

$$\beta_{\mu\nu} = \alpha \cdot 1/2 \cdot (\beta_A + \beta_B) \cdot S_{\mu\nu}$$

where $S_{\mu\nu}$ are the computed overlap integrals.

The set of coordinates of the C, N and H atoms in the molecules used in the calculation, are those obtained by X-ray crystallographic analysis and their values are reported elsewhere /11/.

The parameter α in the expression of the resonance integrals has been chosen in order to fit the optical gap by the lowest allowed transition energy:

$$\Delta E_{ij} = E_j - E_i - I_{ij} + 2K_{ij}$$

E_i (E_j) are the occupied (empty) levels, I_{ij} and K_{ij} are the Coulomb and exchange integrals respectively /6/.

The valence band densities of states have been calculated on the basis of a procedure presented elsewhere /2/ and are reported in figures 3 and 4.

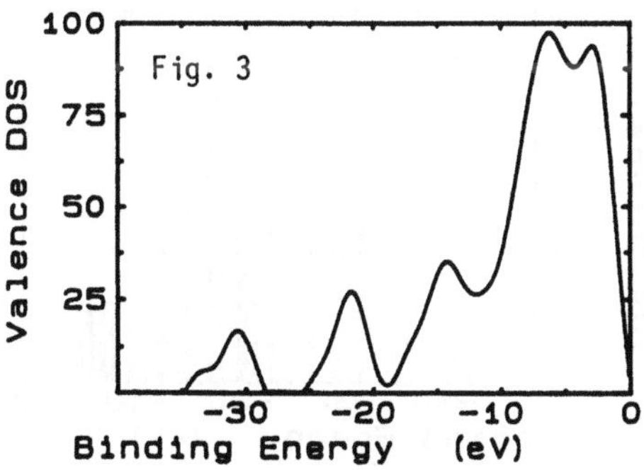

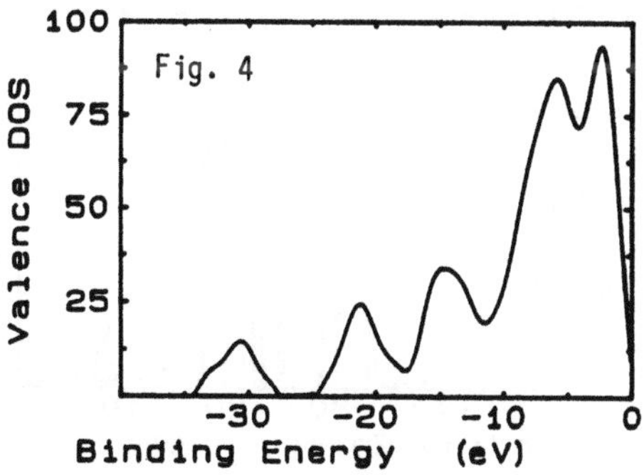

Fig. 3 Valence band density of states of $C_{22}H_{19}N$

Fig. 4 Valence band density of states of $C_{20}H_{15}N$

3. Results and Discussion

In figures 5 and 6 the XPS spectra of $C_{20}H_{15}N$ and $C_{22}H_{19}N$ are presented after a subtraction of the inelastic background due to secondary electrons. As can be seen, the two spectra are characterized by similar groups of peaks with similar relative intensities and spacings. This must be ascribed to the common basic structure of the molecules involved, the differences arising from the chemical substitutions around the main Indole rings.

Starting from the calculated valence band densities of states, an attempt has been done to identify the origin of the observed structures both from the point of view of the atomic species and the electronic symmetries.

In order to obtain a qualitative description of the spectra, we have considered the CNDO-deduced energy levels. Each occupied level E_i is supposed to give a contribution to the photoemission whose intensity is given by the overlap integral

$$I(E_i) = | < \chi \mid \varphi > |^2$$

where $|\chi >$ and $| \varphi >$ are the final (free electron) and initial states of the process.

In the LCAO-MO frame, the i-th molecular orbital, assumed as the initial state, is described by a linear combination of atomic orbitals:

$$|\varphi > = \Sigma_A \Sigma_\lambda C_{A\lambda i} \mid \Phi_{A\lambda}>$$

where A and λ stand for the atomic species and the orbitals symmetry respectively.

The photoemission intensity, in the MO approximation is then /12/:

$$I^{MO}(E_i) = |\Sigma_A \Sigma_\lambda C_{A\lambda i} < \chi | \Phi_{A\lambda} > |^2$$

For each energy E_i, the XPS intensity can be evaluated by substituting each overlap integral $< \chi | \Phi_{A\lambda} >$ with the photoionization cross section of the atomic level $\Phi_{A\lambda}$.

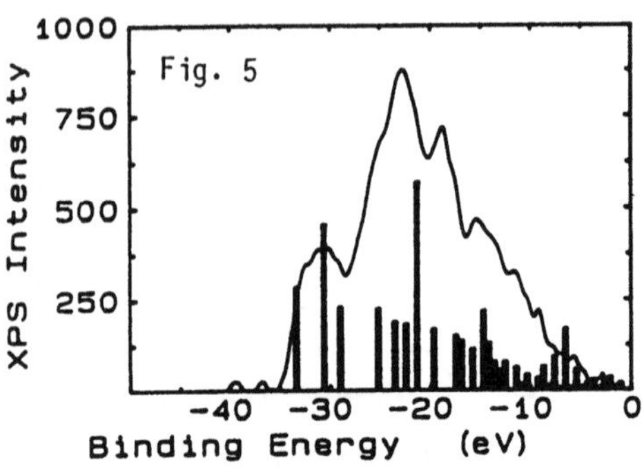

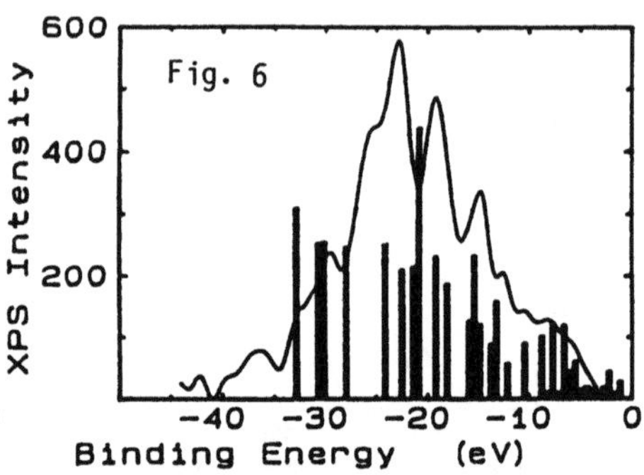

Fig. 5 XPS spectrum of $C_{22}H_{19}N$ (solid line). The histogram refers to the theoretical photoemission yield calculation described in the text

Fig. 6 XPS spectrum of $C_{20}H_{15}N$ (solid line). The histogram refers to the theoretical photoemission yield calculation described in the text

A histogram can be constructed by dividing the energy range of the levels in a certain suitable number of channels and then summing the contributions arising from all the levels falling into a certain channel. The thus deduced histograms are reported in the figures 5 and 6 together with the corresponding spectra. A good qualitative agreement is found between the experimental data and the theoretical intensities, this proving the reliability of our CNDO calculations.

Following the above mentioned method, it was also possible to separate the contributions arising from different atomic species and from different symmetries.

The Carbon s-contribution to the photoemission was found to be mainly responsible for the XPS intensity at least from 6 to 35 eV. The top of the valence band is to be ascribed to p-levels of Carbon and Nitrogen but the corresponding features are negligible with respect to the higher energy Carbon s-contributions.

As a consequence of these considerations also the DOS curves in figures 3 and 4 may be considered as essentially built up by Carbon s-states, thus they may be assumed as local densities of states, useful in interpreting the Carbon KVV Auger spectra.

In figures 7 and 8 the Carbon KVV Auger spectra of $C_{22}H_{19}N$ and $C_{20}H_{15}N$ are reported together with the self convolution of the density of states of figures 1 and 2 respectively. No Coulomb interaction between the two final state holes was introduced due to the band-like profile of the experimental spectrum. The agreement between theory and experiment seems to suggest, also in this case, the reliability of the CNDO method in describing the electronic properties of these materials. In particular we think that the agreement between theory and experiment seems to be stimulating for our future studies of organic compounds and in particular for the investigation of the electronic properties of charge transfer compounds realized utilizing Indoles.

In conclusion no specific difference in the electronic spectra seems to be introduced by the different chemical substituents in the main Indole rings of our compounds. It is our opinion that a detailed analysis of the top of the valence band by means of UPS measurements, may reveal more important differences characterizing this family of Indole derivatives.

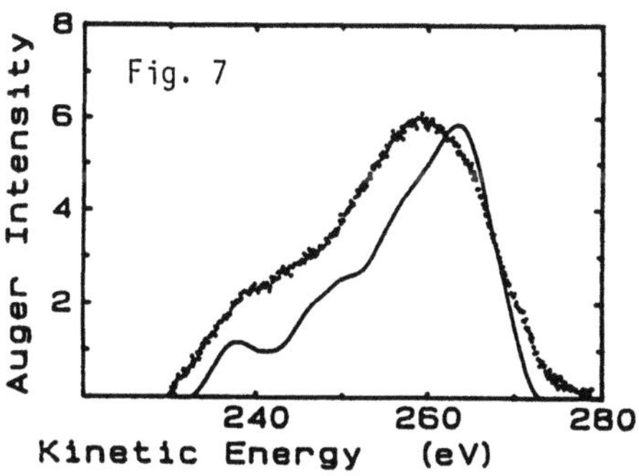

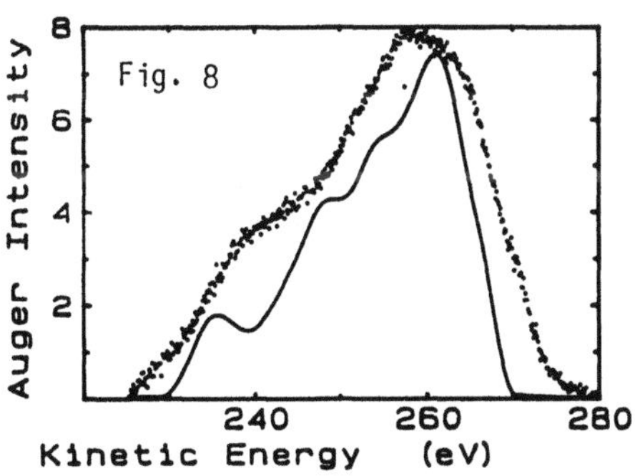

Fig. 7 Auger profile of $C_{22}H_{19}N$ (dotted curve) and the theoretical profile (solid curve) calculated as the selfconvolution of the density of states of figures 3

Fig. 8 Auger profile of $C_{20}H_{15}N$ (dotted curve) and the theoretical profile (solid curve) calculated as the selfconvolution of the density of states of figures 4

References

1 S.F. Lin, W.E. Spicer and B.H. Schechtman: Phys. Rev. B 12, 4184 (1975)
2 R. Girlanda, G. Martino, A.M. Mezzasalma, G. Mondio and G. Saitta: Solid State Commun, 60, 873 (1986)
3 A. Zungher, A. Katzir and A. Halperin: Phys. Rev. B. 13, 5560 (1976)
4 P.R. Underhill and J.A.D. Matthew: J. Phys. C 13, L791 (1980)
5 A. Balzarotti, F. Antonangeli, R. Girlanda and G. Martino: Phys. Rev. B. 29, 5903 (1984)
6 R. Girlanda, G. Martino, A.M. Mezzasalma, G. Mondio, P. Perillo and G. Saitta: II Nuovo Cimento 7 D, 469 (1986)
7 R. Girlanda, G. Martino, A.M. Messasalma, G. Mondio and G. Saitta, II Nuovo Cimento 8 D 229 (1986)
8 J.M. Sichel and M.A. Whitehead: Theor. Chim. Acta 7, 32 (1967), 11, 220 (1968), 11, 239 (1968)
9 N. Mataga: Bull. Chem. Soc. Japan 31, 453 (1958)
10 J.A. Pople, D.P. Santry and G.A. Segal: J. Chem. Phys. 43, S129 (1965), 43, S136 (1965), 44, 3289 (1966)
11 R. Girlanda, G. Martino, G. Mondio, G. Capozzi and G. Valle: To be published
12 U. Gelius and K. Siegbahn: in "X-ray Photoelectron Spectroscopy" edited by T.A. Carlson (Dowden, Hutchinson and Ross Inc., 1978), pp 212-223; Faraday Disc. 54, 257 (1972)

Near Threshold Effects in Ar and CO Auger Spectra

L. Avaldi[1], *G. Stefani*[1], *A. Lahmam-Bennani*[2], *A. Duguet*[2], *I. Harrison*[3], *and G.C. King*[3]

[1]IMAI-CNR, CP 10, Area della Ricerca di Roma, I-00016 Monterotondo, Italy
[2]LCAM, Université Paris-Sud, F-91405 Orsay Cedex, France
[3]Schuster Lab., University of Manchester, Manchester M139PL, UK

1. Introduction

Auger and autoionisation spectra are often interpreted with the assumption that the energy and angular distribution of autoionisation- and Auger-electrons can be precisely predicted from the initial and final states of the target. This is the case only if the creation of the inner-hole, and the subsequent decay can be treated as distinct processes and if the independent free particle approximation is valid. In recent years efforts [1,2] have been made to study collision processes where the projectile and target fragments do not move independently in the Coulomb field of the residual ion. The mutual electrostatic interactions among the free particles may produce a variety of characteristic phenomena, most of which are known under the term of "Post Collision Interactions" (PCI) [3,4] . PCI produce energy and momentum exchanges which manifest themselves as energy shifts and broadening of the Auger and autoionisation peaks [5,6] and distortions in the angular distribution [7] of the reaction products. In the case of an Auger emission following photoionisation PCI can be considered as a threshold process, which vanishes as soon as $h\nu > \epsilon_b + E_A$ [4,5], where ϵ_b is the binding energy of the inner hole and E_A is the energy of the Auger electron.In the case of an Auger emission induced by electron impact ionisation,however it has been recently shown that there remains a finite, asymptotically constant PCI effect, even at the highest impact energies, E_0 [1,6]. In this latter case, which can be schematically written as

$$e(E_0) + A \longrightarrow A^+ + e_s(E_s) + e_{ej}(E_{ej}) \quad (1)$$
$$\qquad\qquad\qquad \hookrightarrow A^{++} + e_A$$

the two electrons (e_s and e_{ej}), undetected by Auger spectroscopy,share the available energy, $E_0-\epsilon_b = E_s+E_{ej}$,in a continuous way so that there is always a finite probability for one of them to have a vanishingly small [1] kinetic energy. This asymptotic PCI effect, though small, is not negligible for today's "state of the art" in electron spectroscopy. It is therefore meaningful to attempt to obtain a better understanding of the PCI effects and their implications on Auger and autoionisation spectroscopy.

This paper reports on an investigation of the Ar $L_3M_{23}M_{23}$ Auger emission and of the satellite Auger decay of $CO(1\sigma_{s_C})(2p\pi)$ $^{1,3}\Pi$ states excited by electron impact close to threshold.

2. Argon: (e,2e) and (e,e'Auger) coincidence experiments

A complete kinematic understanding of the process

$$e(E_0) + Ar \longrightarrow Ar^+ + e_s(E_0-E) + e_{ej}(E-\epsilon_b) \qquad (2)$$
$$\hookrightarrow Ar^{++} + e_A$$

would be achieved by an experiment which detects in coincidence all the three final electrons. As a first step towards this goal we have separately measured the angular correlations of the two pairs of unbound electrons generated in the same ionising event, namely the scattered (e_s) and ejected (e_{ej}) electrons (e,2e) and the scattered and Auger (e_A) electrons (e,e'Auger).The process (2) has been investigated at an incident energy of 8 KeV and for two different energy losses,$E=E_0-E_s$, of 256±5 and 308±5 eV.Since the Ar L_3 ionization energy is 248.5 eV,they correspond to ejected electron energies of 7±5 and 60±5 eV,respectively. The high incident energy and the small scattering angle (θ_s=1.5°) resulted in a momentum transfer ($K=|\vec{K}_0-\vec{K}_s|$ = 0.75 a.u.)close to the minimum kinematically allowed one (≈0.38 a.u.). The vast majority of electron induced ionising events take place upon these kinematics, hence the relevance of achieving a K value as small as possible. A complete description of the experimental apparatus used in these measurements has been reported elsewhere [7].

The Ar $L_{23}M_{23}M_{23}$ Auger spectrum was measured in the range 200-210 eV in coincidence with scattered electrons that had suffered energy losses of 256 and 308 eV (Figs. 1a and b respectively) and had been scattered through an angle of 1.5° . To calibrate the energy scale the non-coincidence Auger spectra were recorded at the same time of the coincidence ones. In fig.1 the broken line is the best fit to the coincidence Auger spectrum, while the full line is the best fit to the non-coincidence one. The fits were achieved by applying a minimised x^2 criterion to a multi-gaussian trial function where widths,heights and positions of the gaussian functions have been used as free parameters. The centroids of the coincidence peaks are clearly shifted towards higher energy values . These shifts, predicted by several authors [4,8,9,10], have been recently observed by other independent coincidence experiments [11,12] and are interpreted as a PCI effect. The present energy resolution is not good enough to allow the determination of the Auger lineshapes, which have been shown to bear further traces of PCI as well as information about interference effects in the final state excitation [11,1].

The interpretation in terms of PCI of the measured energy shifts is supported by the coincidence Auger spectrum measured at an excess energy of 60 eV above the L_3 ionisation threshold . The coincidence spectrum (fig.1b), limited to the region 204-207 eV, is indistinguishable from the coincidence one.This finding is consistent with a semiclassical interpretation of PCI [4] that predicts the extinction of these effects as the energy of the ejected electrons approaches the Auger electron energy.

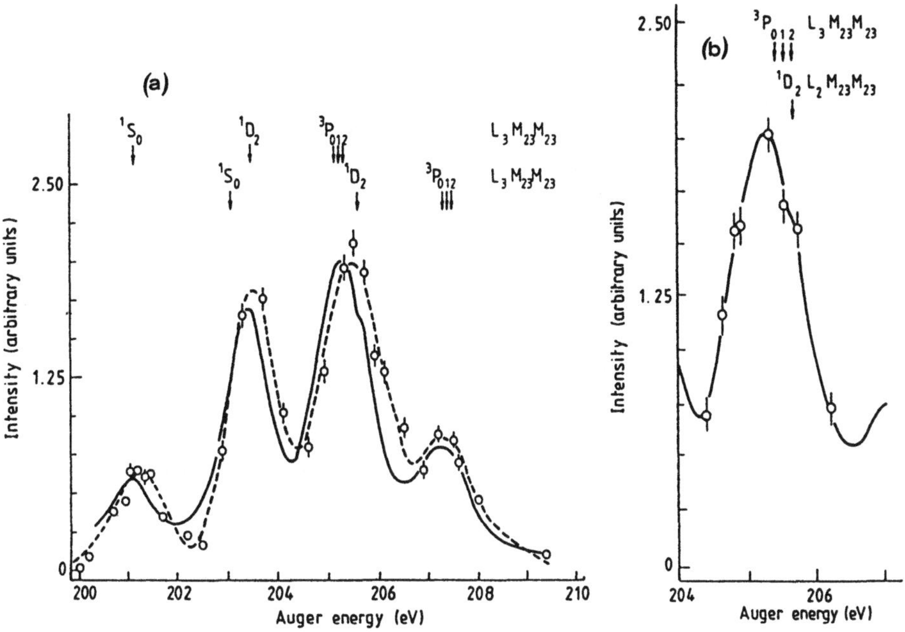

Fig.1 : a) The $L_{23}M_{23}M_{23}$ Auger spectrum (o) measured at θ_A=288 in coincidence with 8000 eV scattered electrons at 1.5° .The energy loss suffered by the primary electrons is 256 eV (i.e. 7 eV excess energy above the L_3-subshell ionisation threshold). b) As figure 1a for 60 eV excess energy.

Assuming the validity of the two step model the coincidence Auger angular distribution is determined by the polarisation state of the intermediate ion[13,14].By neglecting exchange between the scattered and ejected electrons and reducing the ionising interaction to a first order one (1st Born two step model 1B2S) Berezhko et al. [15] showed that the coincident angular distribution in the scattering plane reduces to

$$W(\theta_A) \propto 1 + \beta\left[2 \cos(\theta_A-\psi)\right] \qquad (3)$$

where β ,is the degree of anisotropy of the angular distribution and ψ is the direction of the momentum transfer. Within this picture the first Born approximation implies an axial symmetry around $\hat{K}$ for both the (e,e'Auger) and (e,2e) angular correlation spectra.
The results of the (e,e'Auger) measurements at 1.5° scattering angle and 7 eV excess energy are shown in fig. 2a. The broken curve is the 1B2S model prediction [15], while the chain curves are separate best fits to the experimental data in the two half planes of the angular distribution. The values of β and ψ predicted by the 1B2S theory and those obtained from the best fits are given in table 1. The main feature of fig. 2a is the absence of a common symmetry axis for the angular distributions in the two opposite half

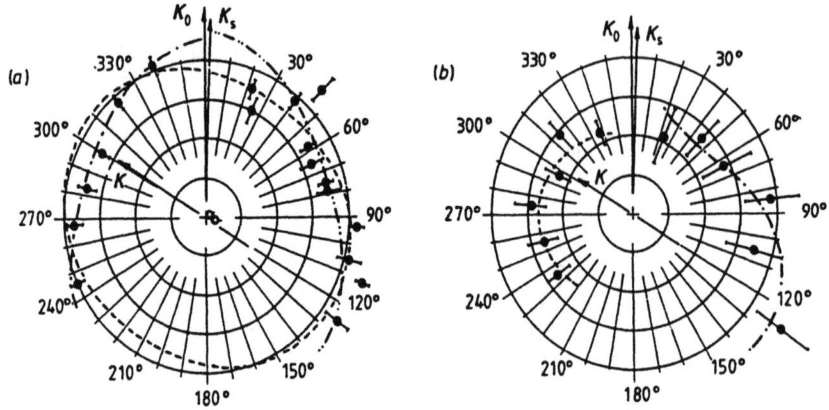

Fig. 2: The (e,e'Auger) angular distribution of the $L_3M_{23}M_{23}(^1S_0)$ transition (a) and the (e,2e) angular distribution (b). The kinematical conditions are E_0=8256 eV, E_s=8000 eV, E_{ej}=7±5 eV and θ_s=1.5°

Table 1: Best fit parameters of the (e,e'Auger) angular distributions compared with 1B2S model predictions.

θ_s(deg)	E_{ej}(eV)	θ_A(deg)	ψ(1B2S)(deg)	ψ(expt)(deg)	β(1B2S)	β (expt)	K(au)
1.5	7±5	<180	301.8	252±4	+0.09	−0.17±0.04	0.75
1.5	7±5	>180	301.8	289±5	+0.09	−0.15±0.04	0.75
1.5	60±5	<180	306.6	290±17	+0.09	+0.08±0.07	0.79

planes. This can be interpreted in terms of PCI. Further support to this interpretation is given by the (e,2e) angular distribution (fig. 2b) measured upon kinematics identical to those of the (e,e'Auger) experiments. The broken and chain curves represent two separate fits to the binary and recoil lobes with the trial function, typical of photoionisation

$$W(\theta_{ej}) \propto 1 + \frac{1}{2}\alpha\left[3\cos^2(\theta_{ej}-\psi)-1\right] \qquad (4)$$

where α and ψ have been used as fitting parameters. This angular distribution confirms the finding of the (e,e'Auger) experiment, i.e. the violation of the symmetry for the two lobes. Similar measurements performed in conditions which are identical, but for the energy of the ejected eletrons (60 instead of 7 eV), have shown for both the (e,2e) and the (e,e'Auger) angular distributions good symmetry around the $\hat{K}$-direction and do not present noticeable disagreement with the 1B2S model.In this latter experiment also the β value agrees both in sign and absolute value with the prediction of the 1B2S theory.

It is therefore straightforward to associate the difference in the behaviour of the two ionising processes to the energy of the ejected electrons and to the interaction of these electrons with the fast Auger ones.

3. Carbon monoxide: satellite Auger spectrum

The process

$$e(E_0) + CO \longrightarrow CO(1\sigma s_C)(2p\pi)^{1,3}\Pi + e(E_0-\epsilon_b) \quad (5)$$
$$\hookrightarrow CO^+ + e_A$$

has been investigated by detecting Auger electrons in the energy range 268-275 eV for various values of incident energy from threshold up to 500 eV. Details of the experimental set up as well as the procedure used in the measurements has been described elsewhere [16,17].A series of satellite Auger spectra obtained at various values of incident energy are shown in fig. 3. The peaks have been assigned to the various transitions following Siegbahn et al. [18] and Ungier and Thomas [19]. Peaks labelled A-a and A-b correspond to

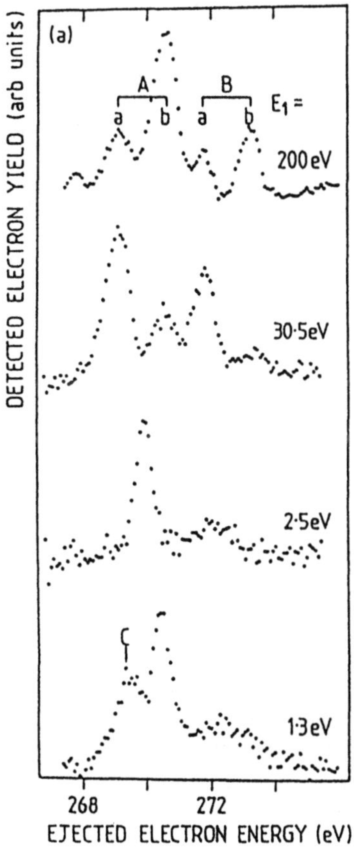

Fig. 3: CO satellite Auger spectra in the ejection energy range 268-275 eV at several values of the excess energy.

transitions from the $CO(1s\sigma_C)(2p\pi)$ $^{1,3}\Pi$ excited states to the $CO^+(1\pi)^{-1}$ ion state, while the ones labelled B-a and B-b correspond to transitions to the $CO^+(3\sigma)^{-1}$ ion state. At the highest value of incident energy used, the multiplet splitting of the singlet and triplet states is particularly evident. The value of 1.5±0.6 eV found for the multiplet splitting is in agreement with previous results from electron energy loss experiments [20] and coincidence scattered-Auger electron measurements [19]. As the incident energy is lowered towards the threshold the peak (labelled A-a in fig. 3) corresponding to the decay of the triplet state increasingly dominates the spectrum. Moreover this peak suffers a shift to higher values of ejection energy. Close to threshold the occurence of an extra peak, marked C in fig. 3, is observed. This peak is attributed to the decay of the $(1s\sigma_C)^{-1}(2p\pi)^2$ inner-shell temporary negative ion of CO [21]. The behaviour of the additional peak was further investigated by measuring excitation functions at different values of the energy loss [16]. The results seem to support the idea that the inner-shell temporary negative ion decays directly to the CO^+ ion rather than through the inner-shell parent state. This finding is in agreement with previous results by Ziesel et al. [22].

The shift, ΔE, in the position of the satellite Auger peak corresponding to the decay of the $(1s\sigma_C)(2p\pi)$ $^3\Pi$ state to the $CO^+(1\pi)^{-1}$ state is plotted against the excess energy, $E_1=E_0-\epsilon_b$, of the incident electron in fig. 4. The values of the shift are measured with respect to the satellite Auger spectrum obtained at a value of E_1 of 200 eV.
The variation of ΔE with respect to E_1, over the range of E_1 from 0.4 to 4.5 eV is well represented by the relationship

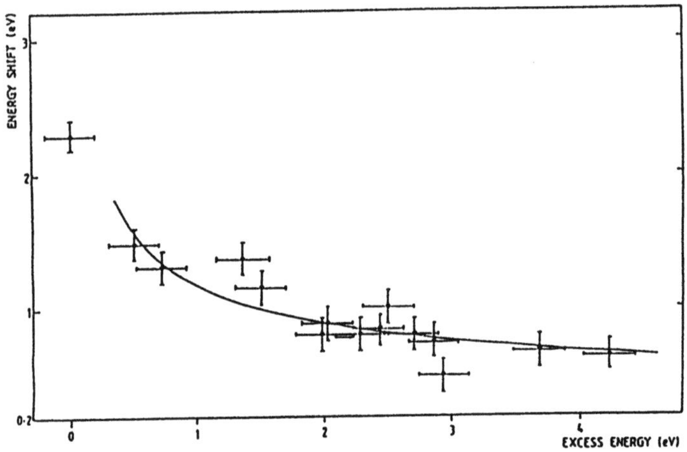

Fig. 4: Energy shift of the satellite Auger line versus the excess energy. The full curve represents relationship (6).

$$\Delta E = (13.8 \pm 0.5)\ \Gamma\ E_1^{-(0.4 \pm 0.04)} \qquad (6)$$

where the coefficient 13.8 and the exponent 0.4 result from a χ^2 minimization procedure.The relationship (6) resembles the scaling law given by the semiclassical PCI theories [3,8] which predict the shift to be proportional to $E_1^{-0.5}$. The observed absolute values of the shift are, however, much larger than predicted by these theories for a value of Γ of 88 meV [20]. The observed shifts might be due to effects other than PCI.The peak under study does not correspond to a single isolated transition, but involves many vibrational levels [18]. The relative intensities of the various transitions will depend on the relative vibrational cross sections,that may change with the incident electron energy. This might also cause a shift in the satellite Auger peak and prevents a direct comparison with PCI lineshape and energy shifts predicted by the semiclassical PCI theories[3,4,8]. Shifts of satellite Auger lines larger than PCI predictions have also been measured by Venezia et al. [23] in a threshold photoionization studies of the L_2 shells of Ge compounds. This result has been interpreted as a screening effect produced by the inner-shell excited electron. Due to the difference in the final state energy levels different screening effects are expected in different molecular systems. Further contribution to the observed shift might also derive from the effect recently investigated by Brown et al. [24] and Armen et al. [25]. These authors showed that directly at threshold the creation of the inner hole and the Auger emission occur in a single process: the Auger resonant Raman effect. This effect, which is expected to be relevant in the excess energy range of a few Γ, also results in a shift of the Auger peak and in a modification of the lineshape.On the basis of the present experimental data a clear distinction among the possible contributions can not be made. However it is evident that in the case of the molecular satellite Auger decay the effect due to the Coulomb interactions between the escaping electrons are mixed with other dynamical and structural effects.

4. Conclusions

We have investigated the correlation between the processes of inner hole creation and the subsequent relaxation via Auger-electron emission.

The energy shift of the coincidence Auger spectrum and the distortion of the (e,2e) and (e,e'Auger) angular distributions, detected at an excess energy of a few eV above the Ar L_3 ionisation threshold, have been attributed to PCI between the slow ejected electron and the Auger electron in the residual ion field. Theoretical two-step models based on first order interactions are inadequate to describe the experimental results. The introduction of distorted waves [27] for the free electrons can account for differences between the symmetry axis of the angular distributions and the $\hat{K}$ direction in the (e,e'Auger) experiments, but not for different symmetry axes in the two half-planes. A correct description of these experiments needs the introduction of at least second order terms in the description of the reaction mechanism, together with PCI effects.

In the case of the experiment on CO the large shift observed in the satellite Auger spectrum is not completely understood. It cannot be explained only in terms of PCI effects. The present results and the findings by Venezia et al. [24] suggest that several different dynamical and structural effects might contribute to the energy shift observed in complex systems that are excited close to threshold.

In the CO experiment an additional structure, attributed to the excitation of an inner-shell temporary negative ion, has been observed for the first time in the electron decay channel.

Acknowledgments

The authors would like to thank CNR-Progetto Strategico "Dinamica Molecolare", SERC and the Italy-France cooperation program for their financial support.

+Permanent address: Dep. Electrical and Electronic Engineering, Un. Nottingham,Nottingham NG7 2RD, England

References

1. W. Sandner in Proceedings XV ICPEAC, Brighton, U.K., 22-28 July 1987, H.B. Gilbody, W.R. Newell, F.H Read and A.C.H. Smith Ed.s North Holland, Amsterdam 1988 and references therein quoted.
2. V. Schmidt J. Phys.(Paris), C9, 401 (1987) and references therein quoted.
3. R. B. Barker and H.W. Berry Phys. Rev. 151, 14 (1966)
4. A. Russek and W. Mehlhorn J. Phys. B:At. Mol. Phys. 19, 27 (1986)
5. M. Borst and V. Schmidt Phys. Rev A33, 4456 (1986)
6. D. Graf and W. Hink J. Phys. B:At. Mol. Phys. 19,L221(1986)
7. G. Stefani, L. Avaldi, A. Lahmam-Bennani and A. Duguet J. Phys. B:At. Mol. Phys. 19,3787 (1986)
8. A. Niehaus J. Phys. B:At. Mol. Phys. 10,1845 (1977)
9. M. Yu. Kuchiev and S.A. Sheinerman Sov. Phys. JEPT 63, 986 (1986)
10. G.B. Armen, J. Tulkki, T. Aberg and B. Crasemann Phys. Rev. A36, 5606 (1987)
11. W. Sandner and M. Völkel J. Phys. B:At. Mol. Phys. 17,1597 (1984)
12. E.C. Sewell and A. Crowe J. Phys. B:At. Mol. Phys. 17,2913 (1984)
13. U. Fano Phys. Rev. 90,577 (1953)
14. A. J. Ferguson Angular Correlation Method in Gamma Ray Spectroscopy , 1965 (New York:Wiley)
15. E. G. Berezhko, N.M. Kabachnik and V.V. Sizov J. Phys. B:At. Mol. 11,1819 (1978)
16. I. Harrison,G.C. King and L. Avaldi to be published in J. Phys. B:At. Mol. Phys. 1988
17. I. Harrison Ph.D. thesis 1986, Victoria University Manchester

18. K. Siegbahn, C. Nordling, G. Johansson, J. Hedman, P.F. Heden, K. Hanner, U. Gelius,I. Benmark, O. Werme, R. Manne and Y. Baer in "ESCA Applied to Free Molecules" (North Holland :Amsterdam) 1969
19. L. Ungier and T.D. Thomas Chem. Phys. Lett. 96, 247 (1983)
20. D. Shaw, G.C. King, D. Cvejanovic and F.H. Read J. Phys. B:At. Mol. Phys. 17,2091 (1984)
21. G.C. King, J.W. McConkey, F.H. Read and B. Dobson J. Phys. B:At. Mol. Phys. 13,4315 (1980)
22. J.P. Ziesel, D. Teillet Billy, L. Bauby and R. Paineau Chem. Phys. Lett. 63,47 (1979)
23. A. Venezia, J. Vayrynen and R.G. Cavell J. Chem. Phys. 87,5195 (1987)
24. G.S. Brown, M.H. Chen, B. Crasemann and G.E. Ice Phys. Rev. Lett. 45,1937 (1980)
25. G.B. Armen, T. Aberg, J.C. Levin, B. Crasemann, M.H. Chen, G. E. Ice and G.S. Brown Phys. Rev. Lett. 54,1142 (1985)
26. E.G. Berezhko and N.M. Kabachnik J. Phys. B:At. Mol. Phys. 15,2075 (1982)

On the Silicon CVV Spectra of Tetramethylsilane and Other Molecules in the Gas Phase

M. Cini[1], *F. Maracci*[2], *and R. Platania*[2]

[1]Dipartimento di Fisica, Università di Roma II, Via O. Raimondo, I-00173 Rome, Italy

[2]Istituto Metodologie Avanzate Inorganiche, Consiglio Nazionale delle Ricerche, Casella Postale 10, I-00016 Rome, Italy

We have measured the Si,C and Cl Auger CVV spectra of the $(CH_3)_nCl_{4-n}Si$ series (n=0,4) with a 1800 eV, 70 μA exciting electron beam. Several spectra were repeated with a 1000 eV beam, and no important change was noticed. The energy resolution for Si was 0.5 eV. Details of the experimental apparatus and procedures have been presented previously /1/. The spectra were also published /2/, with a preliminary analysis of the Tetramethylsilane (TMS) one. Here, we wish to propose a follow-up of that analysis which should be relevant to the spectra of Si containing molecules and to the more general problem of the analysis of the molecular spectra.

Our considerations are based on a full application of the matrix form of the theory of CVV line shapes that was proposed long ago by one of us /3 - 4/. The theory was extended to take into account the interaction of s and p holes; the matrix form is needed here as in most applications, although the tetrahedral symmetry simplifles our task a good deal. We may take the one-hole density of states matrix to consist of two diagonal pieces ϱ_s and ϱ_{1m} for s and p holes respectively.

Using McGUIRE's /5/ and JENNISON's /6/ matrix elements, empirical energy levels /7/ and selfconsistent parentages /7/ we were able to compute the CVV spectrum without adjustable parameters. We computed the repulsion matrix form Hartree-Kock wave functions /8/, and included a common shift <U> of the final state energy levels due to the finite molecular size according to the THOMAS-WEIGHTMAN /9/ prescription; <U> was estimated from molecular geometry to be ~4.9 eV. In Fig.1 we compare the data with an improved version of our previous analysis, in which we corrected a minor error in our computer code (whose consequences, however, can hardly be noticed) and also included the effects of the 0.6 eV Spin-Orbit splitting of the $L_{2,3}$ core level. The overall agreement between theory and experiment is quite encouraging, especially if one considers that the code runs on a personal computer.

We are able to reproduce the positions and intensities of many structures very well, however, it appears that several peaks of the experimental spectrum are missed completely. Discussing our results with P. WEIGHTMAN, we received the clue that the 'missing' peak at ~34 eV binding energy could be due to the production of L_1 holes by the primary beam. They would decay through $L_1L_{2,3}V$-VVV transitions to 3 holes states, which are not contained in our theory. Later, we became aware of experimental results by the DE SOUZA group /10/, which gave us a direct confirmation of this idea. Indeed, their CVV spectra taken with synchrotron radiation of energy close to the $L_{2,3}$ threshold are influenced by threshold effects, with an increased weight of the low binding energy region compared to our data; this complication does not conceal the important fact, that the 'missing' peak is totally absent.

We calculated the relative intensities of the transition in the Coster-Kronig spectrum from TMS in an independent-particle scheme similar to published analysis of elemental Si /11/. The current is found to be, in arbitrary units

Springer Series in Surface Sciences, Vol. 18 Editors: G. Cubiotti · G. Mondio · K. Wandelt

Auger Spectroscopy and Electronic Structure

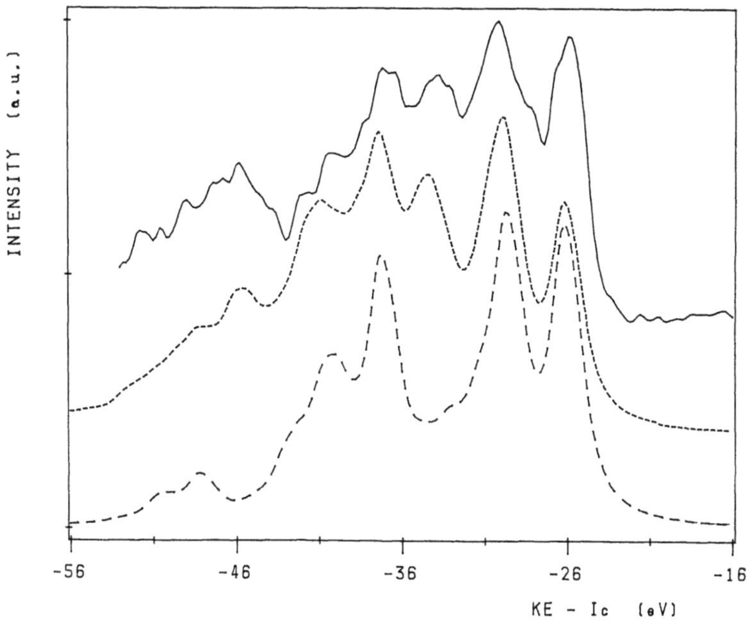

Fig.1 solid line: Si $L_{2,3}VV$ spectrum of TMS on a two-hole binding energy scale; dotted line: calculated line shape including the satellites, as explained below; broken line: line shape including two-hole final states and accounting for the splitting of the $L_{2,3}$ level.

$$J_{L_1,L_{2,3}V} = 12.2\ \varrho_s + 27.77\ \varrho_{1m} \qquad (1)$$

Since the valence band consists of 5 levels, we expect 5 main transitions, where V stands for the $3t_2$,$2t_2$,$2a_1$,$1t_2$ and $1a_1$ energy levels. Using the parentages, the intensities in arbitrary units turned out to be 6.39, 2.5, 3.29, 2.22 and 2.19 respectively. We expect that the interactions should shift them by ~<U> with respect to the independet particle calculations. Taking this into account, the estimated kinetic energies were 35.5, 32, 30.6, 24.4 and 22.1 eV, respectively. Unfortunately such low kinetic energies are very noisy for our apparatus and we were unable to positively identify these transitions experimentally. However this calculation provided us with some argument to say that an important fraction of the Coster-Kronig transitions will leave the valence electrons in the molecular ion at (or near) its ground state, and the ground state should be the most important contribution. Then we can give a qualitative estimate of the 3 holes contribution to our CVV spectrum. Let L(E) be the line shape as obtained by the above theory; if the 3 hole states are shifted by a roughly constant amount S and each of the two hole states generates a satellite which bears a fraction α of its intensity we may expect to see a line shape of the form $L_{tot}(E) = L(E) + \alpha L(E+S)$. The shift S should be given by the balance of electrostatic energy in the final 3 hole state and in the 'initial' state of the $L_{2,3}V$-VVV transition (containing 1 core and 1 valence hole). Since the repulsion energy between one core and one valence hole may be expected to be somewhat larger than that between valence holes, we argue that S will be less than <U> but more than <U>/2, say. We tried a simulation using S and the satellite relative intensity α as parameters. The 'missing' peak then arises as the 3 holes counterpart of its neighbor structure in the 2 holes spectrum, although its intensity turns our to be too low. The presence of important structure in the high binding energy side of the spectrum, which is not yet explained by the 3 holes states, strongly suggests that we have to look for still higher excitations. The simplest idea is perhaps assuming

$$L_{tot}(E) = L(E) + \alpha(1)L(E+S) + \alpha(2)L(E+2S) \qquad (2)$$

This means that we are postulating a second satellite spectrum, shifted by 2S, that could be interpreted as due to 4 holes final states, such as can be produced by shake-off during the 3 holes process. Assuming S=4.4 eV, and the satellite intensities $\alpha(1)=0.48$ and $\alpha(2)=0.40$ we obtain the simulation of Fig.1. Not only the 'missing' peak is now well reproduced, but the whole spectrum now resembles the experimental one. We believe that such a good agreement gives weight to the seemingly bold idea of the 3 and 4 holes peaks; the above treatment of states originating in the Coster-Kronig process, though very simple minded, should be better than their neglect and the '4 holes states' bring intensity where it is needed and simpler processes do not help. We have to recognize that an estimate of $\alpha(1)$ and $\alpha(2)$ parameters is difficult. It would require considering complicated many-body processes also involving the interaction with the primary beam, and we have no detailed theory for doing that at present.

To substantiate the above speculations we may adopt a phenomenological viewpoint and look for other examples of such phenomena in Si containing molecules. We started this analysis with the $SiCl_4$ molecule, which contains Si in the same symmetry as TMS and lends itself to a similar analysis. Unfortunately, we could not find calculated parentages in the literature, and we could not even find a complete information on the levels: we only know the upper 5 of them, with binding energies 12.1, 12.95, 13.44, 15.2 and 17.98 eV /12/. There is a similar lack of informations concerning the other molecules of the series. Thus, we are forced to turn the argument around and see whether we can deduce information concerning the parentages from the experimental spectrum. Although the unknown parentages are 5 the last two turn out to be small and rather uninfluential and an overall factor on the parentages will produce a small effect since the spectrum is rather band like. Thus, the problem may be essentially reduced to a much more manageble two-parameter fit. Thus, it is simple to obtain a simulation like that of Fig.2, where the parentages were taken to be 0.2, 0.05, 0.15, 0.05 and 0.05, respectively. Such parentages do not differ too much from those of TMS /7/. All other data are the same as for TMS, except that $\langle U \rangle$ as obtained by the molecular geomtry is now $\sim$11.3 eV. The computed spectrum reproduces many of the experimental features, which is more significant since this time much of the analysis had to be 'pulled out from the air' due to the lack of input data from the literature. This means that this simple sort of analysis can be widely applied to poorly characterized molecules with useful results.

Presumable, the true CVV transitions are those which are well described, and the analysis serves to recognize them from structures of other origin. Again, we find significant satellite structures at high binding energies and a 'missing' peak, $\sim$7 eV to the left of the main two-holes structure, that we tentatively identify as due to the least bound three hole state. This assignement could be verified by performing an experiments like /10/. We are currently extending the present analysis to other molecules to see if the trend is systematic. For the time being, we conclude that we have found strong (if not compelling) evidence for very prominent 3 and probably also 4 holes structures in the spectra of Si containing molecules. In part, this could have been anticipated due to the presence of competing Coster-Kronig transitions stemming from L_1 holes, but we have no adequate explanation for the large intensity of these satellites. This situation might be fairly general. The real issue for the theoretical analysis of molecular

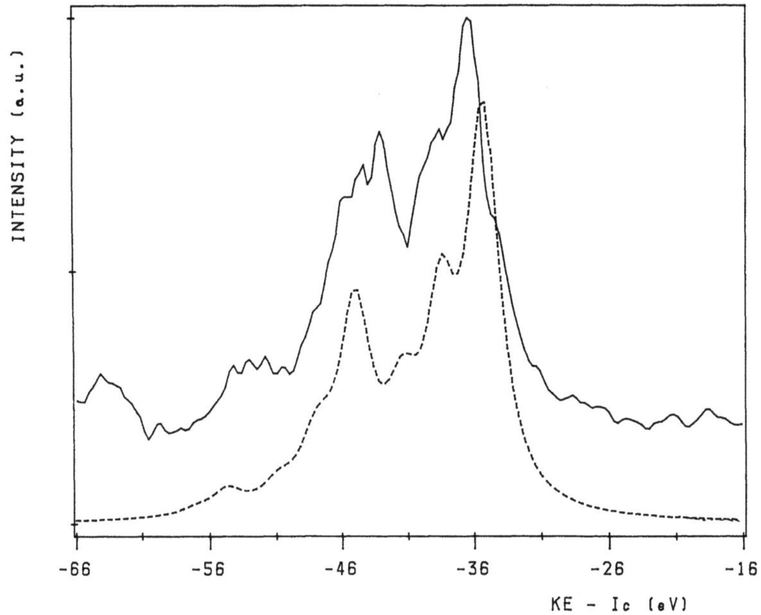

Fig. 2 Solid line: Si $L_{2,3}VV$ spectrum of $SiCl_4$ on a two-hole binding energy scale: dotted line: calculated line shape.

spectra should then be that of including such complex processes into the picture, rather than refining the description of two holes states, that appear to be already fairly well understood by the current theory /3 - 4/.

REFERENCES

1 M. Cini, F. Maracci and R. Platania, J. Electron Spectrosc. Relat. Phenom. 41,37 (1986)
2 M. Cini, F. Maracci and R. Platania, J. Physique C9, suppl. au n.12, Tome 48, 781 (1987)
3 M. Cini, Solid State Commun. 20,605 (1976) and 24,681 (1977)
4 M. Cini, Phys. Rev. B17, 2788 (1978)
5 E. McGuire, Phys. Rev. 185, 1 (1969)
6 D.P. Jennison, Phys. Rev. B18, 6865 (1978)
7 A.E.Jones, G.K. Schweitzer, G.A. Grimm and T.A. Carlson, J. Electron Spectrosc. Relat. Phenom. 1,29 (1972/73)
8 E. Clementi and C. Roetti, Atomic Data and Nuclear Data Tables 14, 177 (1974)
9 T.D. Thomas and P. Weightman, Chem. Phys. Letters 81, 325 (1981)
10 G.G.B. De Souza, P. Morin and I. Nenner, J. Chem. Phys. 83,492 (1985)
11 P.J. Feibelman, E.J. McGuire and K.C. Pandey, Phys. Rev. B15, 2202 (1978)
12 J.C. Green, M.L.H. Green, P.J. Joachin, A.F. Orchad and D.W. Turner, Phyl. Trans. Roy. Soc. 268, 111 (1970)

Part IV

Analytical Aspects of Auger Electron Emission

The Role of Electron Backscattering in AES

A. Jablonski

Institute of Physical Chemistry, Polish Academy of Sciences, ul. Kasprzaka 44/52, PL-01-224 Warszawa, Poland

It has been shown that the correcting parameters for the quantitative analysis by AES can be derived from the characteristics of the elastic and inelastic electron backscattering from surfaces. The theory relating the inelastic mean free path to the elastic backscattering probability extremely well describes the published experimental data. Based on this theory, a method is described for determining the inelastic mean free path. The values of the inelastic mean free path resulting from this method are briefly reviewed. Attention is drawn to the differences in the definitions of terms characterizing the electron transport in solids.

1. INTRODUCTION

The possibility of quantitative analysis by AES has already been suggested in the late sixties /1 /. Until now AES has become a well established quantitative technique for studies of solid surfaces. At present, several monographs extensively review the problem of quantitative analysis /2,3 /. AES as a quantitative tool requires a simple formalism providing the relation between the intensities of Auger transitions observed in the spectra and the concentration of emitting elements. This relation is usually expressed in terms of the matrix dependent parameters, which can be considered as the correcting factors. The mathematical formalism usually involves three correcting parameters /4-6 /:

a) The backscattering factor, r, correcting for the influence of backscattered electrons on the Auger electron yield.

b) The inelastic mean free path, λ, correcting for the variations in the analysed volume.

c) The number of atoms in unit volume, M, correcting for variations in the atomic density.

Thus the concentration of the component A at the surface of the binary alloy AB can be expressed by the formula /4-6 /

$$x_A = \frac{1}{\{(I_B/I_A)/\,[(I_B^o/I_A^o)\,F_{BA}]\} + 1} \qquad (1)$$

where I_A and I_B are the Auger electron intensities recorded for the alloy components A and B, superscript "o" refers to the samples of standards, and F_{BA} is the Auger matrix factor given by /4-6 /

$$F_{BA} = \frac{M_A^o\, r_A^o\, \lambda_A^o}{M_B^o\, r_B^o\, \lambda_B^o}\,\frac{r_B\, \lambda_B}{r_A\, \lambda_A} \qquad (2)$$

where again superscript "o" refers to the standard samples.

Springer Series in Surface Sciences, Vol. 18 Editors: G. Cubiotti · G. Mondio · K. Wandelt
Auger Spectroscopy and Electronic Structure

The backscattering factor accounts for the enhancement of the Auger electron intensity due to the additional ionizations produced by the backscattered electrons on their way out of the solid. The importance of this phenomenon in quantitative analysis has been recognized by BISHOP and RIVIERE /7 /. These authors estimated also numerical values of the backscattering factor for several elements. Considerable attention has been devoted to the problem of determining the backscattering factor. The experimental and theoretical methods published until 1979 were reviewed by JABLONSKI / 8 /. At present, the theoretical approach for calculating the backscattering factor published by SHIMIZU and ICHIMURA / 9 / and ICHIMURA and SHIMIZU /10/ is considered to be the most accurate. A very extensive compilation of the values of the backscattering correction can be found in these reports. Attempts were made to approximate this correction with the simple analytical formulas, which expressed the values of r as a function of the primary electron energy and the atomic number /8,11,12/. These formulas are particularly convenient in routine quantitative analysis. The approximate function proposed by ICHIMURA et al. /12 / seems to be the most accurate at the moment.

The energy spectrum of all electrons backscattered from the solid surface can be roughly divided into three regions /13/:

a) Elastically backscattered electrons (the so-called elastic peak)

b) Inelastically backscattered electrons. These electrons are usually defined as electrons with energy exceeding 50 eV.

c) Secondary electrons, i.e. the electrons with energy up to 50 eV.

The backscattering factor in AES is associated with the inelastically backscattered electrons. In the theoretical models used in calculations of the backscattering factor the elastically backscattered electrons are neglected. It is due to the fact that the current of the elastically backscattered electrons is much smaller than the current of inelastically backscattered electrons. The latter may reach 50 - 60 % of the primary beam current at energies 2 - 3 keV /14,15/, while the elastic backscattering probability in that energy range does not exceed 5 % /13,16/. In general, the elastic backscattering phenomenon is usually not considered in quantitative analysis by AES. The elastic peak intensity is only occasionally monitored during positioning the sample with respect to the cylindrical mirror analyser /17/. However, it turned out that the elastic backscattering effect may also provide important information for the procedure of quantitative analysis. It has been shown recently that the intensity of the elastic peak can be related to the inelastic mean free path of electrons in solid /18-20/, i.e. to the second correcting parameter in the mathematical formalism of AES. The objective of the present paper is to review the recent developments in the theory of elastic backscattering, to compare the theory with the experimental observations, and to discuss the method of determining the inelastic mean free path.

2. THEORY

Relatively limited interest has been devoted to studies of electron elastic backscattering at energies of surface sensitive electron spectroscopies. Several experimental reports were published in the literature /13,16, 20-23 /. Majority of these studies referred to the normal incidence of the electron beam. The process of elastic backscattering can be described by two characteristics:

a) The elastic reflection coefficient. It is the probability that an electron impinging at the surface will leave the solid withouth energy loss.

b) The angular distribution of elastically reflected electrons.

The theoretical models were proposed to describe the elastic electron backscattering. These models are based on the differential elastic scattering cross sections. In the simplest theoretical approach only one large angle elastic scattering is assumed to occur along the electron trajectory length. The reflection coefficient is then expressed by the simple formula /18.,19 /

$$\eta_E = \lambda M \sigma_{eff} \tag{3}$$

where σ_{eff} is given by

$$\sigma_{eff} = \int_{\Delta\Omega} \frac{d\sigma}{d\Omega} \frac{1}{1 - \sec\Theta} d\Omega .$$

Integration in (3) is performed within the solid angle of a given analyser. The theoretical models assuming the multiple elastic collisions along the trajectory length are more realistic. SCHILLING and WEBB / 21/ presented an analytical model which well described the elastic electron backscattering from the liquid mercury at energies up to 500 eV. This model accurately describes first two elastic electron collisions, while further collisions were expressed in terms of the averaged differential elastic scattering cross section. Extensive theoretical analysis of electron backscattering was recently published by TOFTERUP /24/. This analysis was based on the approximate solution of the Boltzmann transport equation (the so-called P_1 approximation). Generally, all the analytical treatments of the elastic backscattering problem involve certain assumptions which simplify the theoretical model describing the electron transport. These assumptions may result in systematic errors which are difficult to estimate.

The alternative theoretical approach frequently used to characterize the electron transport in solids is the Monte Carlo method. Several algorithms simulating the elastic electron backscattering were recently published in the literature /19,20,25,26/. This approach does not involve the simplifying assumptions usually made in the analytical models. The Monte Carlo algorithm may account for the possibly accurate angular deflections after collision, spatial deflections, energy losses, and the particular geometry of experiment (angle of incidence, solid angle of detector, etc.). The disadvantage of this method is the fact that the Monte Carlo simulations usually require considerable amount of the computer time. Furthermore, the parameters monitored during calculations are finally determined with a certain statistical error. However, this error can be controlled by adjusting the time of calculations, and the accuracies of 1 - 2 % are routinely achieved. The Monte Carlo algorithms were frequently used in calculations of the backscattering factor /8-10/. In these calculations the processes of elastic scattering and energy loss were simulated. The Monte Carlo algorithms simulating the proces of elastic backscattering from surface account only for the elastic collisions along the electron trajectory. Thus, the phenomenon of elastic scattering by the scattering centres of the solid should be carefully analysed before constructing the algorithm.

2.1. Elastic Electron Collisions

Majority of the analytical or the Monte Carlo theoretical models of the multiple electron scattering is based on the assumption that the scattering centre of the solid can be approximated by the known potentials characteristic for the isolated atoms. The non-relativistic Schrödinger equation describing the scattering problem has the form

$$[\Delta + K^2 - (2\,m / \hbar)\, V(r)]\; \Psi(\underline{r}) = 0.$$

where K is the wave vector, and V(r) is the scattering potential. All other symbols have the usual meaning. The wave function $\Psi(\underline{r})$ reaches asymptotically the superposition of the plane incident wave and the spherical outgoing wave

$$\Psi(\underline{r}) \xrightarrow[r \to \infty]{} \exp(i\,\underline{K} \cdot \underline{r}) + \frac{1}{r} \exp(i\,K\,r)\, f(\Theta)$$

where $f(\Theta)$ is the scattering amplitude. The Schrödinger equation can be solved within different approximations. According to the first order Born approximation (FBA) the scattering amplitude is given by

$$f(\Theta) = - \frac{2\,m}{\hbar^2} \int_0^{\infty} V(r) \frac{\sin qr}{qr} r^2\, dr$$

where q is the difference between wave vectors before and after elastic collision, i.e. $q = 2\,K \sin \Theta/2$. The accurate solution of the scattering problem obtained within partial wave expansion method (PWEM) provides the scattering amplitude expressed in terms of the so-called phase shifts δ_n

$$f(\Theta) = \frac{1}{2\,i\,K} \sum_{n=0}^{\infty} (2\,n + 1) \left[\exp(2\,i\,\delta_n) - 1 \right] P_n(\cos \Theta)$$

where $P_n(x)$ are the Legendre polynomials. The CALOGERO /27/ method can be effectively used in calculations of the phase shifts at energies of the surface sensitive electron spectroscopies /20,28/. The differential elastic scattering cross section is related to the scattering amplitude according to the simple expression

$$d\sigma/d\Omega = |f(\Theta)|^2$$

The relativistic Dirac equation was also frequently used to calculate the differential elastic scattering cross sections within PWEM /9,10,20, 29/. However, it has been found that the differences between relativistic and non-relativistic cross sections is not very pronounced at energies of surface sensitive electron spectroscopies /19,20/.

ICHIMURA et al. /29/ have drawn attention to the fact that the differential elastic scattering cross-sections calculated within the FBA have limited applicability at energies characteristic for AES. These authors compared the differential elastic scattering cross sections obtained from the relativistic PWEM with the screened Rutherford cross section (which results from the FBA applied to the screened Columb potential), and found considerable discrepancies. Figure 1 shows the angular dependence of the ratio of the differential elastic cross sections calculated within the non-relativistic PWEM and the FBA. The algorithm has been described elsewhere /20,28/. Calculations were made for the Thomas - Fermi - Dirac po-

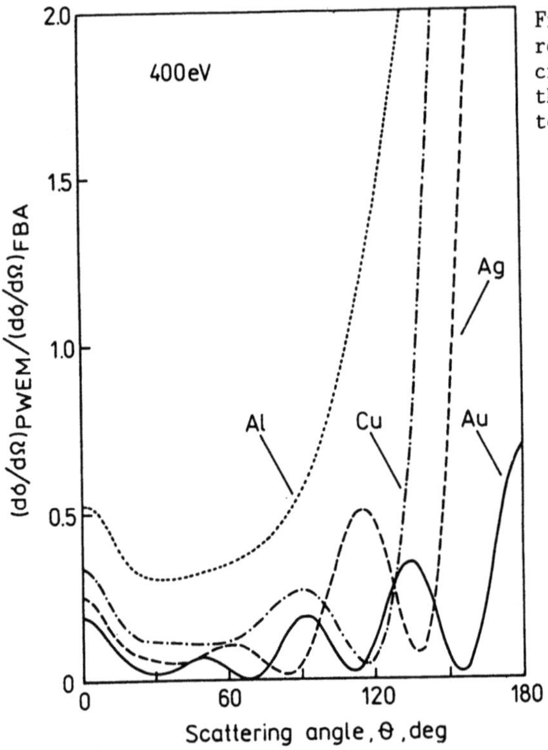

Figure 1. Ratio ot the differential elastic scattering cross sections resulting from the PWEM and the FBA calculated for energy 400 eV.

tential. One can see the complex shape of the plotted dependences. Considerable deviations from unity are observed, even for the lowest atomic number element.

The energy of Auger electrons varies in the range from 50 eV to 2000 eV. Quantitative AES analysis requires knowledge of the values of the inelastic mean free path (IMFP) for the above energy range. Thus, the formalism relating the elastic backscattering intensity to the IMFP should be based on the differential elastic scattering cross sections valid for energies not exceeding 2000 eV. It is evident from Fig. 1 that the accurate relation must be derived from the cross cestions calculated within PWEM.

2.2. The Monte Carlo Algorithm

The most reliable relation between the elastic backscattering intensity and the IMFP seems to be obtained from the Monte Carlo calculations /19, 20,25,26/. Let us discuss here only the basic features of this approach.

The electron trajectory in the solid can be described by the set of the scattering angles corresponding to all scattering events, and by the set of the linear step lenghts between scattering events. The simulated probability density function describing the polar scattering angles has the form

$$W(\Theta) = (d\sigma/d\Omega)\,/\,\sigma_t$$

where σ_t is the total elastic scattering cross section. The non-relativistic PWEM is used in calculations of the elastic scattering cross sections. The azimuthal scattering angles are assumed to follow the uniform distribution. The linear electron path lengths between elastic collisions, L, are described by the exponential distribution

$$F(L) = (1/s_m) \exp(- L/s_m)$$

where s_m is the elastic mean free path. It can be calculated from

$$s_m = (M \sigma_t)^{-1}$$

Electrons are assumed to be attenuated exponentially along the trajectory length. Thus, the contribution to the total elastically backscattered current provided by the i-th electron is given by

$$\Delta I_i = \exp(- x_i / \lambda) \qquad (4)$$

where x_i is the total trajectory length of the i-th electron. Monitoring the intensity contributions, and the corresponding escape angles, one can estimate the characteristics of the elastically backscattered current, i.e. the intensity within a particular solid angle, and the angular distribution.

As mentioned, the probability of elastic backscattering is rather small. For this reason a considerable number of electron trajectories should be generated to obtained a resonable accuracy of the calculated characteristics. It turns out that for the case of the low atomic number electrons the necessary number of trajectories may exceed one million.

3. THE INELASTIC MEAN FREE PATH

3.1. Comparison of Theory of Elastic Backscattering with Experiment

The characteristics of elastically backscattered electrons derived from theory were compared with experimental observations in several reports /19,20/. SCHILLING and WEBB /21/ found reasonable agreement between angular distribution of electrons elastically backscattered from liquid mercury and the theory prediction. Despite simplifications, their theory succesfully related the differential elastic scattering cross sections to the experimental data. There were attempts to use in calculations the differential elastic scattering cross sections resulting from the FBA /19/. Comparison of these results with the experimental values of the reflection coefficient indicates that the multiple elastic scattering model based on the FBA seems to be valid for electron energies exceeding 2000 eV /19/. The angular distribution of elastically backscattered electrons predicted by the theory of TOFTERUP /24/ (P_1 approximation) was found to disagree severely with experimental observations in the energy range from 300 eV to 1200 eV /20/.

Extensive analysis of the validity of the Monte Carlo approach to describe the elastic electron backscattering has been recently published /20/. It turned out that the results of Monte Carlo calculations based on the elastic scattering cross sections provided by the PWEM are in excellent agreement with the available experimental data on the angular distribution of elastically backscattered electrons and the reflection coefficient. Agreement was observed in the energy range from 200 eV to 2400 eV. Exemplary comparison of the theoretical and experimental /23/ angular distributions (normalized to unity at the angle 25° to the surface normal)

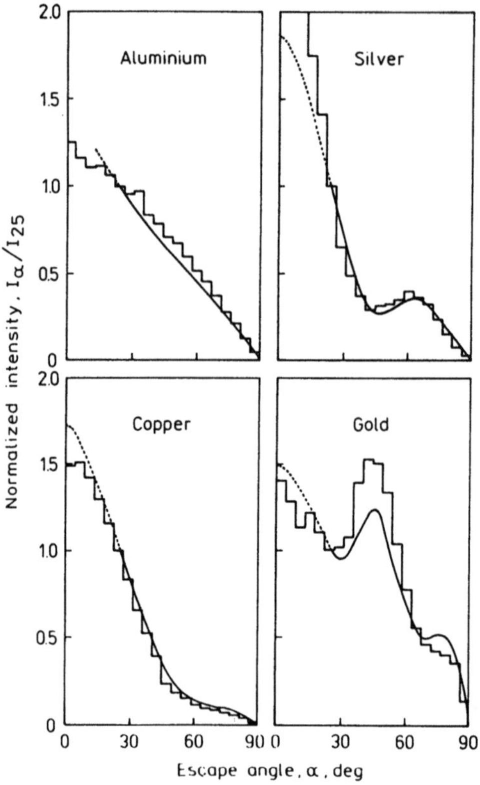

Figure 2. Angular distribution of elastically backscattered electrons from surfaces at energy 400 eV. Histogram: the Monte Carlo calculations with the values of IMFP taken from ASHLEY and TUNG /31/; solid line: experimental distributions reported by BRONSHTEIN and PRONIN /23/

is shown in Fig. 2. Similar agreement was observed for other energies /20/. SCHMID et al. /13/ published the experimental energy dependence of the reflection coefficient for a number of elements. Measurements were made within the solid acceptance angle of the retarding field analyser, i.e. between 6° and 52° to the surface normal. The energy dependences calculated from the Monte Carlo algorithm for the same geometry closely reproduced the experimental data /20/. These energy dependences are shown in Fig. 3. As one cas see, the theory and experiment compare very well.

The Monte Carlo algorithm requires knowledge of the value of IMFP for a given solid (equation (4)). In the reported calculations the input values of IMFP originated from the theoretical expressions published by PENN /30/ and ASHLEY and TUNG /31/. Both expressions provided distinctly different energy dependences of the reflection coefficient, which have similar shape but are somewhat shifted (Fig. 3). The obvious question arises if it would be possible to reverse the situation, and to determine the energy dependence of the IMFP from the experimental values of the elastic backscattering intensity. The reliability of the theoretical model supports such suggestion.

3.2. Determination of the IMFP from the Elastic Peak Intensity

The method by which the IMFP can be measured from the elastic backscattering intensity has been proposed relatively recently /18,19/. In first applications of this method the single elastic scattering model (equation (3)) was used in calculations. The elastic scattering cross sections were then determined within FBA. The theory was extended later by including

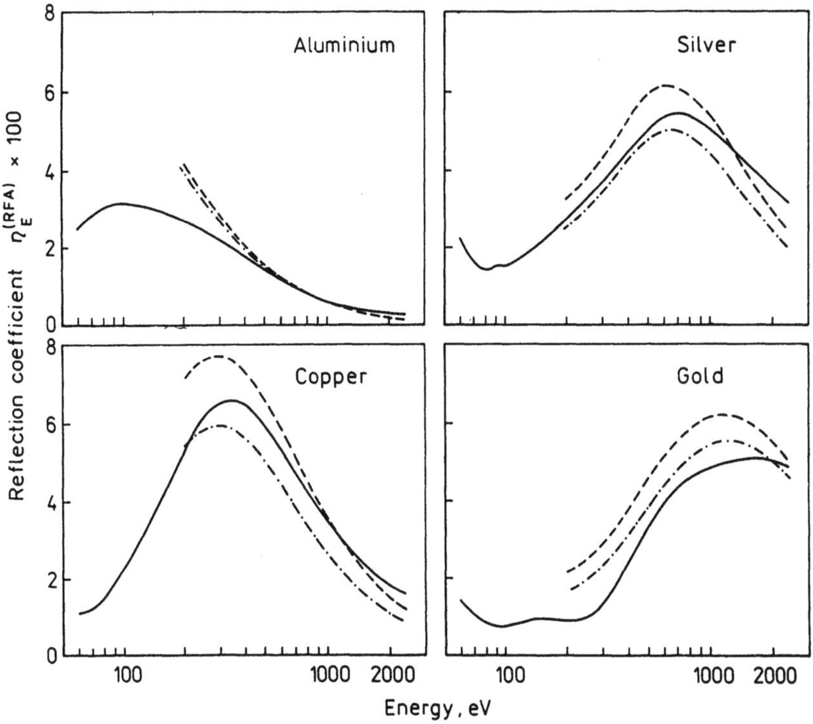

Figure 3. Energy dependence of the elastic reflection coefficient determined within the acceptance angle of the retarding field analyser. Solid line: experimental dependence reported by SCHMID et al. /13/; dashed line: the Monte Carlo calculations with the values of IMFP taken from ASHLEY and TUNG /31/; dot-dashed line: the Monte Carlo calculations with the values of IMFP taken from PENN /30/

multiple electron collisions, and calculating the elastic scattering cross sections within the PWEM /19,25,26,32/. In principle, determination of the IMFP would require experimental measurements of the absolute values of the reflection coefficient for a given geometry. Then, the Monte Carlo calculations would indicate such a value of the IMFP which provides the observed reflection coefficient. Actually, such procedure was used in several reports /19,25,26/. The corresponding experiments can be made in specially adapted systems /13,25,26/. It would be advantageous to modify the experimental procedure so that the measurements can be performed using the AES spectrometer. This requirement satisfies the recently proposed procedure of relative measurements /32,33/. Actually, it is sufficient for the purpose of quantitative AES analysis to determine the ratio of the values of the IMFP rather than the absolute values (equations (1) and (2)). In such a case, a simple experimental procedure can be proposed which involves the measurements of the ratio of elastic peak intensities instead of the absolute elastic backscattering probabilities for each sample separately.

Suppose that the reliable values of the IMFP are known for a certain standard element. This element should have the following properties:

a) Applicability to the theoretical model providing the values of IMFP (the nearly free-electron-like behaviour).

b) Possibly low atomic number (to avoid the influence of electron elastic scattering in overlayer method for determining the IMFP /34/).

c) Reasonable agreement between published experimental and theoretical values of the IMFP.

Aluminium standard seems to satisfy the above properties. The available experimental values of the IMFP very well compare with the theory prediction /33/. The expression of ASHLEY and TUNG /31/ was proposed to provide the standard values of the IMFP for aluminium. However, this metal has certain disadvantages in experimental practice:

a) Difficulties to obtain a clean surface.

b) Rich plasmon structure which may overlap with the elastic peak and thus hinder the integration.

The standard AES spectrometer equipped with the cylindrical mirror analyser can be used for measurements. The elastic peak intensity is recorded for a given sample (I_A) and for the aluminium standard (I_{Al}). Thus, experiment provides the ratio of intensities I_A/I_{Al}. Experiments are accompanied by Monte Carlo calculations in which the intensity I_{Al} is determined using the input value of the IMFP for aluminium (e.g. resulting from the expression of ASHLEY and TUNG /31/). Similar calculations are performed for the element A assuming several different input values of the IMFP. Determination of the IMFP for silver at 1000 eV is illustrated in Fig. 4. The calibration plot provides the value of 12.4 Å which corresponds to experimentally determined ratio of 3.37. As one can see, the calibration plot calculated within single elastic scattering model (equation (3)) is noticeably different, and can be used in determination of the IMFP for possibly low atomic number elements /19/.

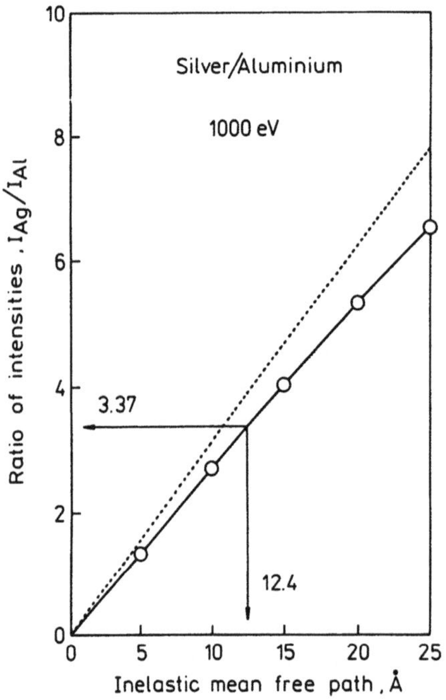

Figure 4. Ratio of intensities I_{Ag}/I_{Al}, calculated for the CMA as a function of the IMFP for silver. Solid line and circles: The Monte Carlo calculations; dashed line: the single elastic scattering model (equation (3))

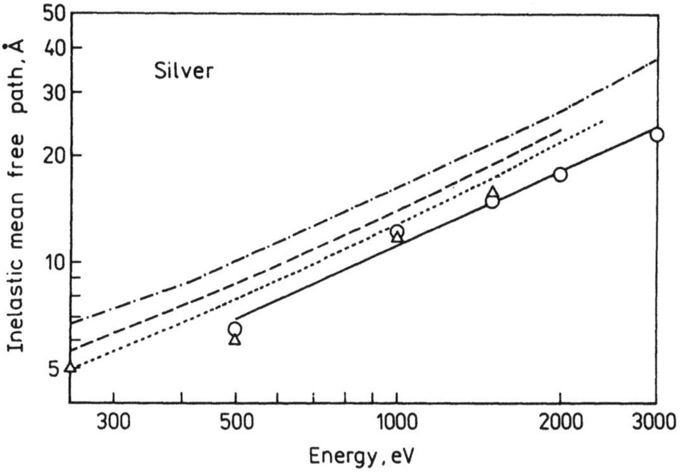

Figure 5. Energy dependence of the IMFP for silver. Solid line and circles: the method of relative measurements /32/: triangles: values derived from the absolute values of the reflection coefficient /25/; dotted line: expression of PENN /30/; dot-dashed line: expression of ASHLEY and TUNG /31/; dashed line: expression of TANUMA et al. /35/

Fig. 5. presents the energy dependence of the IMFP determined for silver /32/ using the aluminium standard. These dependences reasonably well compare with the theoretical dependences published in the literature. ASHLEY and TUNG /31/ proposed to aproximate the energy dependence by expression

$$\lambda = k\ E^{p} \tag{5}$$

The fit of this equation to the values of the IMFP determined for silver provides the constants k = 0.0900 and p = 0.699 /32/, which reasonably well agree with the theoretical values /31/. The solid line shown in Fig. 5 is actually the fitted function. One can see that the fitted dependence well agrees with the values of the IMFP determined by different authors using the method based on elastic peak intensity.

The method based on the aluminium standard can be easily extended for alloys. It has been used to determine the IMFP for the ordered NiAl alloy /33/. Calculations were made within the single elastic scattering approximation. Equation (5) with constants k = 0.0459 and p = 0.845 was found to describe well the energy dependence of the IMFP in the energy range from 500 eV to 3000 eV.

4. DISCUSSION

The experimental values of the IMFP available in the literature originate mainly from the so-called overlayer method /36/. This method consists in depositing the layer of a given material on certain substrate and following the AES or XPS intensity recorded for the overlayer and the substrate as a function of the overlayer thickness. The overlayer method has a number of disadvantages /19,34/:

a) Difficulties in obtaining well defined overlayers.

b) Difficulties in determining the overlayer thickness.

c) Electron backscattering effects (when AES is used in measurements).

d) Effects of elastic electron collisions.

The elastic electron collisions may influence considerably the estimated value of IMFP. POWELL /36,37/ proposed to introduce the term attenuation length (AL) which is defined as a parameter resulting from the overlayer film experiment under the assumption that effects of elastic electron scattering are ignored. It has been found that the difference between the AL and the IMFP may exceed 30 % /28,34/. The E-42 Committee of ASTM proposed two definitions characterizing the electron transport /38/:

a) The IMFP, defined as the average distance travelled by an electron between inelastic collisions.

b) The escape depth (ED), defined as the distance normal to the surface at which the probability of an electron escaping without energy loss drops to 1/e (36.8 %) of its original value.

The relations between the IMFP, the AL, and the ED were recently extensively analysed /34/.

It should be stressed that the parameter resulting from the measurements of the elastic peak intensity corresponds to the IMFP. This method is very convenient in experimental practice. Measurements can accompany the quantitative analysis, even on a given spot on a sample. The method is applicable to amorphous or polycrystalline solids. However, similar limitation applies also to the formalism or the quantitative AES analysis.

The accuracy of the described method is difficult to estimate. The energy dependence determined for several elements well agree with the theoretical predictions. The errors associated with the integration of the elastic peak partially cancel when taking the ratio of intensities. Certain systematic error arises also from imperfections of the theory. The prospects of further developments of the theoretical model describing the elastic backscattering were recently discussed /20/. Another problem which requires further considerations is the determination of the elastic peak intensity. Work is now in progress on the above problems.

5. CONCLUSIONS

Elastic backscattering of electrons from surfaces is very accurately described by the Monte Carlo method based on the elastic scattering cross sections resulting from the partial wave expansion method. Excellent agreement with the experimental data supports the idea to use this theoretical model in determining the inelastic mean free path from the experimentally measured intensities of the elastic peak. The conventional AES spectrometer can be used for measurements. The published values of the inelastic mean free path resulting from this method well compare with the theory predictions.

LITERATURE

1. L. A. Harris: J. Appl. Phys. 39, 1419 (1968)
2. Practical Surface Analysis by Auger and X-ray Photoelectron Spectroscopy, ed. by D. Briggs, M. P. Seah (J. Wiley, Chichester 1983)
3. M. Thompson, M. D. Baker, A. Christie, J. F. Tyson: Auger Electron Spectroscopy (J. Wiley, New York 1985)
4. P. M. Hall, J. M. Morabito: Surface Sci. 83, 391 (1979)

5. A. Jablonski: Surface Interface Anal. 4, 135 (1982)
6. M. P. Seah: In Scanning Microscopy / 1983, ed. by O. Johari, Vol. II, p. 521 (SEM, AMF O'Hare, Chicago, IL 1983)
7. H. E. Bishop, J. C. Riviere: J. Appl. Phys. 40, 1740 (1969)
8. A. Jablonski: Surface Interface Anal. 1, 122 (1979)
9. R. Shimizu, S. Ichimura: Quantitative Analysis by Auger Electron Spectroscopy, Technical Reports No I-006, 76-175 (Toyota Fundation, Tokyo 1980)
10. S. Ichimura, R. Shimizu: Surface Sci. 112, 386 (1981)
11. A. Jablonski: In Proceedings of the Fourth International Conference on Solid Surface and the Third European Conference of Surface Science, Cannes 1980, ed. by D. A. Degras, M. Costa, p. 1307.
12. S. Ichimura, R. Shimizu, J. P. Langeron: Surface Sci. 124, L49 (1983)
13. R. Schmid, K. H. Gaukler, H. Seiler: In Scanning Electron Microscopy / 1983, ed. by O. Johari, Vol. II, p. 501 (SEM, AMF O'Hare, Chicago, IL 1983)
14. P. Sommerkamp: Z. Angew. Physik, 28, 220 (1970)
15. E. H. Darlington, V. E. Cosslett: J. Phys. D, Appl. Phys. 5, 1969 (1972)
16. G. Gergely: Surface Interface Anal. 3, 201 (1981)
17. S. H. Overbury, G. A. Somorjai: J. Chem. Phys. 66, 3181 (1977)
18. A. Jablonski, P. Mrozek, G. Gergely, M. Menyhard, A. Sulyok: Surface Interface Anal. 6, 291 (1984)
19. A. Jablonski: Surface Sci. 151, 166 (1985)
20. A. Jablonski, J. Gryko, J. Kraaer, S. Tougaard: Phys. Rev. B, in press
21. J. S. Schilling, M. B. Webb: Phys. Rev. B2, 1665 (1970)
22. I. M. Bronshtein, V. P. Pronin, V. M. Stozharov: Soviet Phys. - Solid State, 16, 1374 (1975)
23. I. M. Bronshtein, V. P. Pronin: Soviet Phys. - Solid State, 17, 1363 (1976)
24. A. L. Tofterup: Phys. Rev. B32, 2808 (1985)
25. W. Dolinski, S. Mroz, M. Zagorski: Surface Sci. 200, 361 (1988)
26. W. Dolinski, H. Nowicki, S. Mroz: Surface Interface Anal. 11, 229 (1988)
27. F. Calogero: Variable Phase Approach to Potential Scattering (Academic Press, New York 1967)
28. A. Jablonski: Surface Sci. 188, 164 (1987)
29. S. Ichimura, M. Aratama, R. Schimizu: J. Appl. Phys. 51, 2853 (1980)
30. D. R. Penn: J. Electron Spectrosc. Relat. Phenom. 9, 29 (1976)
31. J. C. Ashley, C. J. Tung, Surface Interface Anal. 4, 52 (1982)
32. B. Lesiak, A. Jablonski, G. Gergely: Physica Scripta, submitted
33. P. Mrozek, A. Jablonski, A. Sulyok: Surface Interface Anal. 11, 499 (1988)
34. A. Jablonski, H. Ebel: Surface Interface Anal. 11, 627 (1988)
35. S. Tanuma, C. J. Powell, D. R. Penn: Surface Sci. 192, L849 (1987)
36. C. J. Powell: J. Electron Spectrosc. Relat. Phenom. 47, 197 (1988)
37. C. J. Powell: In Scanning Electron Microscopy / 1984, ed. by O. Johari Vol. IV, p. 1649 (SEM, AMF O'Hare, Chicago, IL 1984)
38. Standard Definitions of Terms Relating to Surface Analysis, Surface Interface Anal. 5, 268 (1983); 10, 48 (1987)

Application of the Pattern Recognition Method in Auger Electron Spectroscopy

B. Lesiak and A. Jablonski

Institute of Physical Chemistry, Polish Academy of Sciences,
PL-01-224 Warszawa, ul. Kasprzaka 44/52, Poland

In numerous reports the chemical state of carbon atoms was correlated with the changes in the carbon KLL spectra. Analysis of variations in the spectra shape would require an objective method for the spectra interpretation. The powerful possibilities of the pattern recognition method were recently proposed for this purpose. Exemplary applications are discussed. The chemical form of carbon atoms can be effectively identified on the basis of the standard set of the spectra. The recognition procedure called the k nearest neighbour rule was found to be reliable and convenient in these applications.

1. Introduction

Chemical applications of the Auger Electron Spectroscopy was extensively reviewed in several recent publications /1-3/. It has been pointed out that the shape of the broad KLL carbon feature is particularly sensitive to the chemical form of carbon atoms /4-15/. This is obviously due to the fact that electrons from the L shell are the valence electrons (2s and 2p). The carbon KLL feature consists of several overlapping transitions extending from 243 eV to 267 eV /1/. Due to the sensitivity to the chemical form of the carbon atom, the KLL spectra can be considered as a "fingerprint". Exemplary spectra are shown in Fig. 1.

Several criteria were proposed to identify chemical forms of the carbon atom at surfaces. MIZOKAWA et al. /14,15/ suggested the peak positions in the fine structure of the XAES and AES spectra of graphite, diamond, silicon carbide and diamond-like carbon. GRAIG et al. /10/ proposed a simple criterion based on the peak-to-peak height within the KLL feature. This criterion was successful in identification of the carbon-carbon and the carbon-metal bonds. GAARENSTROOM /8/ explored the possibility to identify polymers in a series of polyalkyl methacrylate compounds by means of the factor analysis applied to selected frequencies arrising from the Fourier transformation of the spectra. The factor analysis approach is recently continued by KOENIG et al. /16/. LESIAK et al. /12-13/ presented a simple method which visualises clustering of the carbon KLL spectra in a space with dimension equal to the number of channels used in the spectra acquisition. Let us briefly describe this method. The carbon KLL spectra were represented by hundred dimensional vectors corresponding to the content of the hundred channels. The centers of gravity of vectors recorded for two different forms of the carbon atoms were connected by a straight line and all vectors were projected on this line. The projected points can be plotted in form of histogram (frequency vs distance from the arbitrarily chosen origin). Figure 2 shows exemplary histograms illustrating clustering of the spectra recorded for polyacetylene samples with different amount of iron chloride ions. One can see that slight interaction between polymer chain and the iron chloride ions is reflected in the carbon KLL spectra shape.

Springer Series in Surface Sciences, Vol. 18 Auger Spectroscopy and Electronic Structure
Editors: G. Cubiotti · G. Mondio · K. Wandelt

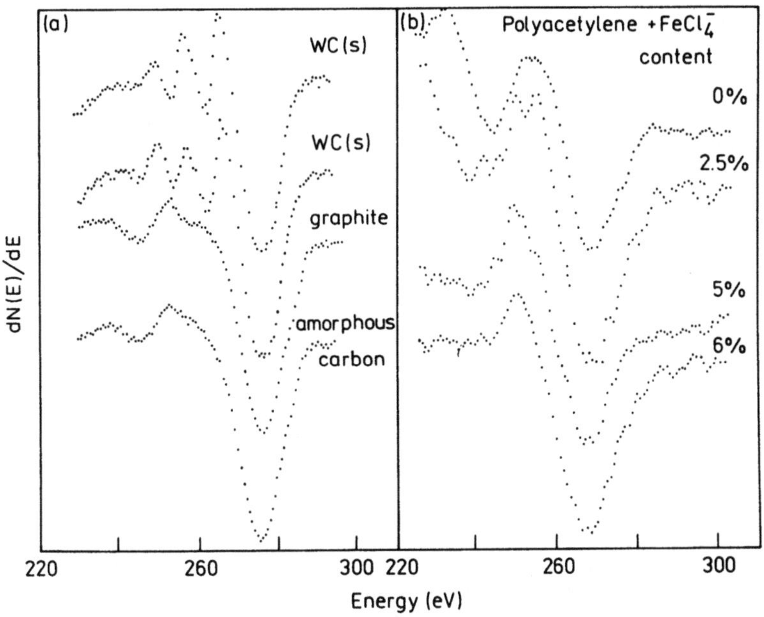

Figure 1. The carbon KLL spectra: (a) basic forms of carbon atoms, (b) polyacetylene doped with different amounts of iron chloride

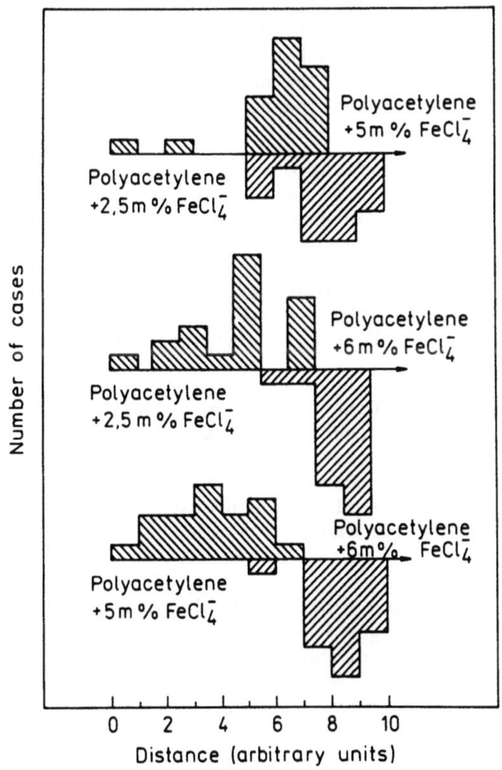

Figure 2. Histograms indicating clustering of vectors corresponding to spectra recorded for polyacetylene samples

2. The Pattern Recognition Method

The literature on the theoretical aspects of the pattern recognition method is extensively reviewed in a monograph of DEVIJVER et al. /17/. Abundant literature is also available on the application of the pattern recognition method in various fields of chemistry /18-20/. The pattern recognition method was recently proved to be extremely sensitive tool for the identification of chemical form of carbon from the shape of the carbon KLL spectra. The application of the pattern recognition method in AES requires several stages of spectra treatment:

1. Spectra acquisition.
   The spectrometer used for this purpose should have the facility of recording the spectra in the digital form, preferably on-line with the computer. The "number crunching" possibilities of IBM PC are sufficient for the calculations.
2. Preliminary treatment.
   The spectra are normalized in order to obtain similar objects for the spectra recognition. Normalization procedure usualy involves modification of intensity (e.g. to obtain the constant p-p height), shift of the energy scale, smoothing etc.
3. Creation of the standard set of spectra to be recognized. This set is called the reference set or the training set, and is necessery as a "computer experience" in the recognition procedure.
4. Selection of the effective recognition procedure. Such procedure is called a classifier.
5. Identification of the unknown spectra using a given classifier.

The spectra corresponding to the training set should cluster in the n-dimensional space. The purpose of the classifier is to ascribe an unknown vector to a certain cluster. If this unknown vector appears inside a well separated cluster than the classification is obvious. Difficulties arrise when clusters overlap or when an unknown vector appears between the clusters. Performance of the classifier is characterized by the so-called misclassification probability, E. This is a probability that the considered vector is ascribed to the wrong cluster (class).

One of the most universal methods of classification is the k-nearest neighbour rule (kNN) /17/. It was extensively used in the pattern recognition method applied in AES. The identification of the unknown spectrum is made by analysing the immediate neighbourhood of the considered vector in the n-dimensional space. The vector is ascribed to the class which is represented by the greatest number of vectors among the k nearest neighbours. Performance of the kNN classifier is schematically outlined in Fig. 3. The number of the nearest neighbours k should be so selected that it minimizes the misclassification probability E.

The performance of the classifier can be further improved by the feature selection procedure /ref. 17 p. 187/. This procedure provides number of channels (features) which are the most informative. In this way one can establish regions of the spectra in which the changes are expected. The feature selection procedure decreases the misclassification probability and shortens the time of computation necessary for the recognition of an unknown spectrum.

It has been found that the best performance exhibit the classifiers in which the recognition is divided into two stages (parallel network of binary classifiers). For each pair of clusters corresponding to different chemical form a separate kNN classifier is determined (binary classifier). In the first stage the unknown vector is processed by all binary classifiers. The final result of classification is obtained in the second stage by "voting" of all the binary classifiers.

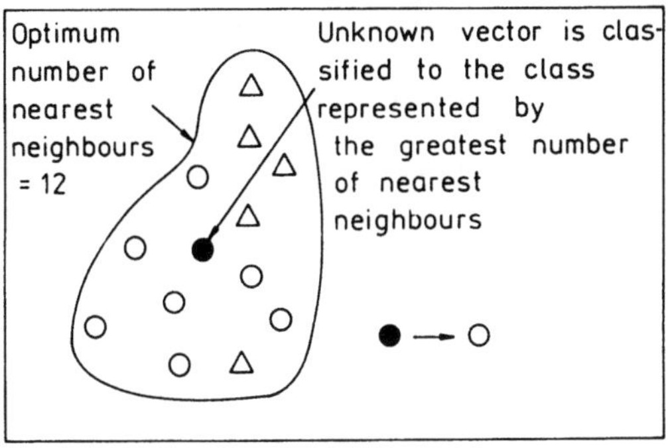

Figure 3. Illustration of the kNN rule

3. Applications of the Pattern Recognition Method

The pattern recognition method in connection with AES has been used in the following problems:

1. Recognition of the basic forms of carbon atoms /11/. The samples of graphite, amorphous carbon and metal carbides were analysed.
2. Identification of different conductive polymers (synthetic metals) containing different dopants /12,13/. The objects to be recognized were the polyacetylene and polypyrrole samples doped with chloride ion and cyanide ion.
3. Aging of conductive polymers in vacuum and the influence of the electron beam on the polymer surface /12/.

To illustrate the pattern recognition method in interpretation of the AES spectra we will discuss here the identification of the conductive polymers containing different amounts of the same dopant. The experiments were performed for the polyacetylene doped with iron chloride. The molar concentration of iron chloride was equal to 0%, 2.5%, 5% and 6%. The spectra were recorded in the differential form with modulation amplitude 4 V p-p. The energy of the primary electron beam was 1000 keV and the current equal to 300 nA. /13/. Preparation procedure of the polymer samples was described elsewere /21/. The samples were analysed as received, since the polymer structure may be profoundly affected by the cleaning procedure. The carbon KLL spectra were recorded in 512 channels and then transferred to the IBM PC computer. Before application of the pattern recognition method, the spectra were smoothed, the dimension was diminished to one hundred, and all the main minima were synchronized. Details of the preliminary treatment of the spectra are available in the literature /12-13/. To create the reference set, 20 spectra were recorded for each polymer. On the basis of the reference set of the spectra three different classifiers were constructed:

A. The kNN rule based on 100 features defining the spectra.
B. The kNN rule after feature selection.
C. The parallel network of kNN binary classifiers which were submitted separately to the feature selection procedure.

The misclassification probabilities characterizing the performance of all three classifiers are listed in Table 1. One can see that the reliability of the classifier C is very good.

The feature selection procedure drastically decreases the number of considered features and simultaneously improves the performance of the classifier. It is sufficient to monitor only certain ranges of the spectra (each feature comprises about 5 channels) for effective identification of a given polymer. The polymers differ only in the amount of the dopant, and

Table 1. Comparison of classifiers for identification of polyacetylene with various dopant concentration. The following numbers are ascribed to the samples: 1 - pure polyacetylene, 2 - 2.5 molar % of iron chloride, 3 - 5 molar % of iron chloride, 4 - 6 molar % of iron chloride.

Classifier	Considered samples	Number of considered features	Misclassification probability, E Pair	Total
A	1,2,3,4	100	-	0.188
B	1,2,3,4	6	-	0.125
C	1,2	2	0	
	1,3	1	0	
	1,4	1	0	0.038
	2,3	5	0.100	
	2,4	2	0	
	2,5	2	0	

the doping affects the carbon KLL spectra only slightly (Fig. 1(b)). The possibile variations in the spectra shape stem only from the changes in interaction between the dopant ions and the polymer chains. Neverthless, the pattern recognition method is sensitive enough to reveal differences in the spectra.

The pattern recognition method was also successful in other applications in AES. The basic chemical forms of carbon atoms were also well identified /11/. Some difficulties were encountered in distinguishing graphite from amorphous carbon (E=0.2). The classifier developed on the basis of this spectra was applied to the process of CO adsorption on polycrystalline nickel at room temperature. The carbide form was distinguished at early stages of adsorption which was followed by the mixture of the graphite and amorphous form. These observations well agree wih the literature data.

Classifier based on the set of the standard spectra obtained for the conductive polymers was used in studies of polymer surface submitted to different processes. The aging of polymers was found to produce distinct changes in the spectra which were probably due to the depletion of the surface region in dopant or to the interaction between the residual gases and the polymer surface. Similarly, the extended interaction with the beam produced noticeable changes in the spectra. On the other hand the polymers after electrochemical reduction (which decreases the dopant content) were not affected by aging and the exposition to the electron beam. This shows that variation in the spectra shape is probably due to the variations in the dopant content.

4. Conclusions

The pattern recognition method can considerably extend the potential of the Auger Electron Spectroscopy in studies of chemical and physical processes at solid surfaces. The classification procedures developed on the basis of the standard set of the spectra offer an objective and convenient method for spectra identification. The pattern recognition method is particularly useful when the slight spectra changes are expected. The fragments of the spectra can be established in which these changes take place. The kNN rule requires considerable amount of calculations necessary for the analysis of the standard set of the spectra (determination of the misclassification probability, calculation of the optimal number of nearest neighbours, etc.), and for the feature selection procedure. However, they are well managed by the personal computer. It is recomended to combine the Auger Electron Spectrometer with the computer for experiments involving applications of the pattern recognition method. In such a case the identification process can be fully automatized.

5. Literature

1. D. Briggs, J.C.Riviere: In Practical Surface Analysis by Auger and Photoelectron Spectroscopy, ed. by D. Briggs and M.P. Seah, J. Wiley and Sons, (Chichester, New York 1983), p. 87
2. M. Thompson, M.D. Baker, A. Christie, J.I. Tyson: Auger Electron Spectroscopy, J. Wiley and Sons, (New York 1985), p. 217
3. B. Carriere, J.P. Deville, P. Humbert: J. Microsc. Spectrosc. Electron., 10, 29 (1985)
4. T.W. Haas, J.T. Grant, G.J. Dooley: J. Appl. Phys. 43, 1853 (1972)
5. R. Ducros, G. Piquard, B. Weber, A. Cassuto: Surface Sci. 54, 519 (1976)
6. M.P. Hooker, J.T. Grant: Surface Sci. 55, 741 (1976)
7. M.A. Smith, S. Sinharoy, L.L. Levenson: J. Vac. Sci. Technol. 16, 462 (1979)
8. S.W. Gaarenstroom: J. Vac. Sci. Technol. 16, 124 (1979)
9. H.P. Bonzel, A.L. Krebs: Surface Sci. 91, 499 (1980)
10. S. Graig, G.L. Harding, R. Payling: Surface Sci. 124, 591 (1983)
11. B. Lesiak, P. Mrozek, A. Jablonski, A. Jozwik: Surface Interface Anal. 8, 121 (1986)
12. B. Lesiak, A. Jablonski, M. Zagorska, A. Jozwik: ibid., 12, 461 (1988)
13. B. Lesiak, A. Jablonski, M. Kisliuk, M. Zagorska, A. Jozwik: Poverkhnost (in russian), in press
14. Y. Mizokawa, T. Miyasata, S. Nakamura, K.M. Geib, G.W. Wilmsen: Surface Sci. 182, 431 (1987)
15. Y. Mizokawa, T. Miyasata, S. Nakamura, K.M. Geib, G.W. Wilmsen: J. Vac. Sci. Technol. A5, 2809 (1987)
16. M.F. Koenig, J.T. Grant: J. Electron Spectrosc. Relat. Phenom. 41, 145 (1986)
17. P.A. Devijver, J. Kittler: Pattern Recognition. A Statistical Aproach, Prentice Hall Intern. INC., (London 1982)
18. P.C. Jurs, T.L. Isenhour: Chemical Applications of Pattern Recognition, ed. by J. Wiley and Sons, (New York 1975)
19. K. Varmuza: Pattern Recognition in Chemistry, ed. by Springer-Verlag (Berlin 1980)
20. Modern Trendes in Analytical Chemistry ed. by E. Pungar and G.E. Veress, Academia Kiado, (Budapest 1984)
21. J.C.W. Chien: Polyacetylene, Chemistry, Physics and Material Approach, Academic Press, (Orlando 1984), p. 24

Chemical Information from Auger Spectra

B. Carrière, J.P. Deville, and P. Humbert

I.P.C.M.S. Groupe Surfaces-Interfaces U.M. no. 380046,
Université Louis Pasteur,
4, rue Blaise Pascal, F-67070 Strasbourg, France

Interpretation of chemical shifts and fine structure modifications in band-like or atomic-like Auger spectra in order to characterize changes in the chemical environment is not always straightforward. The aim of this paper is to present some recent developments on these aspects through some examples: the $SiL_{2,3}VV$ spectrum during silicon oxidation and the OKLL spectra for bulk oxides and metal/oxide interfaces.

1 Introduction

Auger electron spectroscopy has become a standard tool for qualitative and quantitative surface analysis and is widely applied in many fields of materials research. The quantitative approach gives information about surface concentrations, overlayer thicknesses or growth modes through peak intensity measurements. The qualitative approach deals with peak energy shifts and peak shape modifications, attributable to changes in the chemical environment of the surface atoms. The aim of this paper is to precise some aspects of this latter approach.

Chemical effects are expected to be particularly marked for Auger transitions involving at least one valence electron. The corresponding spectra show either valence band structures or quasi-atomic structures. The theoretical background and the main characteristics of these two kinds of spectra are briefly reviewed in the next section. Section 3 focuses on the decomposition of the band-like $SiL_{2,3}VV$ Auger peak according to the different chemical environments of silicon atoms observed by AES during silicon oxidation. Section 4 concerns the effective two-hole final state interaction in quasi-atomic OKLL spectra obtained with X-AES on different oxides. Section 5 deals with the effect of extra-atomic relaxation on the oxygen Auger parameter. In these two last sections, examples from recent metal/oxide interface studies illustrate some relaxation effects and interatomic features observed in oxygen Auger spectra.

2 Band-like and atomic-like Auger spectra

It is well known that Auger spectra obtained from solids show either valence band structures, containing information on the occupied states of the valence band, or quasi-atomic structures which cannot be related to the valence electron distribution [1]. Before recalling how some Auger spectra involving valence electrons may nevertheless be atomic-like, let us first mention the factors which determine the profile of band-like CVV Auger peaks.

For such transitions, electron and hole delocalisation in the valence band leads to an Auger profile which is in first approximation proportional to the self-convolution of the occupied valence states. The Coulomb interaction between the final state holes is negligible due to a nearly complete decoupling of the holes. We could thus think of extracting the density of occupied states from the Auger spectrum by means of a deconvolution technique. However, in such an approach, it is assumed that the valence electron distribution is not distorted in the initial state of the Auger process by the creation of a core hole and that the transition probability is constant over the whole valence states.

Springer Series in Surface Sciences, Vol. 18 **Auger Spectroscopy and Electronic Structure**
Editors: G. Cubiotti · G. Mondio · K. Wandelt

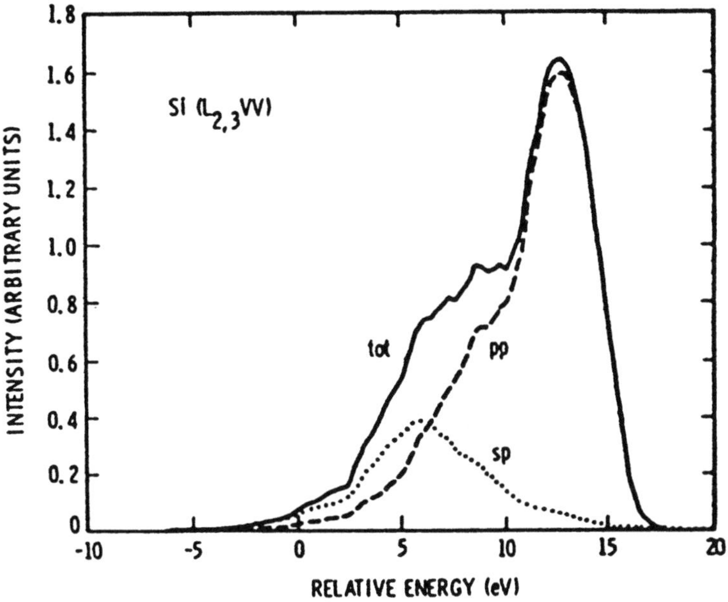

Fig. 1 - Theoretical Si$L_{2,3}$VV Auger spectrum broken into pp and sp components corresponding to the different angular momenta of the final state holes. The ss component is so small that it can be neglected [5].

Theoretical and experimental studies on sp-band materials have shown that initial and final state effects should indeed be taken into account. Auger transition matrix elements depend on the angular momentum of the involved valence electrons and transitions leading to pp final states are favoured relative to sp and especially ss final states [2-6] as it is illustrated in Fig. 1 for the Si$L_{2,3}$VV transition. So the Auger profile is mainly determined by the local charge distribution on the excited atom and only in a lesser extent by the bonding charge distribution. The role of the local density of states has been neatly brought to the fore in the case of GeSe [7]. The density of states worked out from the Ge$M_1M_{4,5}$V and Se$M_1M_{4,5}$V Auger peaks appear to be quite different: in fact they represent local densities on Ge and Se atoms respectively and only their sum accounts for the different features observed in the valence band photoemission spectrum. To add further difficulties, calculations and experimental studies on sp-band metals [6,8] showed that relaxation of the initial core hole involves mainly an s-type screening, modifying in consequence the local charge population.

For some CVV spectra of metals such as Cu, Zn, Ag or Cd, Auger spectra present a fine structure which cannot be related to any valence band density of states, but which is very similar to the spectral features observed for the corresponding free atom. These spectra are therefore termed quasi-atomic and their overall lineshape is determined by the various final state multiplets in the appropriate coupling scheme. The Auger lines are furthermore shifted in energy relative to the position expected for band-like spectra by a quantity denoted U_{eff} which accounts for the effective interaction between the two final state holes. Cini [9] and Sawatzky [10] formulated a simple criterion based on the $U_{eff}/2W$ ratio, where W is the valence band width, to determine when CVV transitions show quasi-atomic or band-like behaviour. Although this criterion allowed for satisfactory predictions of Auger lineshapes as in the case of MgAg alloys [11], the fact that it is strictly valid only for filled bands sets the limits of its applicability.

Atomic-like Auger spectra are also observed for halogens, chalcogens and alkaline elements, either chemisorbed or present in compounds (Fig. 2). Their quasi-

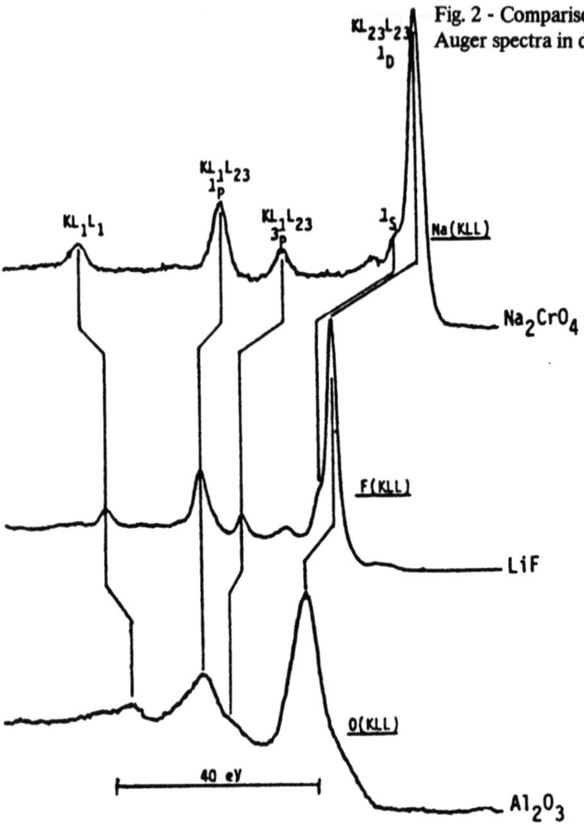

Fig. 2 - Comparison of sodium, fluorine and oxygen KLL Auger spectra in different compounds [12].

atomic character is related to the similarity of their fine structure with that of the nearest rare gas element. Whereas the Cini-Sawatzky model is inappropriate to explain this behaviour, Fuggle et al. [13] pointed out the determining role of screening effects. Extra-atomic relaxation causes an increase of the local density of states on the excited atom and leads thus to a two-hole final state similar to that of the nearest rare gas atom.

Finally, it is worth mentioning the case of interatomic transitions which appear as well for band-like as for atomic-like Auger spectra. These transitions are characterized by a final state where the two holes are no more located on the same atomic site as the initial hole [14]. The consequence of this spatial delocalisation of the holes is the drastic decrease of the two-hole final state interaction energy. These transitions would therefore give rise to band-like features although they appear in atomic-like spectra.

3 The band-like SiL2,3VV Auger spectrum from oxidized silicon

Due to the great sensitivity of their lineshape with chemical environment, band-like spectra are most often used in a fingerprint mode to characterize bonding state modifications at surfaces or interfaces. Deconvolution of this kind of spectra to interpret the fine structure changes is not straightforward. So new features appearing in band-like spectra during some surface treatments are generally connected with well defined reference situations or with results given by other techniques such as photoemission. This procedure is illustrated hereafter in the case of in-situ oxidation of silicon.

We have already reported AES data concerning the initial stages of silicon oxidation at room temperature and low pressure [15,16]. Modifications of the fine structure of the $SiL_{2,3}VV$ and OKVV transitions gave evidence for the formation of a

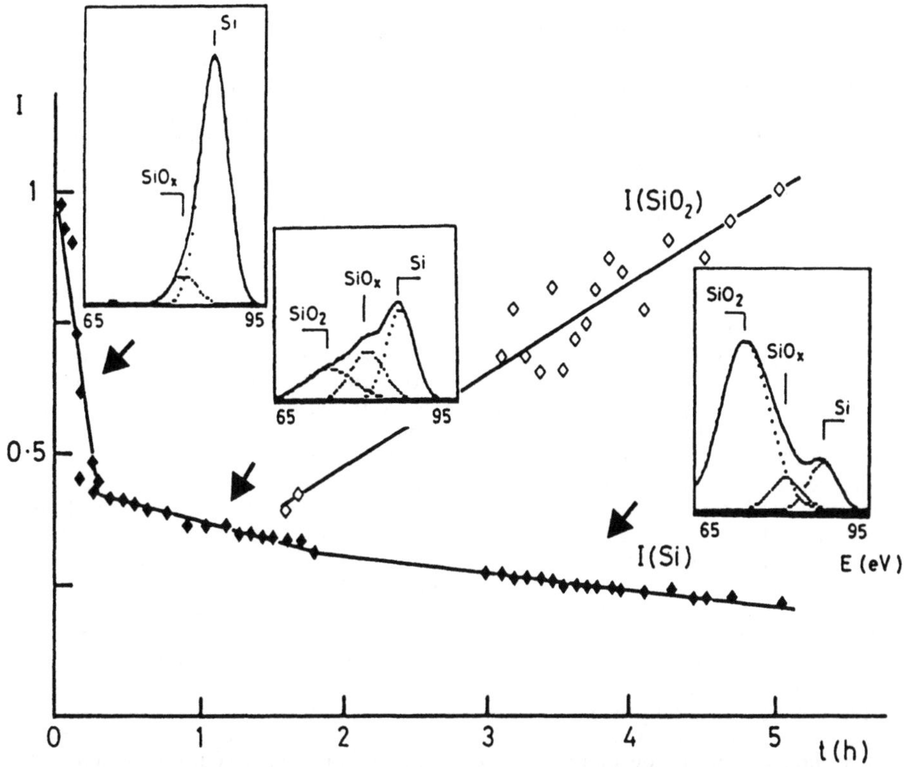

Fig. 3 - Auger signal versus time curve during the electron-beam-induced oxidation of an ordered silicon (111) surface. Curve fitting of silicon $L_{2,3}VV$ Auger spectra in the N(E) mode at different stages of oxidation are shown as insets.

two-dimensional intermediate oxide SiO_x, where $x = 1 \pm \epsilon$, before the formation of SiO_4 tetrahedra. In these studies, oxidations and measurements were performed sequentially, reducing as much as possible sample exposure to the electron beam in order to minimize the beam effects. Next we chose to integrate these effects into our experiments, maintaining continuously the surface under the primary beam [17,18]. A more complete oxidation at room temperature as well as a very detailed description of the evolution of the $SiL_{2,3}VV$ fine structure during this oxidation could therefore be obtained.

Figure 3 illustrates a typical Auger signal versus time curve showing the oxidation of a Si(111) surface. The different segments of the lower curve define three main steps in the oxidation process. The main modifications of the $SiL_{2,3}VV$ spectrum and the corresponding steps are described in reference 18. For the first stage of oxygen incorporation, characterized by the growth of a peak at 84eV, AES by itself does not allow us to be sure that there is only one unique well-defined oxidation state. At this stage, some oxygen atoms would be in a bridging position between two surface silicon atoms and others inserted between the first and the second silicon layers in interstitial positions. However, results obtained by a curve fitting procedure on the SiLVV spectra, shown as insets in Fig. 3, strengthen the physical reality of this first oxidation stage. They show the fast initial growth and after that the levelling of a peak at 81eV attributable to the intermediate oxide. This curve fitting confirms thus that the thin oxide layer grows up first from SiO_x species and later from SiO_4 tetrahedra. But it is clear that further progress in the interpretation of the SiLVV fine structure would require theoretical calculations which could give an account of the experimentally observed features.

4 The atomic-like OKLL Auger spectrum from oxides

As a result of their atomic-like behaviour, the fine structure of quasi-atomic Auger spectra is rather insensitive to bonding state modifications. But, unlike band-like spectra, chemical shifts become consequently meaningful. These shifts arise from changes in electron energies and in extra-atomic relaxation, which are linked to chemical environment differences. However, in the particular case of oxides, Auger line shift measurements are often subjected to uncertainties caused by charging effects or reference level problems. The use of Lang and Williams' [19] or Wagner's [12] Auger parameter is then a convenient way to circumvent these experimental difficulties. The former parameter gives moreover an insight into the screening phenomena of the Auger two-hole final state since it precisely represents U_{eff}, the effective Coulomb interaction energy.

Using the well known three-step model to describe the Auger process, we derived the U_{eff} values of different KLL transitions of oxygen from binding and Auger kinetic energies measured by XPS/XAES, both in transition- and non-transition metal oxides. The Auger OKLL fine structure as well as the multiplet line identification are shown in Fig. 2 in the case of Al_2O_3. The procedure used to work out the results is given in references 20 and 21. The values obtained for the KVV;^{1}D, KL$_1$V;^{1}P and KL$_1$L$_1$;^{1}S transitions are listed in Table 1. Large extra-atomic relaxation effects, attributable to the metal d-band, are observed for transition metal oxides and attest to the greater efficiency of d-band screening over s- or p-band screening. Furthermore, strong evidence emerges that the final state holes are localized in spite of these important relaxation phenomena. However, d-band screening breaks off as the metal d-band is both filled and separated from the actual valence band. A more detailed analysis of the results obtained for non-transition metal oxides revealed that mostly least bound O2p states participate to oxygen Auger transitions involving valence electrons. This means that the corresponding transition probabilities are not uniform over the whole valence band but favour the local density of valence states. Apart from the fundamental aspects of these conclusions, interesting developments are expected from that difference in screening behaviour between transition and non-transition metal oxides. This is illustrated in the next section for the FeSi/SiO_2 interface.

Table 1 - Effective two-hole final state interaction $U_{eff}(kl)$ for Auger OKLL transitions in different oxides. All values are given in eV.

U_{eff}	(VV)	(L_1V)	(L_1L_1)	U_{eff}	(VV)	(L_1V)	(L_1L_1)
Fe0	6.4	10.3	9.1	Al_20_3	14.2	16.3	13.1
Fe_20_3	4.8	9.1	8.0	Be0	11.3	14.2	11.9
$Mn0_2$	5.5	9.5	8.7	$Ge0_2$	10.1	14.4	13.5
Nb_20_5	6.1	8.9	7.2	Mg0	11.3	14.5	11.0
Hg0	11.7	13.5	11.0	$Si0_2$	10.5	14.8	13.1
Zn0	11.6	14.5	11.0	$Ti0_2$	9.6	12.2	9.0

At last, let us focus on interatomic transitions occuring in oxygen Auger spectra. In most of the studied oxides, interatomic features only appear as a slight broadening at high kinetic energy side of the OKVV peak. On the other hand, the spectra of ZnO and HgO present large shoulders as it is shown in Fig. 4. For both oxides, a rough two-component decomposition of the OKVV spectrum leads to a high kinetic energy contribution which can be explained by an OKV(O)V(M) interatomic transition involving the outermost d-band of the metal and a nearly zero Ueff value. With respect to the other oxides, the presence of a filled and localized d-band near the actual valence band seem to favour the probability of interatomic transitions. Similar features were

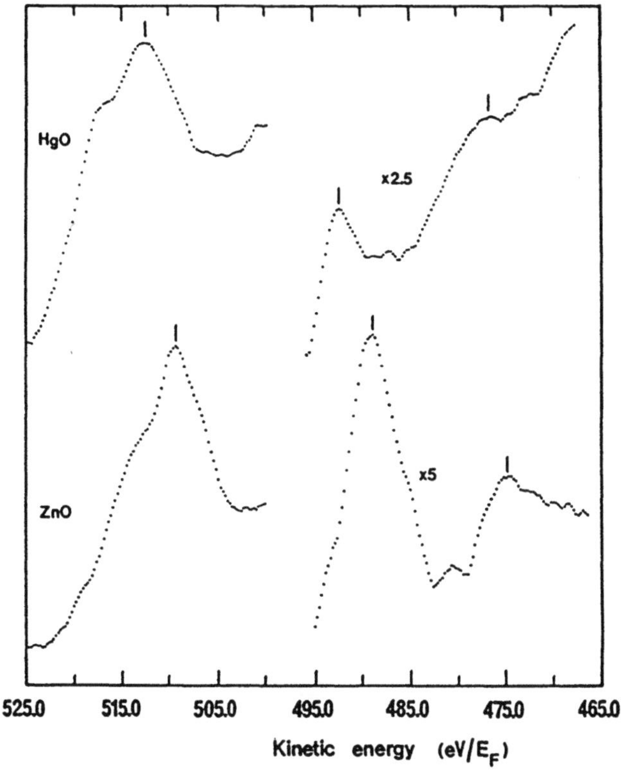

Fig. 4 - Oxygen KLL Auger spectra with expanded multiplet fine structure for HgO and ZnO.

recently observed by Fritsch and Légaré in the case of the Pt/Al_2O_3 interface [22]. The oxygen Auger spectrum (Fig. 5b), obtained after thermal deposition of small platinum particles on an alumina substrate, shows a small peak at 520eV which is absent in the OKVV spectrum of clean Al_2O_3 (Fig. 5a). This feature can be interpreted as an interatomic oxygen Auger transition, where one final state hole is located on a platinum atom, since the corresponding U_{eff} value calculated from experimental data is nearly zero. Although this example shows that useful information can be gained from fine structure analysis of OKLL spectra, the main difficulty to interpret particular features lies in its decomposition which has to take into account the contributions of the different transitions, the plasmon losses and the secondary background.

5 Extra-atomic relaxation and Auger parameter

To illustrate the influence of the presence of transition metal atoms on screening effects and on Wagner's oxygen Auger parameter in silicon oxide, we present some results of an ESCA investigation on the thermal oxidation of an FeSi alloy [23]. The (100) surface of a clean Fe-6 at% Si single crystal was exposed to oxygen at room temperature and low pressure and next annealed at 500°C. Analysis of the Si2p, O1s and $Fe2p_{3/2}$ photoelectron spectra show that iron is reduced whereas oxygen induced segregation leads to the formation of a thin quasi-silica layer at the surface. The oxygen Auger parameter of this oxide layer, composed of SiO_4 tetrahedra, is thus expected to be close to that of bulk silica. However, the obtained value of 1041.1eV is in large discrepancy with the Auger parameter of studied reference situations, i.e. SiO_2, 1039.5eV; Fe_2SiO_4, 1040.2eV; FeO, 1041.8eV; Fe_2O_3, 1042.2eV. The whole set of values is represented in Fig. 6, taken from a publication of Wagner et al. [12].

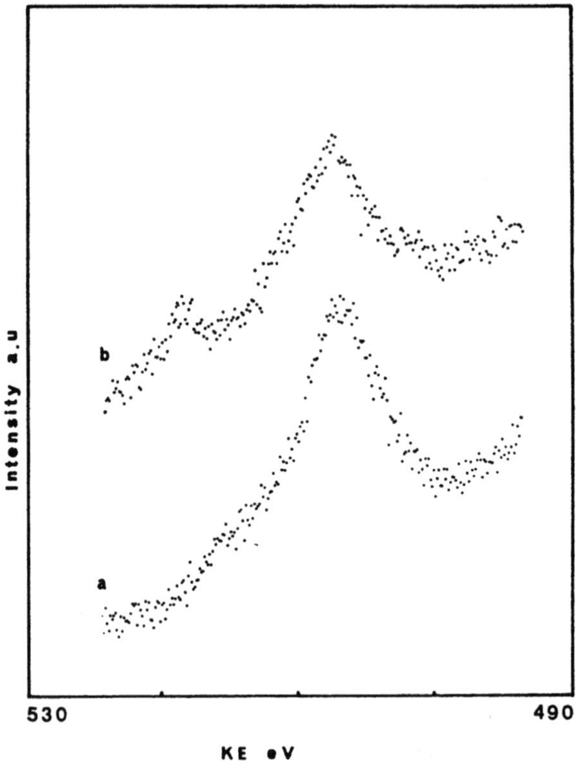

Fig. 5 - Oxygen KVV Auger spectra obtained (a) before and (b) after platinum deposition on Al_2O_3 [22].

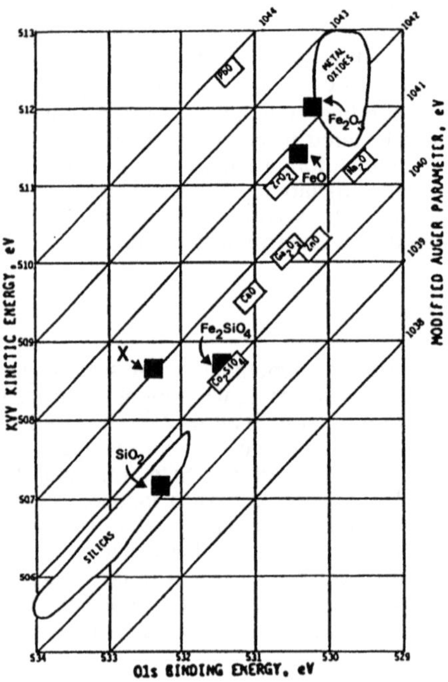

Fig.6 - Oxygen Auger parameter in different oxides after Wagner et al. [12]. The values obtained on reference oxides (FeO, Fe_2O_3, SiO_2 and Fe_2SiO_4) were added on this diagram. The square denoted X corresponds to the value determined for the quasi-silica layer on FeSi.

To throw some light on this problem, we determined the effective Coulomb interaction of the OKL_1L_1 transition for this surface. We found a $U_{eff}(L_1L_1)$ value of 11.2eV, to be compared with the reference values of SiO_2 (13.1eV), Fe_2SiO_4 (10.8eV), FeO (9.1eV) and Fe_2O_3 (8.0eV). According to the conclusions of the preceeding section, the similar value obtained for this surface and bulk Fe_2SiO_4 shows that d-band screening, due to the presence of iron atoms, is involved in the extra-atomic relaxation of the OKL_1L_1 transition. This supposes that iron atoms are close to SiO_4 tetrahedra, as in Fe_2SiO_4 where iron atoms occupy octahedral vacancies of the SiO_4 packing. However, neither the binding energy nor the lineshape of the O1s peak accounts for the presence of this compound. These apparently conflicting results can in fact be explained by an extra-atomic d-band relaxation contribution of the iron atoms at the interface between the metallic substrate and the silicon oxide layer. So, turning back to the measured oxygen Auger parameter, it appears that the shift with respect to SiO_2 in Wagner's diagram can be attributed to extra-atomic relaxation effects due to the metallic substrate. This example thus shows that even in XPS studies, Auger relaxation effects can provide interesting clues for bonding state characterization, in particular in the case of metal/oxide interfaces.

REFERENCES

1. B. Carrière, J.P. Deville, P. Humbert: J. Microsc. Spectrosc. Electron. 10, 29 (1985)
2. C.J. Powell: Phys. Rev. Lett. 30, 1179 (1973)
3. H.H. Madden, J.E. Houston: Solid State Comm. 21, 1081 (1977)
4. P.J. Feibelman, E.J. McGuire: Phys. Rev. B17, 690 (1978)
5. D.R. Jennison: Phys. Rev. B18, 6865 (1978)
6. D.R. Jennison, H.H. Madden, D.M. Zehner: Phys. Rev. B21, 430 (1980)
7. G.D. Davis, P.E. Viljoen, M.G. Lagally: J. Electron Spectrosc. Relat. Phenom. 21, 135 (1980)
8. R. Lässer, J.C. Fuggle: Phys. Rev. B22, 2637 (1980)
9. M. Cini: Solid State Comm. 24, 681 (1977)
10. G.A. Sawatzky: Phys. Rev. Lett. 39, 504 (1977)
11. P. Weightman, P.T. Andrews: J. Phys. C13, 3529 (1979)
12. C.D. Wagner, D.A. Zatko, R.H. Raymond: Anal. Chem. 52, 1445 (1980)
13. J.C. Fuggle, E. Umbach, R. Kakoschke, D. Menzel: J. Electron Spectrosc. Relat. Phenom. 26, 111 (1982)
14. P.H. Citrin, J.E. Rowe, S.B. Christman: Phys. Rev. B14, 2642 (1976)
15. B. Carrière, J.P. Deville: Surf. Sci. 80, 278 (1979)
16. B. Lang, P. Scholler, B. Carrière: Surf. Sci. 99, 103 (1980)
17. B. Carrière, A. Chouiyakh, B. Lang: Surf. Sci. 126, 495 (1983)
18. B. Carrière, J.P. Deville, A. El Maachi: Phil. Mag. B55, 721 (1987)
19. N.D. Lang, A.R. Williams: Phys. Rev. B20, 1369 (1979)
20. P. Humbert, J.P. Deville: J. Phys. C20, 4679 (1987)
21. P. Humbert, J.P. Deville: J. Physique Coll. C9-48, 927 (1987)
22. A. Fritsch, P. Légaré: Surf. Sci. 184, L355 (1987)
23. P. Humbert, A. Mosser, M. Essouni: to be published

Polymer Formation on Nickel Surfaces at Elevated Temperature

B. Dötsch, L. Hammer, and K. Müller

Institut für Angewandte Physik, Lehrstuhl für Festkörperphysik,
Universität Erlangen-Nürnberg, Erwin-Rommel-Str. 1,
D-8520 Erlangen, Fed. Rep. of Germany

Acetylene on Ni(111) has been studied by a combination of Auger electron spectroscopy (AES) and high resolution electron loss spectroscopy (HREELS). Dependent on temperature, different adsorption kinetics and reactions with the substrate are observed. C_2H_2 adsorbs at 240 K moleculary at the surface. Increasing temperature leads to thermal fragmentation of acetylene. In the temperature range between 500 K and 600 K a carbidic overlayer is generated at low C_2H_2 coverage, exposure exceeding 2L leads to the formation of hydrocarbon species on top of the carbidic layer. The adsorption and reaction mechanisms at elevated temperatures are discussed.

1. Introduction

The adsorption and thermal reaction of acetylene on single crystal transition metals is of continued interest. Electron energy loss spectra for C_2H_2 adsorbed on nickel surfaces (see e.g. refs. /1-6, 16/ and LEED measurements of C_2H_2 on nickel surfaces (see e.g. refs. /4,6-8/) have previously been reported. In addition, reports of Auger electron spectra of acetylene adsorbed at higher temperatures on several transition metal surfaces like on Ni(100), polycrystalline nickel, Ta(110), Fe(110) and W(110) (see e.g. refs. / 8-12/) have been done.
In this work the adsorption and reaction behaviour of acetylene on Ni(111) especially in a temperature range from 450 K to 600 K was studied. Several surface sensitive techniques are used with emphasis strongly laid on Auger electron spectroscopy.

2. Experimental

The measurements were carried out in a ultra high vacuum system (base pressure $1 \cdot 10^{-10}$ mbar) equipped for Auger electron spectroscopy, low energy electron diffraction and high resolution electron energy loss spectroscopy. The Ni(111) single crystal was mechanically and electrochemically polished before being mounted on an indirectly heated and liquid nitrogen cooled sample holder. The crystal was cleaned in situ with alternating neon sputtering (3 min, 3 keV, 30 $\mu A/cm^2$) and annealing (5min, 1400 K) until the main contamination sulphur and phosphorus were reduced below the AES detection limit. Carbon contamination could be removed by annealing

Springer Series in Surface Sciences, Vol. 18 **Auger Spectroscopy and Electronic Structure**
Editors: G. Cubiotti · G. Mondio · K. Wandelt

in oxygen (5min, 2 · 10^{-8} mbar oxygen, 600 K) and subsequent heating to 1400 K. Auger electrons were exited by electron impact using a beam current of 1 μA, 2.0 keV defocused to irradiate a 6-7 mm diameter spot on the crystal. In order to check and to avoid electron beam induced damage of the adsorption system electron energy loss spectra were taken before and after the AES measurements. Identical loss spectra indicate that adsorbates remained undisturbed under the influence of the electron beam.

3. Results and discussion

The adsorption and reaction behaviour of acetylene on Ni(111) at different temperatures is given in fig. 1 (temperature increase from bottom to top). The center panel demonstrates the variation of the carbon coverage with acetylene exposure using the peak to peak intensity of the C(KVV) Auger line normalized to the peak to peak intensity of the Ni(M_{23}VV) of the clean nickel surface. High resolution electron energy loss spectra at saturation coverage of C_2H_2 are shown in the left hand panel, carbon Auger lineshapes (first derivative) in the right hand panel. HREELS measurements at temperature lower than 300 K are in good agreement with measurements of Ibach et. al. /3/ and indicate that C_2H_2 adsorbs moleculary at the surface. Increasing temperature leads to thermal fragmentation of acetylene (shown in the left middle panel of fig. 1) square with a change in the C(KVV) Auger lineshape (right middle panel of fig. 1). However, both mechanisms at 240 K and 420 K can be fitted with a first order Langmuir isotherm. Therefore we assume that the rate of adsorption of the C_2H_2 molecules on the surface is proportional to the number of unoccupied sites, and one obtains eq. (1) for the intensity I(A) of acetylene molecules on the surface as a function of the exposure A. Best fit curves are plotted in the center panel of fig. 1 (solid line) together with the experimental points.

$$I(A) = I_0 \cdot \left[1 - \exp(-c(T) \cdot A)\right] \qquad \text{eq.(1)}$$

Here I_0 is the value of I at saturation, A the exposure of acetylene , and c(T) the initial sticking probability.

Starting at 450 K we found a new adsorption respectively reaction behaviour of acetylene on Ni(111). The intensity of the carbon signal initially increases strongly, leading to a plateau and subsequently increases again until a saturation coverage is reached. At 600 K another carbon intensity curve is observed correlated to another change of the carbon Auger lineshape and to the disappearence of hydrocarbon vibrational losses (fig. 1 upper panel).

The following discussion relates to the reaction behaviour of acetylene in the temperature region from 450 K to 550 K. Fig. 2 shows (upper panel) the intensity of the C(KVV) Auger line as a function of the C_2H_2 exposure at 550 K, the insert shows the low exposure part of the curve. Until the plateau is reached we found a C(KVV) Auger lineshape shown in the left panel (fig. 2), additional exposure leads to a change in lineshape which is shown at saturation coverage in the upper right

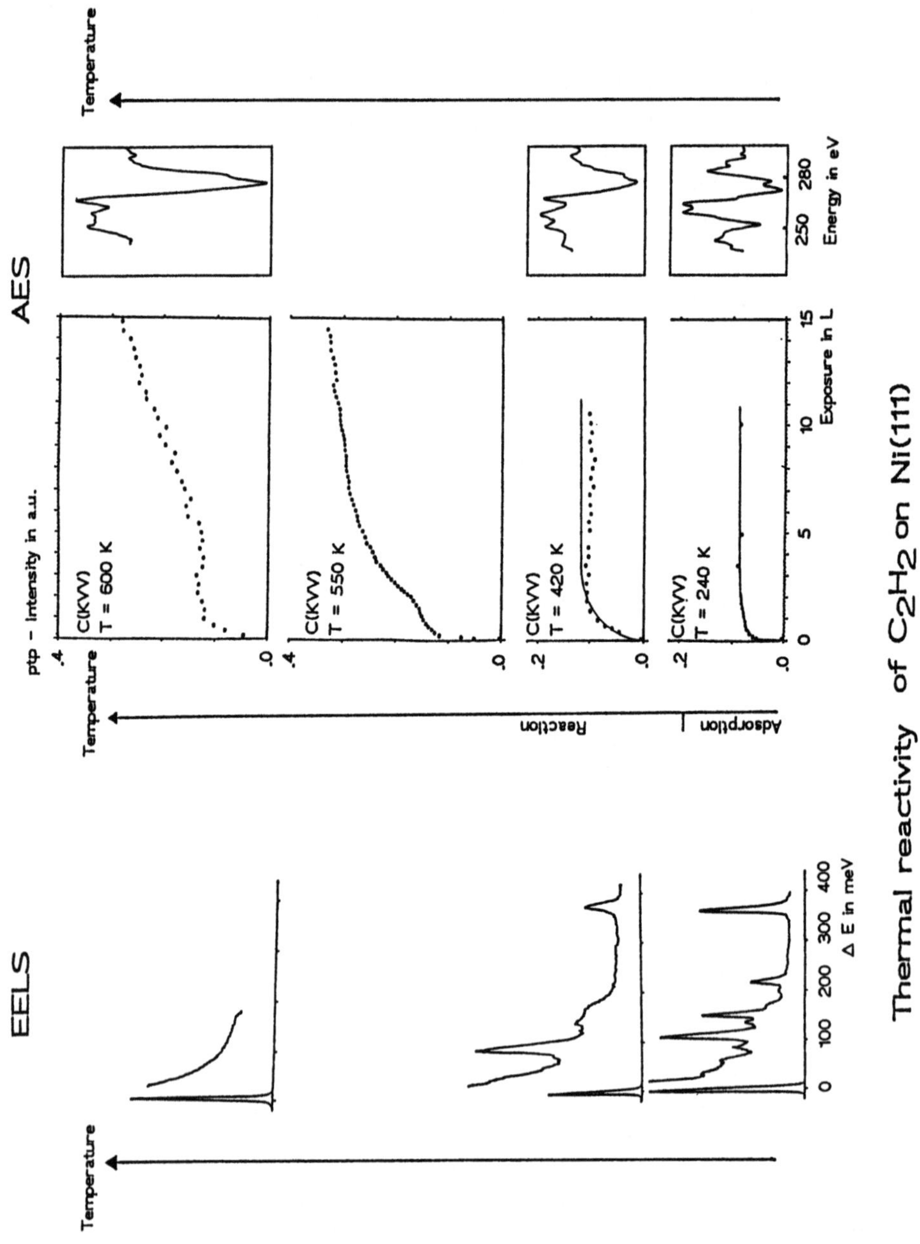

Fig. 1. Thermal reactivity of acetylene on Ni(111). center panel: Intensity variations of C(KVV) Augerline as function of coverage and temperature. Right hand panel: Correlated Carbon Auger lineshape at saturation coverage. Left hand panel: HREELS spectra at saturation coverage.

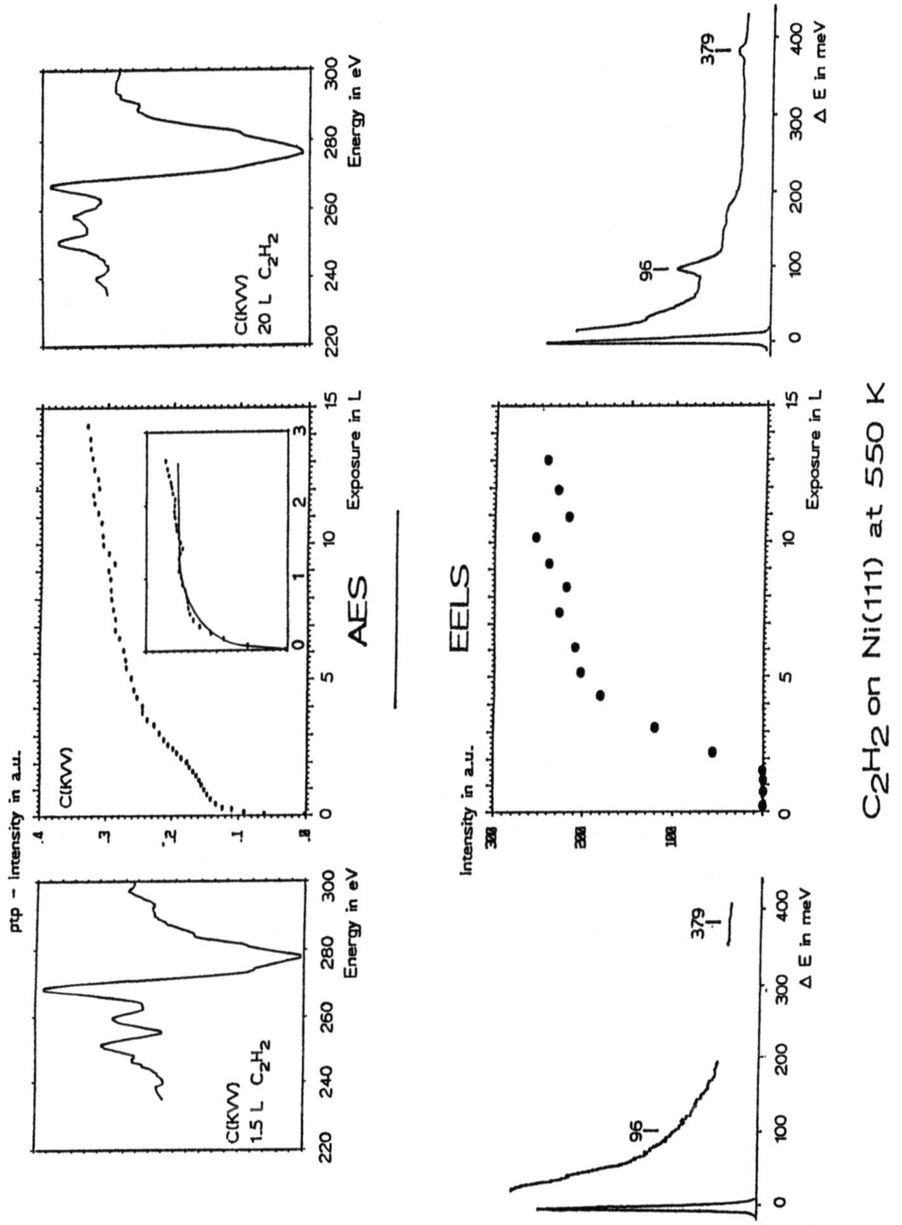

Fig. 2. Acetylene on Ni(111) at 550 K. center panel: Intensity of C(KVV) Augerline and 96 meV loss peak as function of C_2H_2 exposure. Left hand panel: C(KVV) Auger lineshape and HREELS spectra at low exposure (2 L). Right hand panel: C(KVV) Auger lineshape and HREELS spectra at saturation coverage.

of fig. 2. Complementary results were obtained from HREELS and LEED measurements. At the low exposure range no vibrational losses are detected (fig. 2, lower panel). At the same time a complex diffraction pattern arise, which is known from thermal fragmentation of different hydrocarbons on Ni(111) /7/. Bertolini et al. /7/ assign the diffraction pattern to a carbidic carbon superstructure. This is in a good agreement with our Auger lineshape observations which turn out to be carbidic carbon (see e.g. ref. /13/, /14/, /15/). The formation of carbide is terminated by the end of the plateau respectively by the increase of the carbon Auger signal and the beginning change of the C(KVV) Auger line. Correlated to this behaviour are vibrational losses of hydrocarbon species (fig. 2, lower right). In addition spot intensities of the diffraction pattern decrease, before the diffraction pattern finally disappears at saturation coverage. The mechanism of the carbide formation can also be fitted with first order Langmuir kinetics (cp. eq. (1)) (solid line in the middle panel above).

$$I(A) = I_P \cdot \left[1 - \exp(-c(T) \cdot A)\right] \qquad \text{eq. (2)}$$

Here I_P is the value of I at the plateau, A the exposure of C_2H_2 in L, and c(T) the sticking probability.

The further discussion should be limited to the adsorption behaviour after the formation of the carbidic layer. The species built up could be indentified by HREELS measurements. Vibrational losses at 96 meV, 379 meV and a shoulder between 120 meV and 180 meV were dedected (fig. 2, upper right panel). This spectrum is in a good agreement with measurements by Demuth and Ibach/16/, who attached the losses to CH_2 - species respectively to polymers containing CH_2 species. Further references to the formation of a thick polymer layer could be obtained by the investigation of the intensity curves. The intensity of the peak at 96 meV as a function of the acetylene exposure (fig. 2, lower panel,center) can be correlated to the measurement of the carbon intensity Auger curve. Both intensities increase after the formation of the carbidic carbon layer which indicates that a thick hydrocarbon layer was formed. Additional informations were obtained by LEED. The disappearence of the diffraction pattern upon continuous acetylene exposure points also at the growth of a disordered layer of hydrocarbons on the nickelcarbide surface.

4. Conclusion

We suggest that the reaction behaviour of acetylene at temperatures between 400 K and 600 K could be described as follows: At low exposure C_2H_2 adsorbes by first order Langmuir kinetics to form a carbidic overlayer. On top of this overlayer subsequent adsorption of thermal fragments of acetylene probably CH_2 species, takes place. The reaction behaviour at higher exposures favours the conclusion that a netted, thick and amorphous polymer layer of hydrocarbons is formed on the carbide surface.

5. Acknowledgement

The support of the Deutsche Forschungsgemeinschaft is gratefully acknowledged.

6. References

/1/ J.E. Demuth, H. Ibach; Surf. Sci. 85 (1979) 365
/2/ J.C. Bertolini, J. Rousseau; Surf. Sci. 83 (1979) 531
/3/ H. Ibach, S. Lehwald; J. Vac. Sci. Technol. Vol. 18/2 (1981) 625
/4/ L. Hammer, T. Hertlein, K. Müller; Surf. Sci. 178 (1986) 693
/5/ J.A. Stroscio, S.R. Bare, W. Ho; Surf. Sci. 148 (1984) 499
/6/ B.J. Bandy, M.a. Chester, M.e. Pemble, G.S. McDougall, N. Sheppard; Surf. Sci. 139 (1984) 87
/7/ J.C. Bertolini, G. Dalmai-Imelik; in: Structures et Propriètès des Surface des Solides, CNRS, Paris (1969) 135
/8/ G. Casalone, M.G. Catania, M. Simonetta, M. Tescari; Surf. Sci. 62 (1969) 321
/9/ S.R. Kelemen; Appl. Surf. Sci. 28 (1987) 439
/10/ F.A. Londry, A.J. Slavin, P.R. Underhill; Surf. Sci. 140 (1984) 521
/11/ A.K. Bhattacharya, M.A. Chesters, Surf. Sci. 186 (1987) 279
/12/ S.D. Foulias, K.J. Rawlings, B.J. Hopkins; Surf. Sci. 133 (1983) 377
/13/ J.E. Houston, D.E. Peebles, D.W. Goodmann; J. Vac. Sci. Technol. Vol. 1/2 (1983) 995
/14/ M. Salmerón, A.M. Baró, J.M. Rojo; Phys. Rev. B. Vol. 13/10 (1976) 4348
/15/ G. Paolucci, R. Rosei, K.C. Prince, A.M. Bradshaw; Appl. Surf. Sci. 22/23 (1985) 582
/16/ J.E. Demuth, H. Ibach; Surf. Sci. 78 (1978) L 238

Electron Spectroscopic Studies of Acetylene Adsorption on Polycrystalline Iron Films

M. Pirner, D. Borgmann, and G. Wedler

Institut für Physikalische und Theoretische Chemie
der Universität Erlangen-Nürnberg,
Egerlandstr. 3, D-8520 Erlangen, Fed. Rep. of Germany

1. Introduction

Recently both isothermal and temperature programmed reaction mass spectrometry (IRMS and TPRMS, respectively) have been used to study the composition of the gas phase which is in equilibrium with acetylene covered surfaces of thin films of transitions metals, particularly of iron /1,2/. Both TPRMS carried out between 77 and 400 K and IRMS performed at 273 K revealed the formation and desorption of methane, hydrogen, ethane, butane and benzene before a marked partial pressure of acetylene could be detected.

It is the aim of this paper to use electron spectroscopies (HeI UPS, HeII UPS, XPS and AES) and changes in work function to investigate the adsorbate and to correlate the own experiments using evaporated films with literature data preferably obtained with single crystals.

2. Experimental

The experiments were carried out using a Vacuum Generators electron spectrometer of ESCA III-type. The iron films were evaporated from spectroscopically pure iron wires under ultrahigh vacuum conditions, condensed at 80 K on a copper substrate and finally annealed for 60 min at 333 K. The cleanliness of the iron films is as good as that of highly purified iron single crystals. The work function of the clean films used in this work is 4.3 eV, comparable with that of Fe(111) single crystals. Unannealed films exhibit a work function of 4.1 eV. The adsorption properties of these annealed films are, indeed, very similar to those of Fe(111) single crystals, as has been demonstrated e.g. for N_2 adsorption /3,4/ or CO_2 adsorption /5,6/. Details concerning the characterization of such iron films are given elsewhere /1/.

The acetylene (Messer Griesheim, purity 99.7 %) was distilled several times in a separate glass-apparatus under uhv conditions and then filled into ampoules.

The films were discontinuously covered with the purified acetylene at 80 or at 300 K. In the former case the temperature was discontinuously increased in order to investigate the influence of temperature on the behaviour of the adsorption layer.

Springer Series in Surface Sciences, Vol. 18 Auger Spectroscopy and Electronic Structure
Editors: G. Cubiotti · G. Mondio · K. Wandelt

3. Results and Discussion

HeII UP spectra are shown in Fig. 1. They are in very good agreement with the HeI UP spectra, exhibit a slightly worse resolution, give, however, much better information on the position of the signals far away from the Fermi level which are superimposed by the secondary electron cascade in the case of HeI UPS.

When the iron films are exposed at 80 K to increasing doses of C_2H_2 at first signals appear at 5.4, 9.2, and 11.4 eV. Their positions are not influenced by the increase in coverage, the intensities reach a maximum at an exposure to 1.5 L. With further increasing coverage three additional peaks are built up at 6.5, 11.6, and 13.5 eV. When the acetylene pressure in the apparatus is increased to values above the saturation pressure ($4.7x10^{-5}$ Pa) acetylene condenses on the film. This does not influence the separation of the three peaks dominating already at 2.5 L.

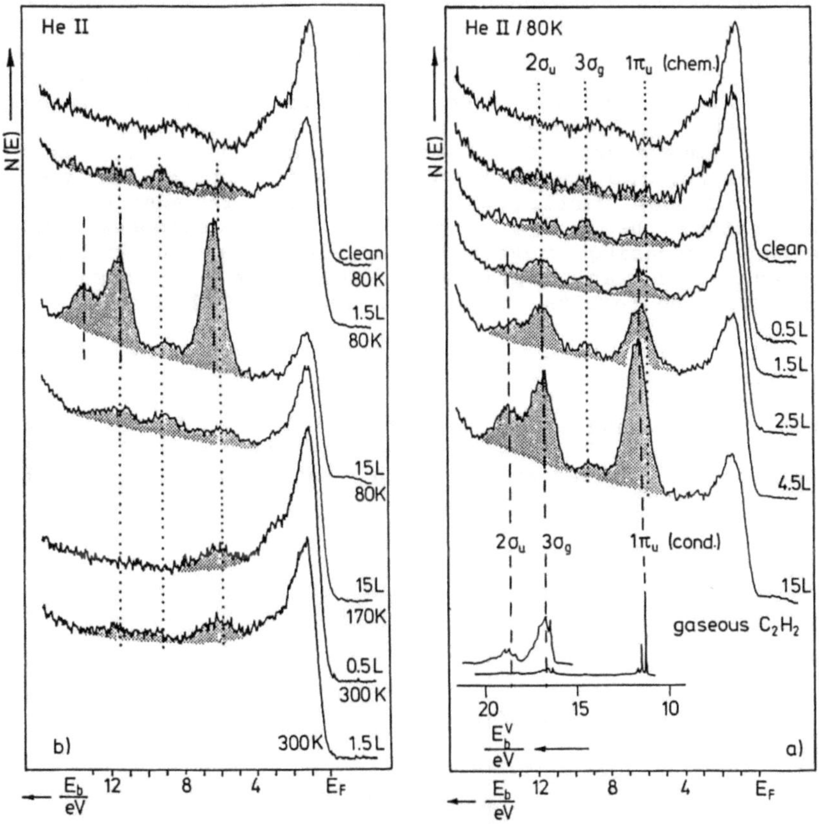

Fig. 1: a) HeII UP spectrum of an Fe film in the clean state and in dependence on exposure to C_2H_2 at 80 K. The energy scale of the spectrum of gaseous C_2H_2 is related to the vacuum level. b) HeII UP spectra of an Fe film after different exposures to C_2H_2 in dependence on temperature

The inset in Fig. 1a shows the gas phase UP spectrum of acetylene which allows to correlate the signals to the $1\pi_u$-, $3\sigma_g$-, and the $2\sigma_u$-orbitals of the acetylene molecule /7/.

When the temperature is increased to 170 K (Fig. 1b) acetylene partially desorbs and the peaks at 6.5, 11.6, and 13.5 eV disappear. The spectrum is now nearly identical with that observed at 80 K after an exposure to 1.5 L C_2H_2.

At 300 K small amounts of adsorbed acetylene produce a spectrum which consists of only one broad peak around 6 eV. However, the adsorption of additional C_2H_2 leads to a spectrum which again is in agreement with the low coverage spectra observed at lower temperatures.

A first qualitative interpretation of the spectra of Figs. 1a and b is the following:

At 80 K and low coverages acetylene is chemisorbed on iron represented by signals at 5.4, 9.2, and 11.4 eV (dotted line). C_2H_2 can be adsorbed in multilayers at high coverages. This is confirmed by volumetric measurements performed earlier /1,2/ and indicated by the appearance of additional signals at 6.5, 11.6, and 13.5 eV (dashed line). At high C_2H_2 pressures (starting with the spectrum taken at 15 L) condensation occurs. Therefore the 15 L spectrum exhibits a superposition of the signals due to chemisorbed and condensed C_2H_2. With increasing temperature the adsorbed and the condensed acetylene desorb, the chemisorbed gas remains on the surface, however, decomposes partially. When the gas is adsorbed at room temperature the initially admitted doses decompose, and only with increasing coverage acetylene is molecularly chemisorbed. Such an explanation would be in complete agreement with the results of previous studies /1,2/.

There are some more facts which support the given interpretation:

In Fig. 2 the changes in work function are plotted as a function of exposure. They were determined from the shift of the onset of the secondary electron cascade in the HeI UP

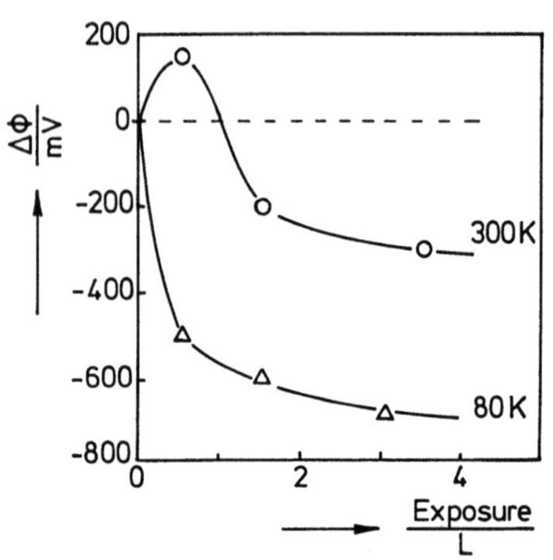

Fig. 2: Changes in work function of an Fe film as a function of exposure to C_2H_2 at 80 and 300 K

spectra. The curve at 80 K was also measured by means of the photoelectric method after Fowler /1/. There is complete agreement. The shape of the curve is characteristic of the adsorption of only one kind of species, whereas the curve obtained with the adsorption at 300 K clearly shows that the processes occuring at low coverages are quite different from those dominating at high coverages. There is strong evidence for decomposition followed by molecular adsorption.

Such an interpretation is supported by the C1s XP spectra (Fig. 3). At 80 K and - in the case of high coverages - also at 300 K the position of the C1s signal is observed at 284.0 eV below the Fermi level. This is in excellent agreement with the values reported for acetylene adsorption in molecular form on Fe(110) (283.9 eV) /8/, on Fe(111) (283.2 eV) /9/ and on an iron sheet (283.9 eV) /10/. In the low coverage range, however, a shift of the C1s peak to 283.2 eV is observed at 300 K. This energetic position lies in the regime (282.3 to 283.3 eV) characteristic of "carbidic" carbon, i.e. carbon bound to the metal /9,11,12/ and strongly points to breakage of the C-C bond.

Another hint as to the state of the carbon atom is the shape of the differentiated Auger signal. Fig. 4 shows that there is no marked change in the form of the KLL signal when the temperature is increased from 80 to 170 K. This is in agreement with the UPS results. At 300 K, however, the signals look like a superposition of the signals a) or b) with a signal characteristic of "carbidic" carbon, particularly in the case of spectrum c) which was obtained after lower exposures than spectrum d) and e).

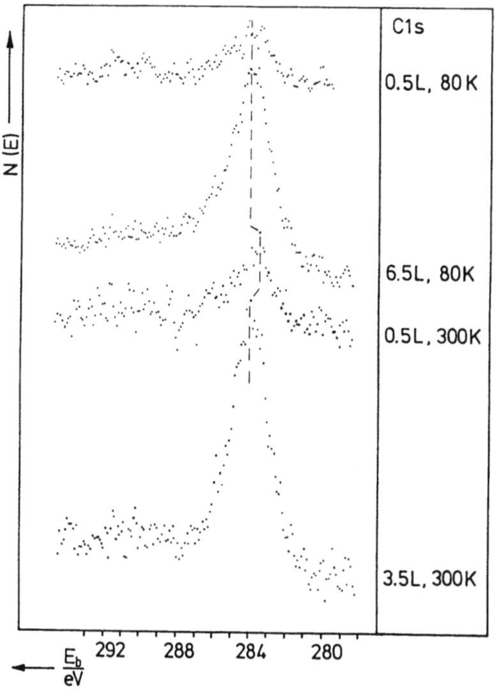

__Fig. 3:__ C1s XP spectra of an Fe film after different exposures to C_2H_2 at 80 and 300 K

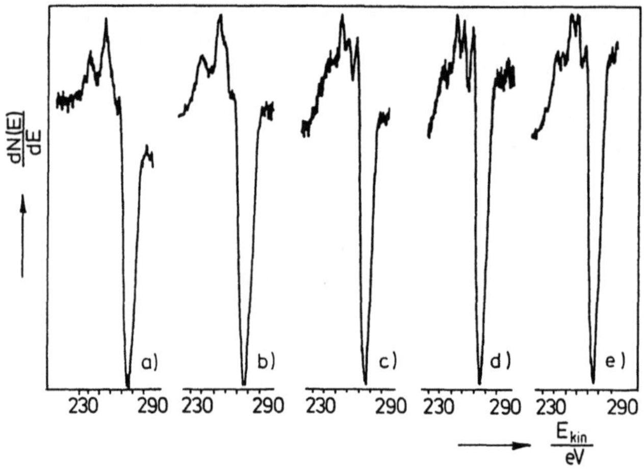

Fig. 4: C_{KLL} Auger signals of Fe/C_2H_2 at different exposures and temperatures. a) 80 K, b) 170 K, c) 3.5 L; 300 K, d) 7.5 L 300 K, e) 20 L; 300 K

These electron spectroscopic results agree with the conclusions drawn from measurements using IRMS and TPRMS. These latter methods are not appropriate for a study of the species present in the adsorption layer, i.e. of the nature of the species observed at 80 K (Fig. 1). In Fig. 5 the energetic position of the peaks of the $1\pi_u$-, $3\sigma_g$-, and $2\sigma_u$-emissions related to the vacuum level are plotted for acetylene in the gas phase, in the chemisorbed and in the condensed state. The position of the energy of the non-bonding $3\sigma_g$-orbital is taken as a reference energy. It is evident that the separation of the three orbital energies is the same for gas phase acetylene and condensed acetylene. Taking into account the work function of the annealed film (4.3 eV) and a change in work function by -0.7 eV due to C_2H_2 adsorption (see Fig. 2) there remains a shift of 1.7 eV which has to be traced back to relaxation. In the case of chemisorbed acetylene (change in work function by -0.5 eV) the relaxation shift is even 3.5 eV. These values are relatively high, are, however, in agreement with literature data. YU et al. /13/ report a relaxation shift of 3.2 eV for

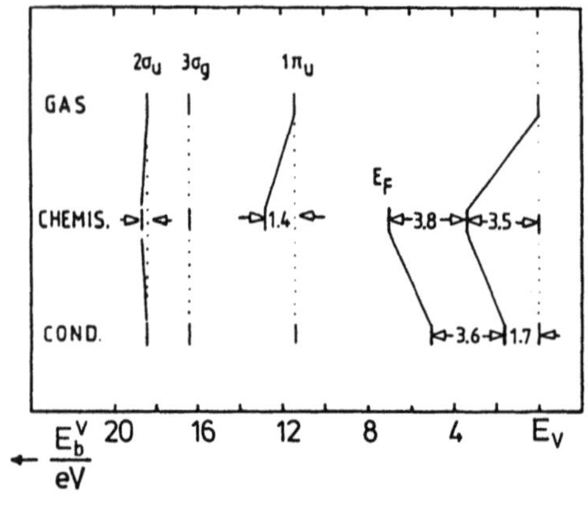

Fig. 5: Relative binding energies of gaseous, chemisorbed and condensed C_2H_2 related to the vacuum level. The position of the $3\sigma_g$-orbital is taken as reference

acetylene adsorption on polycrystalline Ni films, 4.0 eV for adsorption on Cu. The strong interaction of the chemisorbed acetylene with the iron film is indicated by the strong shift of the $1\pi_u$-emission (1.4 eV) relative to the $3\sigma_g$-emission. Even the $2\sigma_u$-orbital energy is slightly shifted.

4. Acknowledgements

This work was financially supported by the "Deutsche Forschungsgemeinschaft" and the "Fonds der Chemischen Industrie".

5. References

1. D. Borgmann, H. Behner, M. Pirner, and G. Wedler: Langmuir 2, 261 (1986)
2. W. Alter: Thesis, University of Erlangen-Nürnberg, 1988
3. G. Wedler, D. Borgmann, W. Alter, and K. Witan: Ber. Bunsenges. Phys. Chem. 90, 235 (1986)
4. D. Borgmann, F. Kroninger, P. Steidl,$and G. Wedler: Surface Sci. 189/190, 485 (1987)
5. R. Bauer, H. Behner, D. Borgmann, M. Pirner, W. Spiess, and G. Wedler: J. Vac. Sci. Techn. A5, 1110 (1987)
6. M. Pirner, R. Bauer, D. Borgmann, and G. Wedler: Surface Sci. 189/190, 147 (1987)
7. D.W. Turner:In Molecular Photoelectron Spectroscopy (Wiley and Sons, Interscience, London 1970)
8. H.P. Bonzel and H.J. Krebs: Surface Sci. 91, 499 (1980)
9. A. Mason and M. Textor: Proc. Roy. Soc. (London) Ser. A 356, 47 (1977)
10. H.P. Bonzel, H.J. Krebs, and W. Schwarting: Chem. Phys. Lett. 72, 165 (1980)
11. G.Wedler and J. Ertel: Ber. Bunsenges. Phys. Chem. 87, 469 (1983)
12. J. Benziger and R.J. Madix: Surface Sci. 94, 119 (1980)
13. K.Y. Yu, W.E. Spicer, J. Lindau, and P. Pianetta: Surface Sci. 57, 157 (1976)

Photoelectron and Auger Electron Emission from Physisorbed Xenon

K. Wandelt

Institut für Physikalische und Theoretische Chemie, Universität Bonn, Wegelerstr. 12, D-5300 Bonn, Fed. Rep. of Germany

1. Introduction

In this book electron spectroscopic techniques have been applied to elucidate the electronic structure of solid surfaces. Emphasis has been put on Auger electron spectroscopy as a very powerful tool to probe electron correlation effects taking advantage of the two-hole final state of the Auger emission process. Similarly important is the investigation of the electronic structure of adsorbed atoms and molecules, because their electronic properties govern the elementary processes of chemical reactions with and catalytic reactions at solid surfaces; of thin film growth; of Schottky barrier formation etc. Also for this problem Photoelectron Spectroscopy (PES) and Auger Electron Spectroscopy (AES) are the techniques which are most widely used in order to answer the basic question: How does the electronic structure, e.g. the electronic binding energies, of the molecules change between the gas phase and the adsorbed phase? This change, in fact, is a complex composite of several contributions, namely of physical and chemical initial state (ΔE_{init}) as well as of final state relaxation (ΔE_{relax}) effects. Firstly, the electron binding energies of adsorbed molecules are measured with respect to the Fermi level E_F of the substrate (E_B^F(ad)) while the respective ionization potentials in the gas phase are referred to the vacuum level E_V (E_B^V (gas)). This makes a correction by the work function $\varphi = E_V - E_F$ necessary. Secondly, upon approaching the surface the electrostatic potential acting on all charges of the molecules changes already in the inital state (prior to ionization) due to the polarizable charge in the substrate. This causes corresponding level shifts ΔE_{pot}. Thirdly, individual orbitals of the adsorbed particles are involved in the chemical bond with the surface leading to orbital specific shifts ΔE_{chem} of the respective electronic states. If this chemical bond is associated with some net charge transfer from or to the substrate, this extra charge will also affect all other electron energy states of the molecule which are not directly involved in the chemical bonding, namely the core levels. Finally, the experimentally determined difference $\Delta E_{exp} = E_B^V(gas) - E_B^F(ad)$ is also influenced by screening or final state relaxation effects (ΔE^{1+}_{relax}) of the one hole state created by photoionization. The total difference between measured adsorbate and gas phase electron binding energies is therefore given by:

$$\begin{aligned} \Delta E_{exp} &= E_B^V(gas) - E_B^F(ad) = \\ &= \varphi + \Delta E_{pot} + \Delta E_{chem} + \Delta E^{1+}_{relax} = \qquad (1) \\ &= \Delta E_{init} + \Delta E^{1+}_{relax} \end{aligned}$$

Springer Series in Surface Sciences, Vol. 18 Auger Spectroscopy and Electronic Structure
Editors: G. Cubiotti · G. Mondio · K. Wandelt

This break-up in the various contributions is purely conceptional. In practice, it is not possible to determine them separately. It is, for instance, still unclear whether φ corresponds to the work function of the bare substrate (prior to adsorption) or whether φ includes the adsorbate induced work function change and, if yes, to what extent. Another fundamental problem inherent in the interpretation of electronic spectra from adsorbates is the assessment of the relative size of the substrate induced extra-atomic relaxation energy ΔE^{1+}_{relax} of the photo-ionized final state. Adsorbed (physisorbed) rare gases have always been considered to be good test-cases to study this question /1,2/, because their initial state bonding interaction with any kind of substrate is very weak ($\Delta E_{chem} \approx 0$). Furthermore final state relaxation of these adsorbates is therefore also dominated by simple image polarization effects /3/ rather than by more complex charge transfer processes. Hence, adsorbed rare gas atoms may be considered as probes which enable to study the existence and the magnitude of the physical contributions ΔE_{pot} and ΔE^{1+}_{relax}. In principle, these contributions are present in any adsorbate system. However, in adsorbates with a strong chemisorption bond ΔE_{chem} may dominate the whole picture.

The present paper is an attempt to demonstrate that for adsorbed rare gases, in particular xenon, the sum $\Delta E_{pot} + \Delta E_{chem} + \Delta E^{1+}_{relax}$ has a constant value, largely independent of the substrate, and that any remaining substrate dependent variation of E^F_B(ad) and, hence, of ΔE_{exp} (see equ. (1)) is purely due to differences $\Delta\varphi$ in work function between different (bare) substrates. These conclusions are based on the interpretation of combined photoemission (one hole final state) and Auger electron (two hole final state) spectra of adsorbed xenon, but they should apply to other physisorbed (also molecular) adsorbates likewise.

2. Photoemission from adsorbed Xenon

At temperatures around 50 K the noble gas xenon can be adsorbed on any solid surface. As an overview the UV-photoemission spectrum (excited with He I radiation of $h\nu$ = 21,2 eV) of a Ru(0001) surface covered with one monolayer of Xenon is shown in Fig. 1. The structure between E_F and $\sim$4 eV binding energy corresponds to emission from the valence band of the Ru(0001) substrate. The intense emission between $\sim$10 eV and $\sim$16 eV arises from unspecific secondary electrons. Emission from the Xe overlayer appears between $\sim$4 eV and $\sim$8 eV superimposed on the substrate background emission (dotted line), and consists of two sharp peaks arising from the $5p_{3/2}$ and $5p_{1/2}$ photoemission final (hole) states of adsorbed Xe atoms. Fig. 2 shows just the spectral range of the $5p_{3/2,1/2}$ emission of a monolayer of Xe adsorbed on a Ru(0001) and a Ag(111) surface together with the $5p_{3/2,1/2}$ spectrum of gaseous xenon. Note, that in order to be able to compare both spectra of adsorbed xenon with the one of gaseous xenon, the energy scale in Fig. 2 is referred to the vacuum level (E^V_B) by adding the work function φ of the respective (bare) metal surface to the Xe(5p) binding energy E^F_B(ad) measured with respect to the Fermi level of Ru and Ag (see Fig. 1):

$$E^V_B(\text{ad}) = E^F_B + \varphi \; . \qquad (2)$$

Comparison of the gas phase spectrum with both adsorbate spectra reveals three interesting differences. Firstly the adsorbate signals are significant-

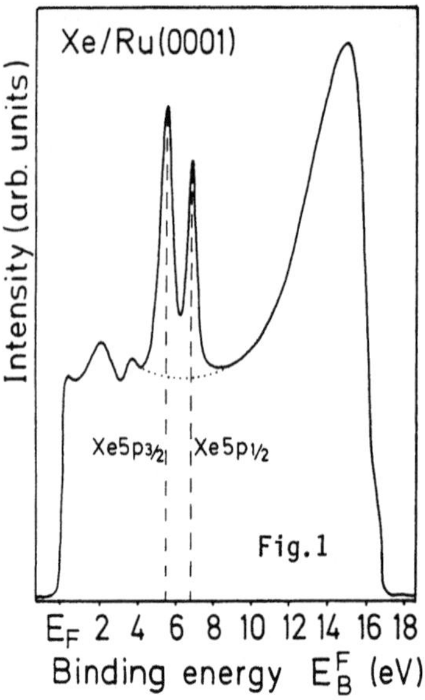

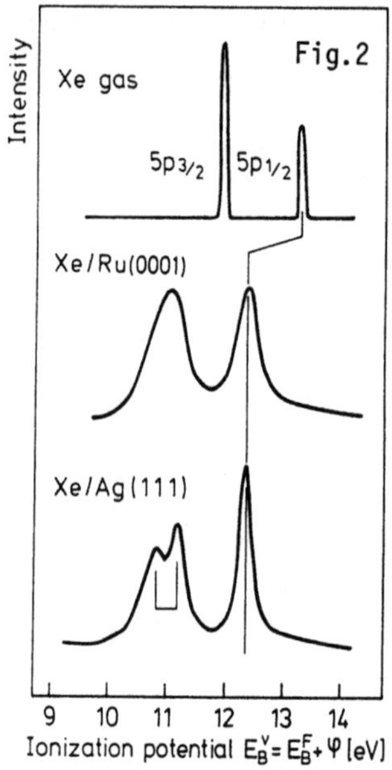

Fig. 1 Photoelectron spectrum (excited with He I-radiation of 21.2 eV) of the valence band region of a Ru(0001) surface covered with one monolayer (ML) of xenon.

Fig. 2 $5p_{3/2,1/2}$ spectra from gaseous xenon and xenon adsorbed on Ru(0001) and Ag(111). The noticeable changes in peak shape, relative intensities and ionization potentials are discussed in the text. Note the same ionization potential $E_V^B(5p_{1/2})_{ad}$ on both metals.

ly broader. Secondly, in the adsorbed phase the $5p_{3/2}$ signal is broader (on Ru(0001)) or even split (on Ag(111)) compared to the $5p_{1/2}$ peak in contrast to the gas phase spectrum. Thirdly, $E_B^V(5p_{1/2})$ of adsorbed Xe is ~1.1 eV lower than for xenon gas. Of course, the $5p_{3/2}$ signals are shifted by the same amount, but this is less easy to recognize, because their structure is more complicated. In brief, the overall line broadening caused by adsorption is due to a combination of inhomogeneous broadening /e.g. 4,5/, phonon-broadening /6/ and dispersion effects /7-9/, while the selective further broadening or splitting of the $5p_{3/2}$ line results from a lifting of the energetic degeneracy of the $5p_{3/2}$, $m_j = \pm 3/2$ and $5p_{3/2}$, $m_j = \pm 1/2$ substates into two separated lines due to interaction with the substrate /10-12/ and lateral interactions accompanied by two-dimensional band structure formation within the adsorbed xenon layer /7-9/. For more details concerning this peak broadening and splitting the interested reader is referred to the original literature. In the context of the present work we consider only the third observation, namely the fact, that

$$\Delta E^V_{exp} = E_B^V(5p_{1/2})_{gas} - E_B^V(5p_{1/2})_{ad} \approx 1.1 \text{ eV} \qquad (3)$$

This difference of ~ 1.1 eV, which according to equs. (1) and (2) is due to a combination of initial (ΔE_{init}) and final (ΔE^{1+}_{relax}) state shifts, has been shown to hold (on average) for a large number of different substrate surfaces as manifested by the compilation in Tables 1 and 3. Since these lists include so different materials such as transition metals, noble metals, alkali metals, semiconductors and oxides, the (nearly) substrate independent value of $E_B^V(5p_{1/2})_{ad}$ (being the reason for the nearly constant shift $\Delta E^V_{exp} \approx 1.1$ eV, because $E_B^V(5p_{1/2})$ is a constant, namely 13.4 eV) can not just be fortuitious, but points to a general physical principle underlying the interaction of xenon atoms with these solid surfaces. This principle becomes more obvious if one reads the entries in Table 1 from right to left, that is the difference between the measured value $E_B^F(5p_{1/2})_{ad}$ (from any of

Table 1 Compilation of work functions φ and electron binding energies E_B^F (with respect to the Fermi level) of the $5p_{1/2}$ level of xenon adsorbed on transition-, sp- and s-metals. All work function values are determined consistently from the width of the UPS (He I) spectrum of the respective surface. Note the constancy of the values $E_B^V = E_B^F + \varphi$. ΔE^{1+}_{relax} corresponds to the extra atomic relaxation shift between gas and adsorbate phase.

Sample	φ [eV]	E_B^F [eV]	$E_B^V = E_B^F + \varphi$ [eV]
Pd(110)	5.20	7.03	12.23
Pd(100)	5.65	6.75	12.40
Pd(111)	5.95	6.47	12.42
Pt(111)	6.40	5.90	12.30
Ir(100)(1×1)	6.15	6.24	12.39
Ir(100)(5×1)	6.00	6.38	12.38
Ru(0001)	5.52	6.76	12.28
W(001)	4.50	7.90	12.40
W(110)	5.10	7.15	12.25
Ni(110)	4.65	7.75	12.40
Ni(100)	5.30	6.90	12.20
Ni(111)	5.40	6.80	12.20
Cu(110)	4.48	7.80	12.28
Cu(111)	4.90	7.10	12.0
Ag(111)	4.72	7.40	12.12
Ag(100)	4.65	7.68	12.33
Au(111)	5.68	6.72	12.40
Al(111)	4.48	7.74	12.22
K(poly)	2.20	10.21	12.41
Cs(poly)	1.80	10.50	12.30
Xe gas			13.40
ΔE^{+1}_{relax}	= const	=	1.1 eV

the various surfaces) and the constant (largely substrate independent) value $E_B^V(5p_{1/2})$ yields the work function φ of the substrate (see equ. (2)). In other words the difference between the $E_B^F(5p_{1/2})$ values from two different surfaces of Table 1 corresponds to the difference in work function between these two samples:

$$\Delta E_B^F(5p_{1/2}) \simeq - \Delta \varphi \ , \qquad (4)$$

as follows also from equ. (2) and Fig. 2. This correlation may be visualized as if the potential well of an adsorbed Xe atom "floats" with the vacuum level E_V outside the respective surface.

This mere floating of the Xe potential well has two physical reasons, namely the weakness of the "physisorptive" interaction between Xe and any substrate, and the large diameter of Xe atoms.

Fig. 3 shows a potential energy diagram for a Xe atom being adsorbed on the surface of a metal 1, with $E_{V,1}$ being the vacuum potential near this surface, E_F the Fermi level and φ_1 the work function of the (bare) surface. The curve $\Phi_1(Z)$, which connects $E_{V,1}$ outside the surface with the average electrostatic potential $\Phi(-\infty)$ inside the metal corresponds to the variation of the electrostatic potential (without image effects) due to the existence of the surface dipole layer. $E_{V,1} - \Phi_1(-\infty) = \Delta\Phi_1$ is the surface contribution to the total work function φ_1. In the case of a metal the work function is given by the sum of the chemical potential $\bar{\mu}$ and the surface potential barrier $\Delta\Phi$ /13/:

$$\varphi_{metal} = \Delta\Phi - \bar{\mu} \ . \qquad (5)$$

As regards the interaction of Xe atoms with this surface it is crucial to note, that $\Phi_1(Z)$ varies very steeply reaching its saturation value $E_{V,1}$

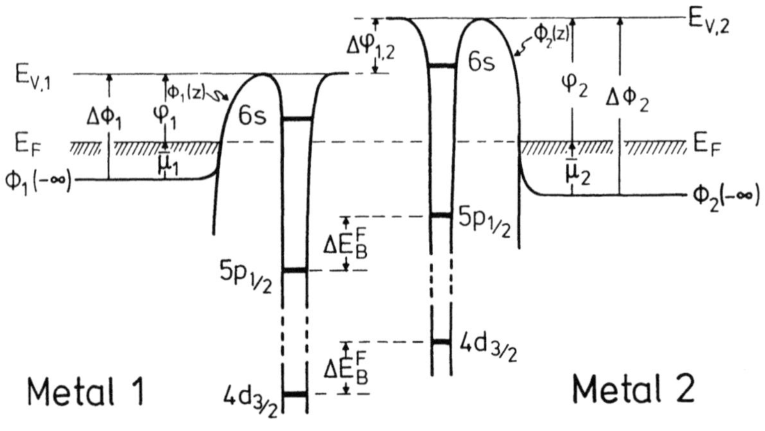

Fig. 3 Potential energy diagram for a Xe atom adsorbed on two extended homogeneous metal surfaces with different work function φ_1 and φ_2. $\bar{\mu}$ is the chemical potential and $\Delta\Phi$ the surface dipole barrier. The potential well of an adsorbed Xe atom as a whole "floats" with the vacuum potential E_V in front of the surface. (Reproduced from Ref. 15).

already ~2 Å outside the surface /13,14/. Typical Xe-metal interatomic spacings as derived from LEED (Low Energy Electron Diffraction) studies are, however, larger than ~3 Å. Consequently, the center of an adsorbed Xe atom sits outside of the surface potential barrier as depicted schematically in Fig. 3 for Xe(1). This picture has received strong theoretical support by self-consistent charge density calculations for a single Xe atom on a jellium surface /14/. The same calculations also demonstrated that the Xe-metal interaction which includes bonding and polarization of the Xe atom leads to absolute initial state shifts of the Xe(5p) valence levels of less than 0.3 eV, in accordance with Xe adsorption energies ranging only between 20 and 40 kJ/mole. Consequently, changes of this small initial state shift when going from one metal substrate to another are even much smaller, namely of the order of $\lesssim 0.1$ eV which is within the scatter of the E_B^V-values in Table 1.

All the above arguments hold likewise for a Xe atom (2) which is adsorbed on the metal 2 which has a work function φ_2 (see right half of Fig. 3). Since the Fermi levels E_F of both metals are aligned through their contact to the spectrometer, the potential well of the Xe(2) atom is shifted up together with $E_{V,2}$ by the amount $\Delta\varphi_{1,2}$. Consequently, because of the constancy of $E_B^V(5p_{1/2})_{ad}$, the electron binding energy $E_B^F(5p_{1/2})$ changes precisely by the same amount with respect to the Fermi level.

This simple "floating model" has received further support by several quite different independent experiments.

First of all, deeper lying core levels of adsorbed xenon should not only behave in the same way as the outer 5p levels, but they should comply even better with equ. (2), because they are more contracted than the 5p orbitals, and, hence, are located even further outside of the surface dipole barrier. This, in fact, proved true for the Xe(4d) states /15/ as well as for the Xe(3d) states /e.g. 5/. Some results for the Xe(4d) states, measured with synchrotron radiation and similar resolution as the 5p data from Table 1, are compiled in Table 2. The $E_B^V(4d_{5/2})$ values are virtually identical for Xe

Table 2 Xe(4d) core level binding energies E_B^F and $E_B^V = E_B^F + \varphi$ for Xe adsorbed on four different metal surfaces. Note the almost identical values of E_B^F, although the work functions deviate by up to 2.6 eV. ΔE_{relax}^{1+} is the extra atomic relaxation shift of the 4d states between gas and adsorbate phase.

Sample	φ [eV]	$4d_{5/2}^F$ [eV]	$4d_{3/2}^F$ [eV]	$4d_{5/2}^V$ [eV]	$4d_{3/2}^V$ [eV]
Pd(111)	5.95	60.22		66.17	
Pd(100)	5.65	60.5	-	66.15	-
Gd(0001)	3.3	62.8	-	66.10	-
Al(111)	4.48			66.00	68.00
Xe gas				67.55	69.5
ΔE_{relax}^{+1}				1.4eV	1.5eV

adsorbed on the four given surfaces although the respective $E_B^F(4d_{5/2})$ and φ values deviate by as much as 2.6 eV /15/.

So far all the results reported were obtained with metal substrates. By contrast, semiconductor surfaces are usually expected to exhibit quite different bonding, polarization and screening properties. Therefore experiments with semiconductor surfaces appear to be a critical test for the simple "floating model" illustrated in Fig. 3. Table 3 includes some $Xe(5p_{1/2})$ results obtained from different semiconductor surfaces, and interesting enough the $E_B^V(5p_{1/2})$ values, again obtained as the sum of $E_B^F(5p_{1/2})$ and φ, are nearly constant for the surfaces studied so far, and are, moreover, the same as determined for metals (see Table 1). Particularly convincing are the results obtained for n- and p-doped Si(111) surfaces. In these cases the Fermi level is shifted within the band gap of the semiconductor only as a function of the dopant, while the properties of the very surface, e.g. the 7x7-reconstruction of the Si(111) surface are unaffected. The shift of the Fermi level through the band gap results in a corresponding change of the work function of the respective surface which again manifests itself in a corresponding shift of the Xe(5p) levels (with respect to the Fermi level determined by the spectrometer Fermi level) /16/. Fig. 4 shows valence band spectra from an n- and a p-doped Si(111) 7x7 surface covered with a monolayer of xenon. The Xe(5p) spectra from both surfaces are again shifted with respect to each other by the difference in work function between these two particular surfaces, namely by 0.9 eV /17,18/.

Table 3 Compilation of work functions φ and electron binding energies (with respect to the Fermi level) of the $5p_{1/2}$ level of xenon adsorbed on various semiconductor surfaces. The work function values are measured photoelectrically (from the width of the corresponding UPS (He I) spectrum) at low temperatures (= Xe adsorption temperature). Under these conditions the surfaces exhibit no band bending, but flat band conditions due to the surface voltaic effect (see text and Ref. 16).

ZnO (000$\bar{1}$)O	4.35	8.00	12.35
ZnO(0001)Zn	3.75	8.70	12.45
ZnO (10$\bar{1}$0)	4.05	8.25	12.30
n-Si(111)	4.50	7.70	12.20
p-Si(111)	5.40	6.80	12.20
n-Si(100)	4.60	7.70	12.30
Xe gas			13.40
ΔE_{relax}^{+1}	= const	=	1.1 eV

Although the experimental evidence presented in Tables 1-3 is already convincing support for the "floating model" suggested in Fig. 3, it still leaves the possibility open, that the constancy of $\Delta E_{exp}^V = E_B^V(gas) - E_B^V(ad) \approx 1.1$ eV is only due to a (fortuitous or systematic) cancellation of substrate dependent differences in initial state shifts ΔE_{init} and final state shifts ΔE_{relax}^{1+}. For instance, one could speculate that

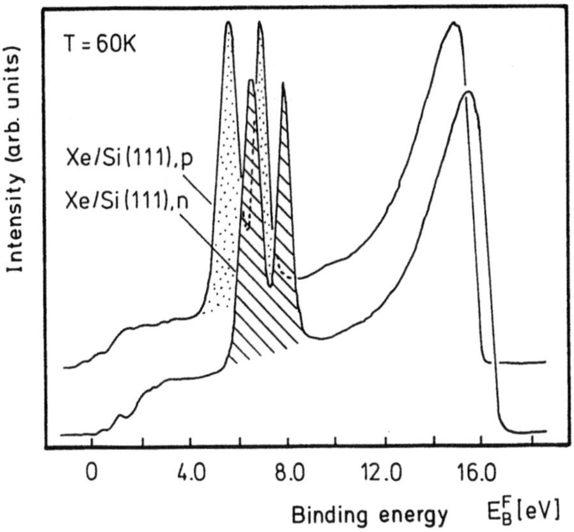

Fig. 4 UV-photoelectron spectra from Xe-covered n- and p-doped Si(111)7x7 surfaces, detected at 60 K. Both spectra including the Xe(5p) signals are shifted with respect to each other by the difference in work function of both surfaces at this low temperature (see Ref. 16).

$\Delta E_{pot} + \Delta E_{chem}$ on one hand and ΔE^{1+}_{relax} on the other hand are different from substrate to substrate, but that $\Delta E_{pot} + \Delta E_{chem} \approx - \Delta E^{1+}_{relax}$. The theoretical calculations for Xe on jellium by Lang and Williams /14/, however, suggested that $\Delta E_{pot} + \Delta E_{chem} \approx 0$ (see above), which would result in ΔE^{1+}_{relax} also being largely substrate independent. In fact, ΔE^{1+}_{relax} would be the only remaining physical origin of $\Delta E^{V}_{exp} \approx 1.1$ eV.

There is an independent experimental way to prove that ΔE^{1+}_{relax} is, indeed, substrate independent, namely by means of Auger electron spectroscopy of Xe adsorbed on substrate surfaces of very different physical nature, as described in the following section.

3. Auger electron emission from Xe adsorbed on Ru(0001) and potassium

The idea behind comparing Xe(5p) photoemission and (NOO) Auger electron spectra for Xe adsorbed on two very different metals is the following. As mentioned in the introduction final state relaxation of photoionized adsorbed rare gases is dominantly ascribed to image polarization effects rather than to charge transfer processes. The relaxation energy due to image polarization is given by the Coulomb potential between the ionized hole state and its induced image charge in the substrate surface /3/:

$$\Delta E_{relax} = - \frac{(z \cdot e)^2}{4\, r_o} \qquad (6)$$

with e = electronic charge and r_o = distance from the image plane. The photoemission final state is a singly ionized (z=1) Xe^{1+} ion (one 5p hole), while the Auger final state corresponds to a doubly ionized (z=2) Xe^{2+} ion (two 5p holes). Taking, furthermore, relaxation of the initial ion state of the (NOO) Auger transition (one hole in the N-level) into account theory shows /19/, that to a first approximation ΔE^{2+}_{relax} (Auger) should be three times larger than ΔE^{1+}_{relax} (photoemission). Hence, also any difference in the Auger (two hole) final state relaxation for Xe adsorbed on two different metals should be enhanced by this factor of three compared to photoemission (one hole). On the other hand, if ΔE^{1+}_{relax} is substrate independent as suggested in the previous section, then also ΔE^{2+}_{relax} may be expected to be very similar for various metals (however, three times larger than the corresponding ΔE^{1+}_{relax} values). In this case the kinetic energy of Auger electrons arising from an (NOO) Auger transition of Xe adsorbed on two different metals should again only be shifted by the difference in work function between the two metals because ΔE_{pot} and ΔE_{chem} (see equ. (1)) in the (neutral) initial state are, of course, the same in the Auger and in the photoemission experiment. In summary, if ΔE^{1+}_{relax} and ΔE^{2+}_{relax} are substrate independent we may expect (based on equs. (1) and (4)):

$$\Delta_{1,2}\left[E_{kin}(\text{Auger})\right] \approx \Delta_{1,2}\left[E^F_B(\text{Photo})\right] \approx \Delta\varphi_{1,2} \qquad (7)$$

were 1 and 2 denote two different metals as in Fig. 3. In other words, we may expect that both the difference in (photoemission) electron binding energy and the difference in (NOO) Auger kinetic energy measured for Xenon adsorbed on two substrates should be equal to the difference in work function of the two substrates. The previous section suggested this relationship to hold for the photoemission case. The actual experiment to prove this relationship also for the Auger case was carried out as follows /20,21/.

The clean Ru(0001) surface was positioned in front of a CMA Auger spectrometer and electron induced low energy Auger spectra were recorded from the bare as well as the Xe covered Ru(0001) surface. Unfortunately, as can be seen in Fig. 5 (A,1) and 5 (A,2) most of the low energy Auger transitions of Xenon overlap with the $Ru(N_{23}N_{45}N_{45})$ transition at 37 eV. Still, the $Xe(N_{45}O_{23}O_{23})$ peak is clearly detected at ~42 eV. Within the accuracy of the measurement (± 0.2 eV) its position (peak minimum) with respect to that of the Ru peak was constant for coverages $\leq$ 1 monolayer of Xe. The bare Ru surface was then covered with a thick and clean potassium film. The low energy Auger spectrum from Fig. 5 (B,1) shows no more Ru contribution. Xe on this film produced the low energy Auger spectra shown in Fig. 5 (B,2) and 5 (B,3) for approximately a half and a complete Xe monolayer. Desorption of the xenon restored the clean potassium Auger spectrum (Fig. 5 (B,1)). Finally, the Ru sample was heated to 1500 K in order to desorb the potassium without changing the position of the sample in front of the CMA /22/. This restored the clean Ru Auger spectrum as shown in Figs. 5 (A,1) and 5 (B,4). Using this $Ru(N_{23}N_{45}N_{45})$ Auger peak as the common reference in both Fig. 5 A and 5 B one determines a shift of 3.8 ± 0.2 eV, which, indeed, is rather similar to the difference in work function between the two surfaces, $\Delta\varphi$= 3,32 ± 0.1 eV, and to the difference $\Delta E^F_B(5p_{1/2})$ = 3,45 ± 0.1 eV measured for Xe adsorption on the same two surfaces (see Table 1).

4. Conclusions

In agreement with the expectation expressed by equ. (7) the difference in kinetic energy measured for the $Xe(N_{45}O_{23}O_{23})$ Auger transition on Ru(0001) and on potassium is very similar to the difference of the UPS electron binding energies of the $Xe(5p_{1/2})$ level measured on the same two surfaces. Furthermore, both differences agree closely (within the accuracy of the mea-

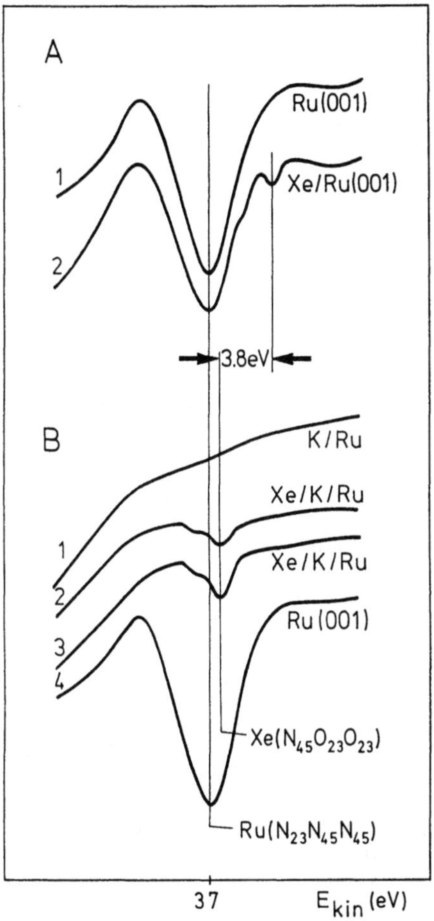

Fig. 5 Electron induced low energy Auger spectra of differently covered Ru(0001) surfaces. Spectra A1 and B4 correspond to the Ru($N_{23}N_{45}N_{45}$) transition of the clean Ru(0001) surface; A2 corresponds to one monolayer of Xe on the Ru(0001) surface; B1 was taken from the Ru surface covered with a thick film of potassium, and B2 and B3 correspond to ~0.5 and ~1.0 monolayer of Xe on the thick potassium film. Note that the Xe($N_{45}O_{23}O_{23}$) transition is shifted by 3.8 ± 0.2 eV between the clean Ru(0001) surface and the potassium surface (see text and Ref. 22).

surement) with the difference of the work functions of both (bare) surfaces, namely $\Delta_{1,2}[E_{kin}(\text{Auger})] = 3.8 \pm 0.2$ eV, $\Delta_{1,2}\, E_B^F(5p_{1/2}) = 3.45 \pm 0.1$ eV and $\Delta\varphi_{1,2} = 3.32 \pm 0.1$ eV. This agreement strongly suggests, that, in fact, both $\Delta_{1,2}[E_{kin}(\text{Auger})]$ and $\Delta_{1,2}\, E_B^F(5p_{1/2})$ are mainly due to the work function difference $\Delta\varphi$, and that, in turn, the difference in final state relaxation on the two surfaces is very small (despite their very different physical properties). In fact, we can roughly estimate that the difference in photoemission final state relaxation between the two surfaces is $\Delta_{1,2}[\Delta E^{1+}_{relax}] = 1/3\ \Delta_{1,2}[E_{kin}(\text{Auger})] \approx 0.1$ eV. This value is very small, even though both surfaces are of quite different nature, and it is well within the scatter of the $E_B^V(5p_{1/2}) = E_B^F(5p_{1/2}) + \varphi$ values of Tables 1 and 3.

Therefore we are led to conclude, that the same "floating model" according to which the potential well of an adsorbed xenon atom floats up and down with the vacuum potential in front of a surface holds not only for the photoemission (one hole) final state (see Fig. 3) but also for the Auger (two hole) final state situation. Of course, to a first approximation the two hole relaxation energy is three times larger than the one hole relaxation energy. This effect, however, cancels when taking the respective Auger and photoemission energy differences detected for xenon on two metals. Any remaining difference between the Auger or photoemission energies measured for Xe on two different metals is due to the difference in work function between the two metal (or semiconductor, see Table 3) surfaces. According to equ. (1) this is an initial state shift (of the neutral Xe atom prior to photoionization). Changes in final state relaxation when adsorbing xenon on one or another metal, namely Ru and potassium, do not play an important role.

This sensitivity of $E_B^F(5p_{1/2}$ of adsorbed Xe atoms towards the substrate work function (see equs. (2) and (4) and Fig. 3) has been shown to hold also for Xe adsorption on heterogeneous surfaces, that is on surfaces which consist of patches of different local surface potential ("local work function"), and has made Photoemission of Adsorbed Xenon (PAX) a powerful technique for the characterization of heterogeneous (e.g. bimetallic) surfaces on an almost atomic scale (Fig. 6). As an example Fig. 7a reproduces $Xe(5p_{3/2,1/2})$ spectra for Xe adsorbed on a Ru(0001) surface, which prior to adsorption of increasing amounts of Xe, was covered with 0.1 monolayer of potassium. It is clearly seen that the multipeaked spectra (in comparison to Figs. 1 and 2) are a superposition of Xe(5p) spectra arising from Xe atoms sitting at inequivalent surface sites. A quantitative analysis of the highest spectrum in Fig. 7a, which corresponds to a full Xe monolayer over the whole 0.1 ML K/Ru(0001) surface, is shown in Fig. 7b. The information retrieved from this Fig. 7b is threefold. 1) The assignment of the three superimposed $5p_{1/2}$ peaks refers to Xe atoms which are adsorbed on free Ru-patches [Xe(Ru)], next to individual potassium atoms [Xe(K/Ru)] and ontop of individual K-atoms [Xe(K)] . 2) The percentages determined by the Lorentzian fits are a quantitative measure of the relative abundance of the respective Xe adsorption site, and 3) the φ_{loc}-values derived from the respective $5p_{1/2}$ binding energy using equ. (2) are the local surface potentials at the respective surface sites. For more details the interested reader is referred to Refs. 4,15,23,24 and references therein.

$$E_B^F = \text{const.} - \varphi_{loc}$$

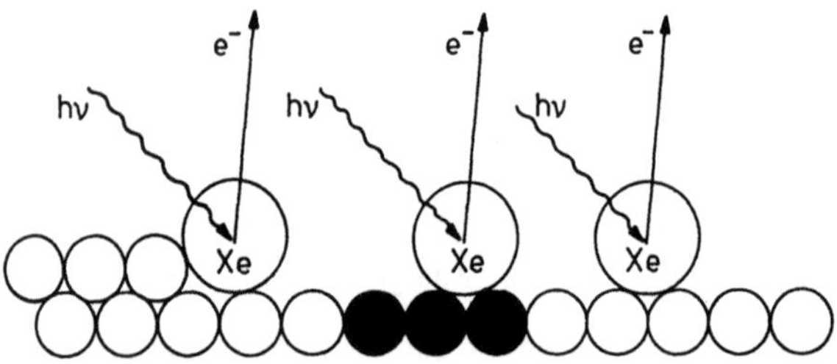

Fig. 6 Schematic representation of the fact that the photoemission from adsorbed xenon (PAX) is sensitive to atomic scale surface heterogeneities such as steps, hetero-atoms, etc.

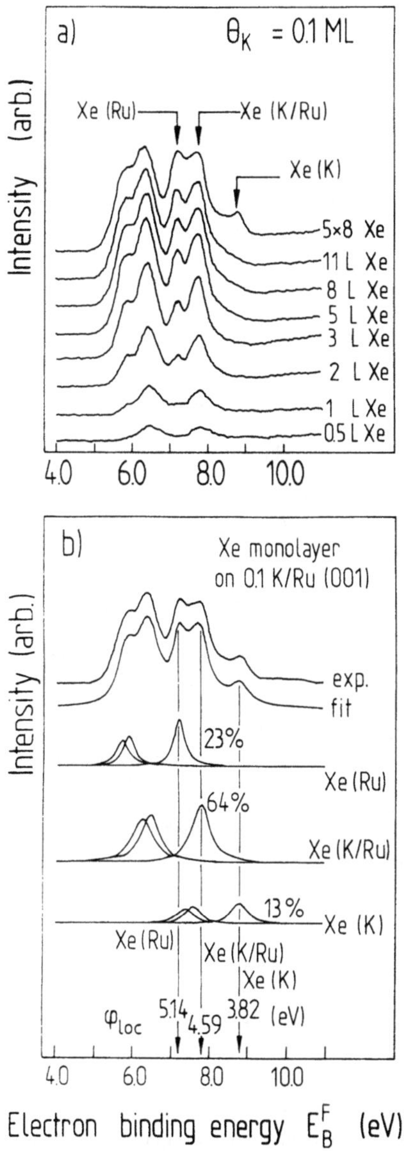

Fig. 7 a) $5p_{3/2,1/2}$ spectra of xenon adsorbed on a Ru(0001) surface precovered with 0.1 ML potassium. Note the increasing number of peaks as the Xe coverage θ increase. The highest spectrum corresponds to a complete Xe monolayer, and the three $5p_{1/2}$-peaks are assigned to Xe atoms at free Ru-sites [Xe(Ru)] , next to a K-atom [Xe(K/Ru)] and ontop of a K-atom [Xe(K)] . This spectrum is decomposed (see Ref. 15) into the three Xe states in panel b), the $5p_{1/2}$ signals of which provide a measure of the relative abundance of the respective kind of surface site (percentages) and its local work function φ_{loc} (see text).

References

1. B.J. Waclawski, J.W. Gadzuk and J.F. Herbst, Phys. Rev. Lett. 41 (1978) 583
2. H.P. Citrin and D.R. Hamann, Phys. Rev. B10 (1974) 4948
3. J.W. Gadzuk, Phys. Rev. B14 (1976) 2267
4. K. Wandelt and J. Hulse, J. Chem. Phys. 89 (1984) 1340
5. R.J. Behm, C.R. Brundle and K. Wandelt, J. Chem. Phys. 85 (1986) 1061
6. J.W. Gadzuk, S. Holloway, C. Mariani and K. Horn, Phys. Rev. Lett. 48 (1982) 1288
7. K. Horn, M. Scheffler and A.M. Bradshaw, Phys. Rev. Lett. 44 (1978) 822
8. M. Scheffler, K. Horn, A.M. Bradshaw and K. Kambe, Surf. Sci. 80 (1979) 69
9. K. Kambe, Surf. Sci. 105 (1981) 95
10. B.J. Waclawski and J.F. Herbst, Phys. Rev. Lett. 35 (1975) 1594
11. R.P. Antoniewicz, Phys. Rev. Lett. 38 (1977) 374
12. S. I. Ishi and Y. Ohno, J. Electron Spectrosc. 33 (1984) 85
13. N.D. Lang and W. Kohn, Phys. Rev. B3 (1971) 1215
14. N.D. Lang and A.R. Williams, Phys. Rev. B25 (1982) 2940
15. K. Wandelt, in: Studies in Surface Science and Catalysis, Vol. 32: Thin Metal Films and Gas Chemisorption, Ed. P. Wissmann (Elsevier, Amsterdam, 1987) p. 280
16. Due to photon irrediation from a UV-lamp at low temperatures ($\lesssim$ 50 K) band bending at semiconductor surfaces may be compensated (J.E. Demuth, W.J. Thompson, H.J. DiNardo and R. Imbihl, Phys. Rev. Lett. 56 (1986) 1408) resulting in flat band conditions and a corresponding temperature/photon induced work function change (surface voltaic effect). Since Xe adsorbs only at low temperatures ($\lesssim$ 50 K) on these surfaces Table 3 includes only the low temperature φ -values. At room-temperature the work function of both the n- and the p-doped Si(111)7x7 surface is φ = 4.9 eV. Furthermore both the Si(111) and the Si(100) surface are topographically heterogeneous due to their surface reconstruction. Therefore the $E_B^F(5p_{1/2})$ values in Table 3 correspond to the center of gravity of the total $5p_{1/2}$ intensity at Xe monolayer saturation (for more details see e.g. **17**.
17. K.Markert, PhD-Thesis, Berlin, 1988
18. K.Markert, P.Pervan, W.Heichler and K.Wandelt, J.Vac.Sci.Technol. submitted
19. G.Kaindl, T.-C.Chiang, D.E.Eastman and F.J.Himpsel, Phys.Rev.Lett. 45 (1980) 1808
20. K.Markert, Diplom-Thesis, München, 1985
21. K.Markert and K.Wandelt, to be published
22. This care was exercised because changes of the position of the sample in front of a CMA may result in energy shifts due to different entrance angles and transmission trajectories of the detected electrons.
23. K.Wandelt, in: Alkali Adsorption on Metals and Semiconductors, H.P. Bonzel, A.M. Bradshaw and G. Ertl (Eds.) Elsevier, Amsterdam, 1989
24. K.Wandelt, J.W.Niemantsverdriet, P.Dolle and K.Markert, Surf.Sci. 213 (1989) 612

An Auger Electron Spectroscopy Study of the Influence of Sputter Ambient on the Electrical Characteristics of PtSi/Si Contacts

P. Lanza, O. Viscuso, F. La Via, C. Magro, and G. Ferla

ST SGS-Thomson Transistors and Standard Circuit Division, R & D Department, 50 Stradale Primosole Catania, Italy

1. Introduction

Metal-semiconductor contacts have been investigated /1-6/ for their importance in integrated circuits as Schottky and ohmic contacts. Over the last few years these studies are increased because of the importance of low contact resistance for circuit performances.

A great improvement in metal-semiconductor contacts has been made by the use of silicides, which reduce the influence of surface contamination and surface damage on the electric characteristics. But the use of these metal-semiconductor contacts has created new problems. Previous works /7/ have shown a great influence of the oxygen partial pressure on platinum diffusivity. It has been seen that an increasing of the O_2 partial pressure during the evaporation of platinum decreases the Pt diffusivities in both Pt_2Si and PtSi. Besides when the oxygen partial pressure is greater than 10^{-7} torr a rough PtSi-Si interface has been seen. On the contrary this interface is extremely flat when the oxygen partial pressure is lower.

In this work we have investigated the influence of the oxygen partial pressure on the electric characteristics of platinum silicide Schottky diodes by using Auger Electron Spectroscopy. We have measured the indepth distribution of the oxygen after the annealing process to obtain PtSi and found that, a high oxygen concentration stops the reaction between Pt and Si.

2. Experimental Details

The diodes were prepared on a low resistivity (5-20 10^{-3} Ω cm) Si(111) substrat, 520 μm thick, with an epitaxial layer 4 μm thick and of 0.5 Ω cm resistivity. A thermal silicon oxide 1.2 μm thick, was grown on the surface and the contact areas were opened by a photolithographic process. Before platinum deposition the surface was cleaned by an argon sputter etch in situ.

The oxygen partial pressure introduced in the sputter chamber has been controlled by a Residual Gas Analyzer. A typical R.G.A. spectrum of the ambient is shown in Fig. 1. From this analysis it is possible to observe that the residual gases present in the vacuum chamber, are essentially: hydrogen, water, nitrogen and oxygen.

The aim of this work is to study the influence of these gases on the platinum silicide formation and on the electrical characteristics of the Schottky diode PtSi/Si. Then we have prepared four groups of samples under different vacuum conditions. In Fig. 2 these partial pressures of nitrogen, H_2O and oxygen have been plotted for the different groups of samples. All the samples were subsequently annealed at 520°C for 30 min; under a nitrogen (95 %) and oxygen (5 %) flow. Part of the platinum reacted with silicon to form PtSi; the unreacted platinum was etched by acqua regia solution at 85°C. The front metallization was completed by a subsequent deposition of Ti-W and aluminium films.

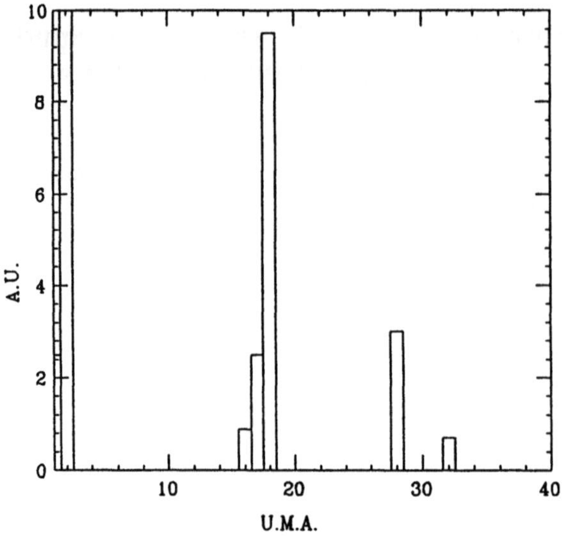

Fig. 1 A typical residual gas spectrum of the sputter chamber

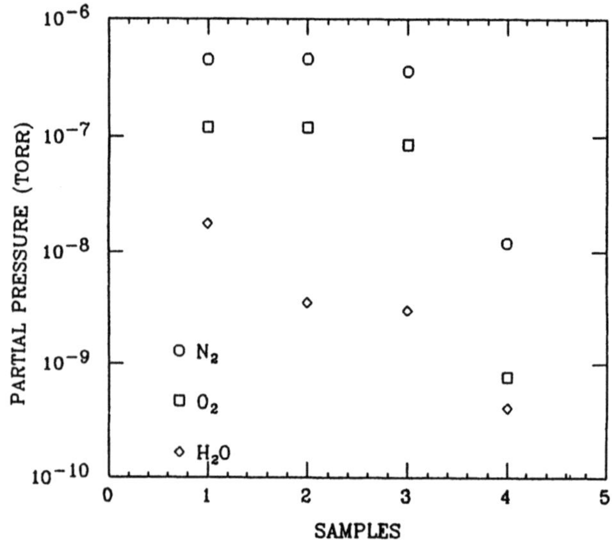

Fig. 2 Partial pressures of N_2, H_2O and O_2 for the different sample preparations

3. Experimental Results

The samples were first investigated by Rutherford Back Scattering to see the different phases that are present after the annealing. From the literature it is well known that the first phase formed at the Pt-Si interface is Pt_2Si. Only when all the Pt is reacted to form Pt_2Si does PtSi begin to grow from the Pt_2Si-Si interface. This growth continues until all Pt_2Si is consumed. The R.B.S. spectra are shown in Fig. 3. From these analyses we can observe that the reaction has not started in the first sample, in fact the platinum peak has the same height as the unreacted sample. In the second sample the reaction is started but after the annealing it has been obtained the phase

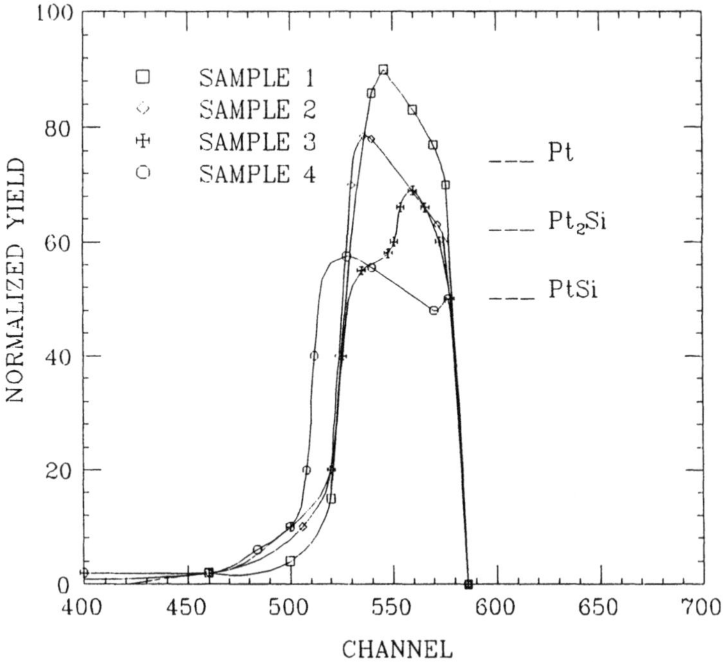

Fig. 3 R.B.S. spectra of the platinum silicide obtained after an annealing for 30' at 520°C

Pt_2Si. In the third sample the phase Pt_2Si is present near the surface while PtSi is beginning to start near the silicon interface. For the last sample the R.B.S. spectrum shows that only the PtSi phase is present.

To explain these different behaviours some Auger spectra have been measured. In fact, also from the literature it is unclear which mechanism slowes or completely stops the reaction between Pt and Si due to contamination of the sputtered platinum film as seen with the first sample.

Fig. 4 shows the indepth profile of the first sample. As already seen from the R.B.S. spectra, the sample has reacted only a little at the interface with silicon. The oxygen contamination is very high across the whole film and a little peak at 76 eV, characteristic of silicon dioxide is observed at the interface. This thin oxide layer is a diffusion barrier for the further platinum diffusion and causes the reaction to stop immediately.

In the other samples (Fig. 5, Fig. 6) the oxygen contamination is lower and is not enough to stop the reaction. The silicidation is only slowed proportionally to the impurity content. In Fig. 6 it is possible to observe also a different level of oxygen contamination between the two phases Pt_2Si and PtSi. This effect has been observed before and it has been ascribed to the swept out of the oxygen during the reaction. But not all the oxygen segregates at the surface, a little amount remains in the PtSi and influences also the PtSi reaction.

The last sample (Fig. 7) shows a very low oxygen contamination and a complete reaction between Pt and Si.

The oxygen contamination of the Pt film produces not only a variation of the silicide formation kinetics but, at the same time, it has also a great influence on the electrical characteristics of the Schottky diode. In Fig.

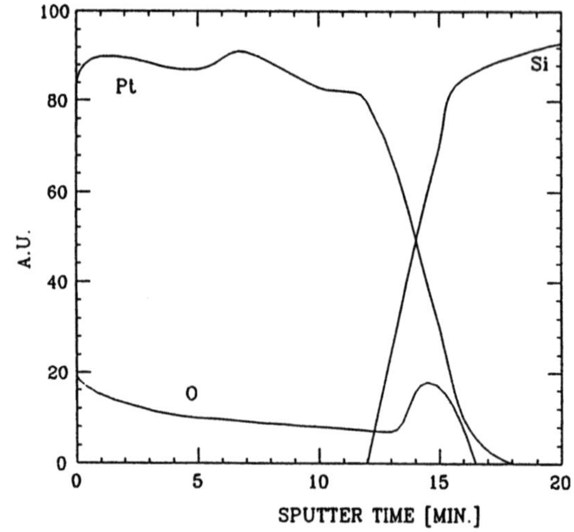

Fig. 4 Auger depth profile of sample 1

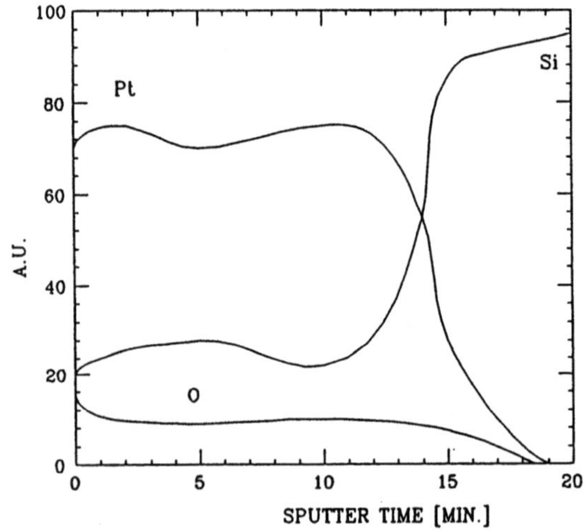

Fig. 5 Auger depth profile of sample 2

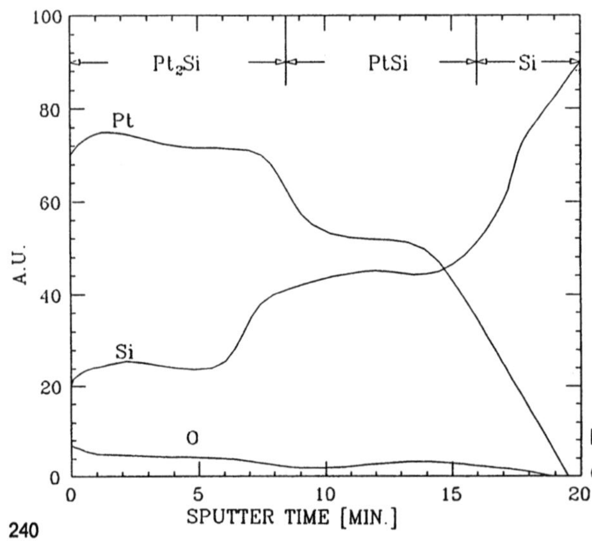

Fig. 6 Auger depth profile of sample 3

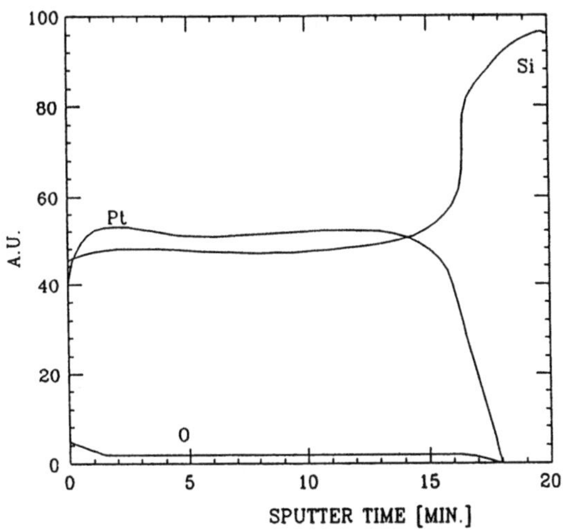

Fig. 7 Auger depth profile of sample 4

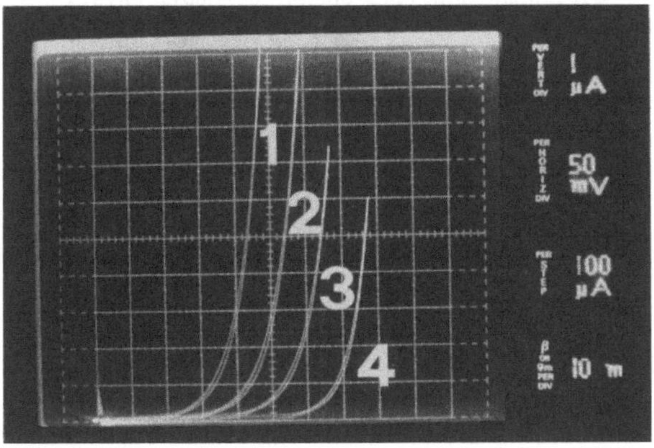

Fig. 8 I-V direct characteristics of the Schottky diodes

8 these I-V characteristics are shown. From this photo it is possible to observe that the sample 4 has the highest forward bias Vf measured at a current of 1 µA (Vf = 345 mV). This value corresponds with the ideal Schottky barrier height (Φ_B= 0.87 eV) of the PtSi/Si contact. The other samples have a lower Schottky barrier height and a greater dispersion of the characteristics.

4. Conclusion

In this work we have studied the influence of the vacuum conditions on the platinum silicide formation kinetics and on the I-V characteristics of the Schottky diode.

The Auger analyses show that only the oxygen is present in the reacted film and that this contamination increases with the deterioration of the vacuum conditions. This oxygen contamination in Pt decreases the speed of the reaction and, for a partial pressure of $1 \cdot 10^{-7}$ torr, the silicidation stops immediatly because of the formation of a thin SiO_2 layer at the interface. Furthermore the Auger profiles have shown that not all the oxygen is swept out during the formation of Pt_2Si, but that a little amount remains in the PtSi and so slows down also this pase formation.

The I-V characteristics of the Schottky diode is strongly influenced by the vacuum conditions during Pt deposition. In fact, while the sample, that has been sputtered in good vacuum conditions, has an ideal characteristic, the other ones show a lower Schottky barrier height and a greater dispersion around the mean value.

This behaviour is not easy to explain with the generally used theories of the Schottky diodes. Further work can be interesting and can give a great improvement in our understanding of the metal-semiconductor contacts.

5. References

1. H.K. Henish: In Rectifying Semiconductor Contact (Oxford 1957)
2. C.R. Crowell and S.M. Sze: Solid State Electron., 9, 1035 (1966)
3. F.A. Padovani and R. Stratton: Solid State Electron., 9, 695 (1966)
4. C.R. Crowell and S.M. Sze: Solid State Electron., 8, 979 (1965)
5. C.R. Crowell and S.M. Sze: J. Appl. Phys., 37, 2683 (1966)
6. R.M. Anderson and T.M. Reith: J. Electrochem. Soc., 122, 1337 (1975)
7. C.A. Crider, J.M. Poate, J.E. Rowe and T.T. Sheng: J. Appl. Phys., 52, 2860 (1981)

Temperature and Core Hole Dependence of EXFAS (Extended Fine Auger Structure) of Ag(111) and Cu

M. De Crescenzi[1], *T. Tylsizczak*[2], *and A.P. Hitchcock*[2]

[1]Dipartimento di Fisica, Università di Roma II, Tor Vergata, Rome, Italy
[2]Institute for Materials Research, McMaster University, Hamilton, Canada

This work reports a comparative study of the temperature dependence of the Ag $N_{23}\varepsilon V$ and Ag $M_{45}\varepsilon V$ EXFAS (Extended Fine Auger Structure) above the $N_{23}VV$ and the $M_{45}VV$ Auger transitions,respectively. In addition the temperature dependence of the $M_{23}\varepsilon V$ EXFAS of polycristalline Cu is studied in detail and compared to the corresponding Cu M_{23} EELFS (Extended Energy Loss Fine Structure) and to band calculation of p->εd extending up to 200 eV above the Fermi level. These results provide strong support for an EXAFS-like origin of the EXFAS signal extending in this way the capability of the Auger technique to give also surface structural results.

1.Introduction

During the last few years a number of experimental techniques have been developed with the purpose to obtain EXAFS-like structural information without resorting to the synchrotron radiation /1/. Aim of the present paper is to show some recent experimental results concerning one of the most recent of these techniques which basically uses electrons instead of X-rays to excite the investigated system. This technique has been called Extended Fine Auger Structure,with the achronim EXFAS /2/,and it refers to the study of the features which appear in the secondary electron distribution above the discrete the CVV Auger line of metals.The proposed mechanism /2-4/ to explain the presence of well defined structures in the background of the Auger spectrum has been schematized in Fig.1 together with the EELFS and Auger processes.

The EXFAS mechanism invokes a one step excitation-decay process in which the core-ionized electron (which has been excited to a final state lying Δ eV above E_F) is involved in the decay of the core hole releasing its excess energy to eject a valence band electron.We emphasize that the core-ionized electron fill in a continuous way the final states Δ above E_F due to the energy loss mechanism with the primary excitation beam E_P.

In this way a valence band electron can be emitted from the sample with a kinetic energy $E_{k'}$ (greater than the kinetic energy E_k of the corresponding CVV Auger transition) gaining an energy Δ according to the following equation:

$$E_{k'} = E_k + \Delta + E\,(V_1) \qquad (1)$$

As the excited Δ states of the core electron are EXAFS-like modulated in energy,then the

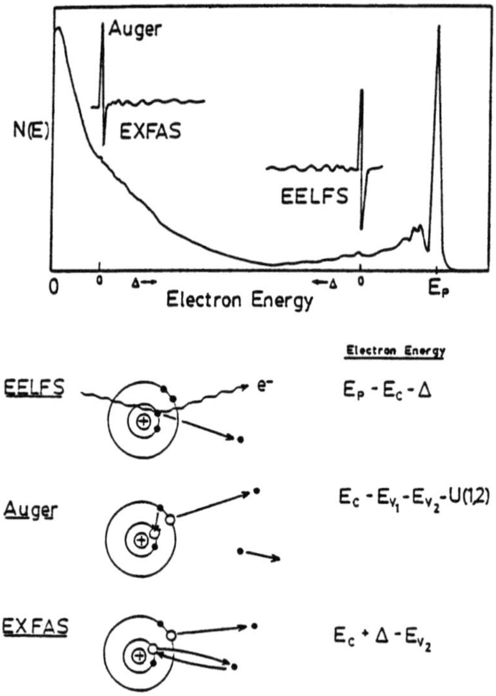

Fig.1 - *Schematic of the energy distribution N(E) of the electrons with a primary beam energy E_p reflected from a surface.The initial creation of the core hole is detected as the extended energy loss fine structure (EELFS) /3/. In the conventional Auger process the core hole is filled by an electron from an occupied level of the core-excited atom and another valence electron is ejected. In the proposed extended fine Auger structure (EXFAS) mechanism,the core-ionized electron takes part in the core hole decay as the electron that fills the core hole.The equations give the relationship between the energy of the detected electron and the energy levels of the target for each spectroscopy. U(1,2) is the Coulomb interaction of the two holes in the Auger final state while Δ is the amount by which the the energy of the ionized electron exceeds the ionization threshold.In the case of EXFAS the Coulomb interaction U(1,2) should be negligeable because the two holes are not located on the same valence band.*

measured intensity of the ejected valence band electron will contain entirely the EXAFS structures.

In essence the proposed mechanism is an autoionization process and thus the EXFAS should be,in principle,a mirror image of the extended energy loss fine structure (EELFS) (see Fig.1),reflecting the same pattern arising from the interference between the outgoing and the backscattered components in the final state of the ionized core electron.

The interference of scattered wavelets returning to the central atom backscattering from atoms in each shell with the outgoing wavelet from the central atom gives rise to band structure-like effects as widely demonstrated by Müller and coworkers /5,6/. Thus any electronic technique which probe the unoccupied density of states can be used as a structural tool because it is strictly related to the local geometry (lattice parameter,coordination number,thermal and disorder effects...) as in the EXAFS spectroscopy.

Early interpretation /7,8/ of the EXFAS ,as well as a recent review /9/ have proposed,instead, diffraction effects experienced by the secondary electrons to explain the presence of features at kinetic energy higher that an Auger line.In order to clarify better the origin of the signal and to distinguish between a secondary-electron diffraction and an EXAFS-like process we report a complete investigation of the temperature dependence of the EXFAS features on a Cu polycrystalline sample and a core hole dependence on a single crystal Ag(111) /10,11/.

2.Experimental

The spectra were acquired with a retarding grid LEED/Auger analyzer with normal incidence excitation which integrates the ejected electrons over >150° solid angle /10/.The sample temperature was controlled by a combination of liquid nitrogen cooling and electron beam from the back of the sample.The spectra have been collected as function of the first derivative of the N(E) yield to enhance the signal to noise ratio and to eliminate the background of the inelastic electrons.The EXAFS oscillating signals $\chi'(k)$ are obtained from the original N'(E) spectra after subtraction of a smooth background. The samples were cleaned by Argon ion bombardment and annealed prior to EXFAS measurements.The Ag(111) sample showed a good LEED pattern.

3.Results and discussion

The EXFAS spectra above the $M_{23}VV$ Auger (60 eV) of polycrystalline Cu are presented in Fig.2 (a) at a number of temperatures between 150 K and 560 K. The spectral intensity has been normalized to the Cu M_1VV Auger intensity. The origin of the k wavevector scale was taken at the minimum of the main Auger N'(E) line,according to Eq.1 for $\Delta=0$.

The Fourier transforms of the EXFAS are presented in Fig.2 (b).It is clear that the signal decreases at all R values but that from more distant atoms is more strongly suppressed relative to the nearest neighbour signal at elevated temperature. The selective damping of higher R signal is a behaviour expected for an EXAFS-like structural origin of the EXFAS since the increased thermal motion has a larger damping effect on larger distances involving more weakly correlated motion,as it has been shown in EXAFS studies of many systems,including the K-shell of Cu /12,13/. The Fourier filtered first shell signals,along with the amplitude functions are reported in Fig.3 in order to demonstrate the similarity of these signal at various temperature.We observed a systematic decrease with increasing temperature of the overall amplitude and a sizeable shift towards low k-values of the maximum amplitude,as indicated by arrows in Fig.3. All these effects are expected for an EXAFS-like Debye-Waller thermal

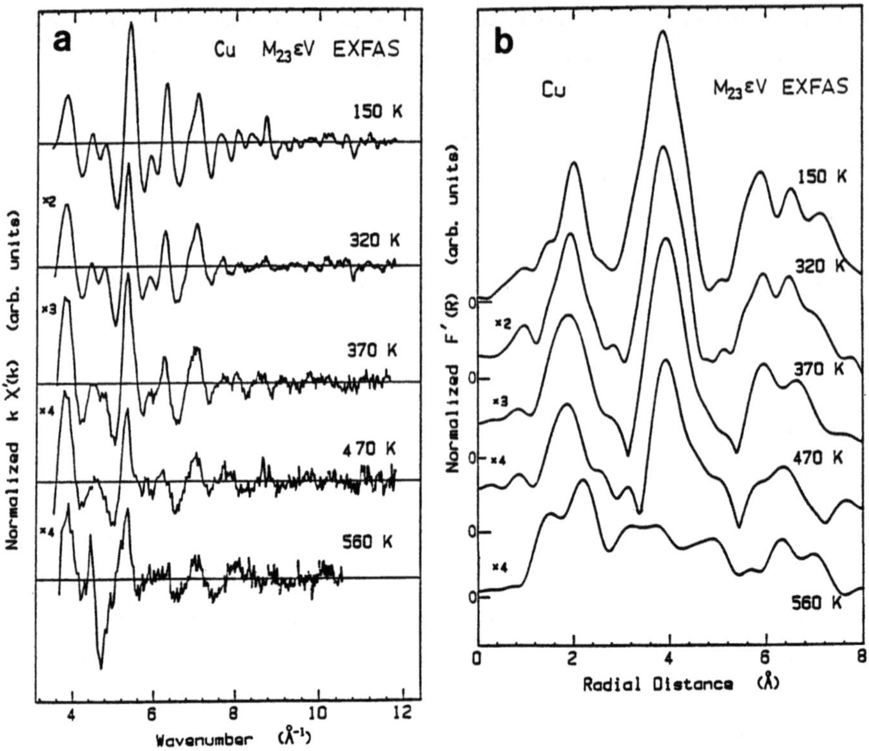

Fig.2 - *(a) First derivative of the Cu $M_{23}\varepsilon V$ EXFAS of polycrystalline Cu at the indicated temperatures /11/. The origin of the k(Å^{-1}) scale was taken at the minimum of the $M_{23}VV$ transition. (b) Fourier transforms of the EXAFS spectra of (a).*

motion damping.The logarithm ratio of these amplitude functions provided an evaluation of the Debye-Waller factors which strongly incresed on passing from 150 K to 560 K.These values are similar in magnitude to previous EXAFS results for the temperature dependence /12,13/. These observations are in contrast with the secondary electron diffraction interpretation /7-9/, because in this case one should measure a temperature dependence similar to that for LEED,with a damping with increasing temperature which is both less rapid and more uniform over all k values scale /14/ than for EXAFS.

Fig.4 (a) shows the EXAFS spectra of clean Ag(111) in the region above the $N_{23}VV$ Auger line /11/. Also in this case a dramatic variation of the intensity of the structure between 2 and 7 Å^{-1} (which correspond to approximately 60-200 eV above the Auger line) is observed as a function of the temperature.The magnitude of the Fourier transforms are presented in Fig.4 (b). As expected from the raw data ,the radial distribution curves change considerably with temperature indicating a structural origin of these features while the preferential damping of the high-R components with increasing temperature is consistent with an EXAFS-like rather than a diffraction origin.

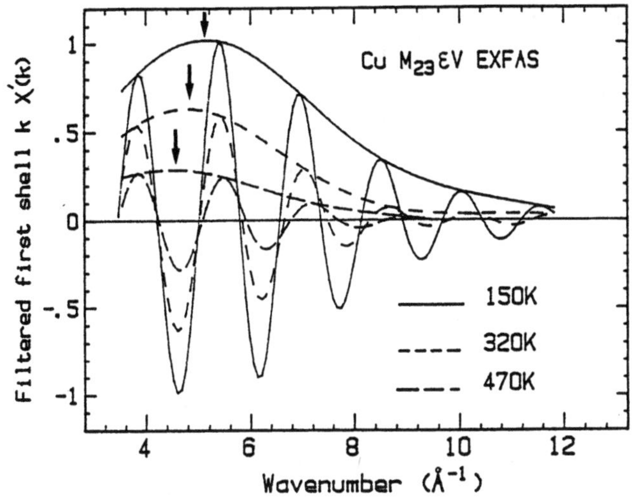

Fig.3 - *Fourier filtered first shell signal of the Cu $M_{23}\varepsilon V$ EXFAS of polycrystalline Cu as a function of the temperature.The solid and dashed lines are the envelopes functions derived from the Fourier filtering analysis.*

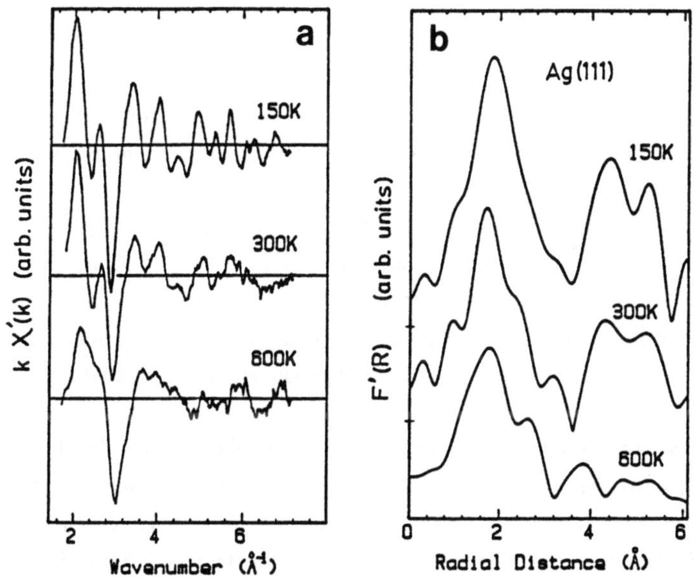

Fig.4 - *(a) EXFAS first derivative spectra of Ag(111) above the $N_{23}VV$ line recorded at the indicated temperatures /10,11/. (b) Fourier transforms of the Ag $N_{23}\varepsilon V$ EXFAS as a function of temperature.*

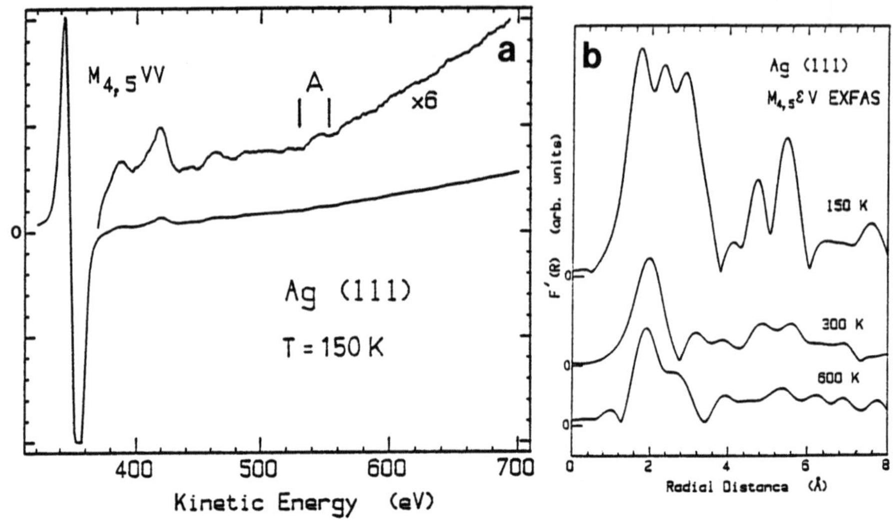

Fig.5 - *(a) First derivative Auger spectrum of Ag(111) at 150 K in the region above the Ag $M_{45}\varepsilon V$ region /11/.The lines marked A indicate weak,conventional $M_2N_{23}V$ and $M_3N_{23}V$ Auger lines. (b) Magnitude of the Fourier transforms of the Ag $M_{45}\varepsilon V$ EXFAS at the indicated temperatures.*

Fig.5 (a) shows the spectrum of Ag(111) in the region above the $M_{45}VV$ Auger transition.All the structures above 370 eV kinetic energy are strongly temperature dependent as clearly shown by the Fourier transforms reported in Fig.5 (b).

This is the first report of EXFAS associated with a relatively deep core level. All previous EXFAS /2-4/ have been connected with core levels below 100 eV binding energy. Our observation of similar EXFAS signals from two edges at widely separated energies emphasizes the local-sensitive EXAFS-like origin of the signal.If the signal is dependent primarily on local geometry then a very similar nearest neighbour signal is expected at each edge.

The Ag $M_{45}\varepsilon V$ EXFAS signal is at least an order of magnitude weaker than the Ag $N_{23}\varepsilon V$ EXFAS reported in Fig.4 (b).One possible reason is the influence of core hole lifetime on the EXFAS mechanism. If the average core hole lifetime greatly exceeds the period in which there is an appreciable continuum state amplitude at the core-excited atom then the EXFAS signal may not occur. Thus the EXFAS could be observable only with rather short lived core hole states such as the M_{23} and the N_{23} states of 3d and 4d transition metals which have short lifetimes on account of the rapid Coster-Kronig process .This may be the explanation for the much weaker M_{45} than the N_{23} EXFAS signal for Ag and also the inability up to now of observing EXFAS associated with L_{23} core states of 3d transition metals.

Also crystalline materials with broad s-p hybridized valence bands (Si,Ge,C) should not exhibit sizeable EXFAS features, whereas relatively strong signals should persist if the

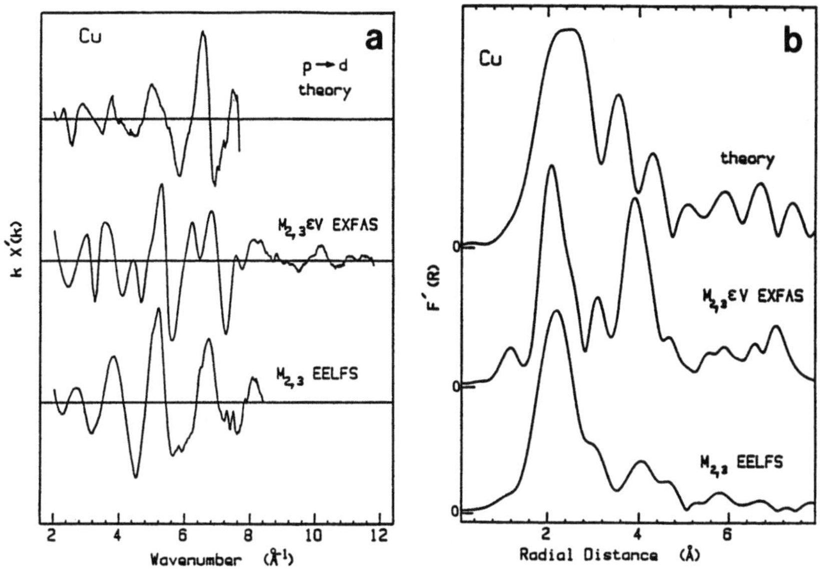

Fig.6 - *(a) Comparison of the Cu L_2 (p->εd) EXAFS calculated by band structure technique /6/,with the EXFAS detected above the $M_{23}VV$ Auger line and with the EELFS above the M_{23} core edge of a thick Cu film /15/.The experimental N'(E) curves have been integrated. (b) Fourier transforms of the above spectra. Note that Cu $M_{23}\epsilon V$ EXFAS has a stronger high frequency components and extends to higher k-values than the M_{23} EELFS signal.*

diffraction mechanism is valid.The Auger spectrum of Si (111) does not show significant structure above the $L_{23}VV$ line /11/ no matter how well the surface region is crystallized.

Fig.6(a) compares the room temperature Cu $M_{23}\epsilon V$ EXFAS to normal incidence EELFS /15/ and to band calculation of the L_2 EXAFS of Cu /6/.

Note that the EXFAS was integrated in this case, in contrast to all other reported spectra which were derived from the recorded N'(E) spectra without integration.

The integration removes a π/2 phase shift as well a sligth enhancement of high-R components that occurs when differentiated spectra are analyzed. All three spectra show similar first shell periodicities and Fourier transforms (Fig.6(b)),but do not match exactly in phase or in amplitude envelope. In our opinion this is not surprising since,relative to EELFS,the proposed origin of EXFAS involves an additional electron (the outgoing d-band electron that is ejected to conserve energy and which is actually detected in the experiment) and thus a different phase shift.

In the case of the $M_{23}\epsilon V$ and $N_{23}\epsilon V$ EXFAS both p->εd and d->εf phase shifts need to be considered.Unfortunately calculated d->εf phase shifts are not yet available.The deviation of R from the true nearest neighbour distances obtained,when calculated M_{23} or N_{23} phase shifts are used,can thus attributed mainly to an additional phase associated with the d->εf transition.

4.Conclusions

A deeper analysis of the electron inelastic scattering yield opens the way to a local geometry determination at surfaces and interfaces. Using fine structure detected in kinetic energy above the CVV Auger lines is now possible to obtain EXAFS-like structural information by means of a very simple experimental apparatus (any Auger UHV equipment) and not sophisticated data analysis following the procedure of the EXAFS spectroscopy.

Research in this direction is in progress and much efforts should be devoted towards several problems which are still open:

-phase shifts computations for shallow core levels,in order to obtain structural results without using standard samples,

-overlapping of various Auger lines,which reduces the k-range available for the Fourier transform,

-size and narrowing of the valence band,

-variation of the extended features as a function of the collection angle,to disciminate better the diffraction long-range effects from a local geometry origin,

-surface structural sensitivity,which should be comparable to the mean free path of the corresponding Auger lines.

These points represent a wide and interesting program to pursue before the EXFAS technique can be used with confidence as a structural tool in surface science laboratories.

References

/1/ E.A.Stern,J.de Physique **47**-C8,3(1986) and references therein.

/2/ M.De Crescenzi,E.Chainet,J.Derrien,Solid State Commun.**57**,487(1986).

/3/ J.Derrien,E.Chainet,M.De Crescenzi,C.Noguera,Surf.Sci.**189**,590(1987).

/4/ E.Chainet, J.Derrien,R.C.Cinti,T.T.A.Nguyen and M.De Crescenzi, J.de Physique **47**,C8-209(1986).

/5/ J.E.Müller and J.W.Wilkins,Phys.Rev.B**29**,4331(1984).

/6/ R.C.Albers,A.K.McMahan and J.E.Müller,Phys.Rev.B**31**,3435(1985).

/7/ L.McDonnel,B.D.Powell and D.P.Woodruff,Surf.Sci.**40**,669(1973).

/8/ G.E.Becker and H.D.Hagstrum,J.Vac.Sci.Technol.**11**,284(1974).

/9/ D.P.Woodruff,Surf.Sci.**189**,64(1987).

/10/ T.Tyliszczak,A.P.Hitchcock and M.De Crescenzi,Phys.Rev.B**38**,5768(1988).

/11/ M.De Crescenzi,T.Tyliszczak and A.P.Hitchcock,Phys.Rev.B15 submitted.

/12/ R.B.Greegor and F.W.Lytle,Phys.Rev.B**20**,4902(1979).

/13/ E.Sevillano,H.Meuth and J.J.Rehr,Phys.Rev.B**20**,4908(1979).

/14/ J.B.Pendry,in Low Energy Electron Diffraction ,Academic Press,London,1974.

/15/ T.Tyliszczak,M.De Crescenzi and A.P.Hitchcock,Phys.Rev.B**37**,10664(1988).

A Technique for Performing Angle-Resolved Auger Electron Spectroscopy (ARAES) and Angle-Resolved Electron Energy Loss Spectroscopy (ARELS)

G.B. Hoflund

Department of Chemical Engineering, University of Florida, Gainesville, FL 32611, USA

1. INTRODUCTION

Auger electron spectroscopy (AES) is the most widely used surface analytical technique. It is used in most surface studies at least for monitoring surface composition even when the study does not focus on AES. A difficulty with interpreting AES spectra is that AES is not highly surface sensitive but actually probes fairly deeply beneath the surface. This makes it difficult or impossible to determine where a given species lies with respect to the surface and how it is distributed in the direction normal to the surface. Another closely related problem is that AES may not be particularly sensitive to changes in the outermost layer of surface atoms. This point is lucidly illustrated in a study of strong-metal-support interaction (SMSI) behavior of platinized titania by HOFLUND et al. /1/. Encapsulation of the platinum by reduced titania species formed during a reduction in H_2 at 400°C is clearly evident in ion scattering spectroscopy (ISS) data but barely visible in the corresponding AES spectra.

A method has been developed recently for increasing the surface sensitivity of AES, and initial results have been reported by HOFLUND et al. /2/. In this method, referred to as angle-resolved AES (ARAES), both the incident angle of the primary electron beam and the collection angle of the Auger electrons are varied independently. Both angles strongly influence the surface sensitivity of AES. The ARAES data taken thus far suggest that the ultimate surface sensitivity approaches that of ISS and is greater than that of angle-resolved electron spectroscopy for chemical analysis (ARESCA), but these generalizations may not hold in all situations. Recently, a new method for collecting high-energy-resolution AES data rapidly has been presented by GILBERT et al. /3/. Combining this method with ARAES (HRARAES) provides depth-sensitive chemical state information. An important aspect of the experimental configuration used for ARAES is that it can also be used to perform angle-relsolved electron energy loss spectroscopy (ARELS) as discussed by HOFLUND and CORALLO /4,5/. This present paper describes several different aspects of ARAES, HRARAES and ARELS, illustrates the powerful nature of these techniques for obtaining new types of information, and discusses future development of these techniques.

2. EXPERIMENTAL METHODS

Figure 1a shows the geometrical configuration commonly used to collect AES spectra, and figure 1b shows the configuration used to collect ARAES, HRARAES and ARELS spectra. The energy analyzer used in these studies was a Perkin-Elmer PHI Model 25270AR double-pass cylindrical mirror analyzer which contained an internal, coaxial electron gun and a movable angularly resolving 90° aperture. An off-axis electron gun was also contained in the system to perform glancing-incidence-angle experiments. With this experimental arrangement the incidence angle could be varied from 0 to 80° off sample normal and the collection angle could be varied from 0 to 80° off sample normal.

3. RESULTS AND DISCUSSION

3.1 ARAES

The previous study /2/ shows that collecting AES spectra in the typical manner shown in figure 1a nearly maximizes the bulk sensitivity of the technique. The dramatic increase in surface sensitivity which can be attained using the experimental configuration shown in figure 1b is illustrated by the ARAES spectra shown in figure 2a and b. These spectra were taken from a sputter-cleaned polycrystalline Ag foil which contained Li as an impurity. Annealing in oxygen drives the Li to the near surface region where Li_2O is formed. The ARAES spectrum shown in figure 2a was taken using the internal electron gun by tilting the sample normal 32° off the CMA axis and collecting emitted electrons over the full 360° acceptance cone (angle resolving aperture withdrawn). The three primary features are due to Li, Ag and O. The ARAES spectrum shown in figure 2b was taken from the same surface using the same experimental conditions except that only electrons emitted approximately 74° off the sample normal were collected using the angle-resolving movable aperture in the CMA. The Li-to-O peak-height ratio remains unchanged from that shown in figure 2a, but the Ag peak height is greatly reduced with respect to the Li and O peaks. These spectra demonstrate that Li_2O forms a layer over

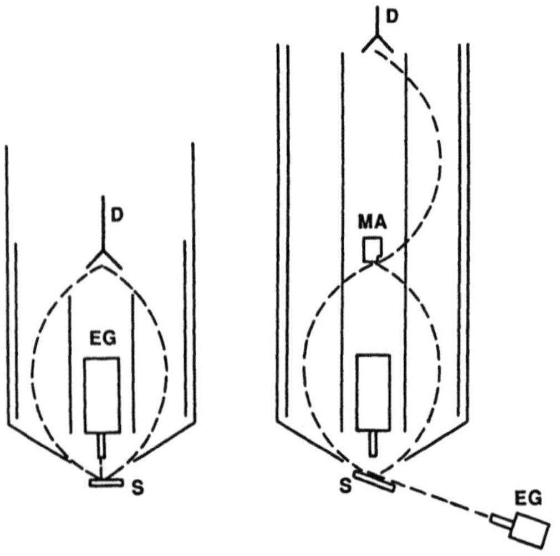

Figure 1. (a) The most commonly used technique for performing AES is to place the sample (S) normal to the CMA axis. A coaxial electron gun (EG) provides the primary electrons, and electrons emitted in a cone about 42° off normal are collected by the detector (D). This arrangement yields AES spectra for which the bulk sensitivity of the technique is nearly maximized. (b) In the angle-resolving configuration, the sample is tilted with respect to the CMA axis and either an internal electron gun or a grazing incidence electron gun can be used. The collection (emission) angle is selected using the 90°-slotted movable aperture.

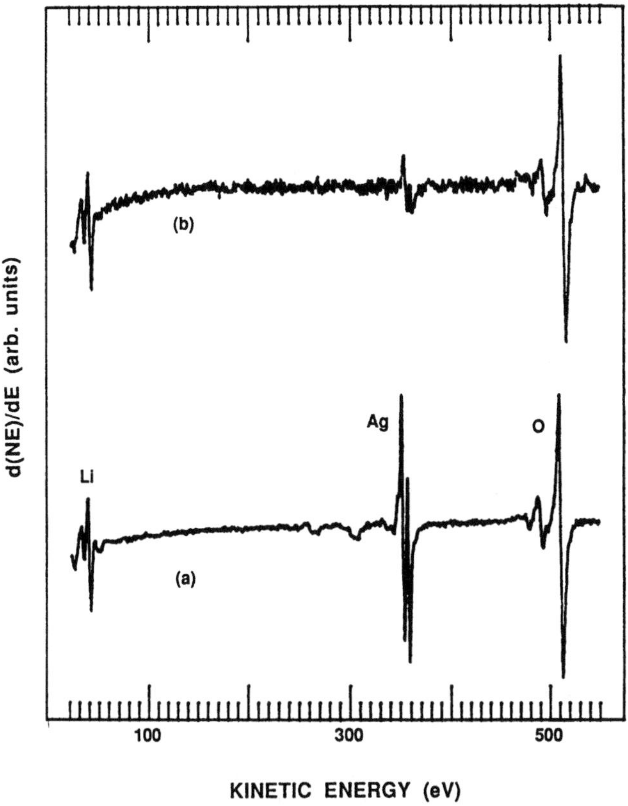

Figure 2. ARAES spectra taken from a polycrystalline, Li-contaminated Ag foil after sputter cleaning and heating in oxygen. (a) was taken by tilting the sample 32° and collecting electrons emitted over the full 360° acceptance cone. (b) was taken by collecting electrons emitted about 74° off the same normal thereby increasing the surface sensitivity.

the Ag and, more importantly, illustrates the increased surface sensitivity attainable using ARAES.

Determining the ultimate surface sensitivity of ARAES is an important topic which has not been fully addressed. This can be done by using more grazing incidence and collection angles. It should be possible to increase both of these angles to more than 80° with respect to the sample normal. The increased surface sensitivity of the ARAES spectrum shown in figure 2b is primarily due to increasing the collection angle from the standard 42° to 74°. A further increase in sensitivity would be gained by using a more grazing incidence angle /2/.

3.2 HRARAES

Since high-resolution AES yields chemical state information /6/, it is possible to obtain depth-sensitive, chemical-state information using angle-resolved techniques (HRARAES). This point is illustrated in figure 3 which is reproduced from a study by ASBURY and HOFLUND /7/ of a polycrystalline Pt_3Sn surface after vacuum annealing or oxygen exposure. Curve a shows the high-resolution N(E) Sn AES peaks taken from a cleaned and annealed Pt_3Sn surface. Curves b were taken from the same

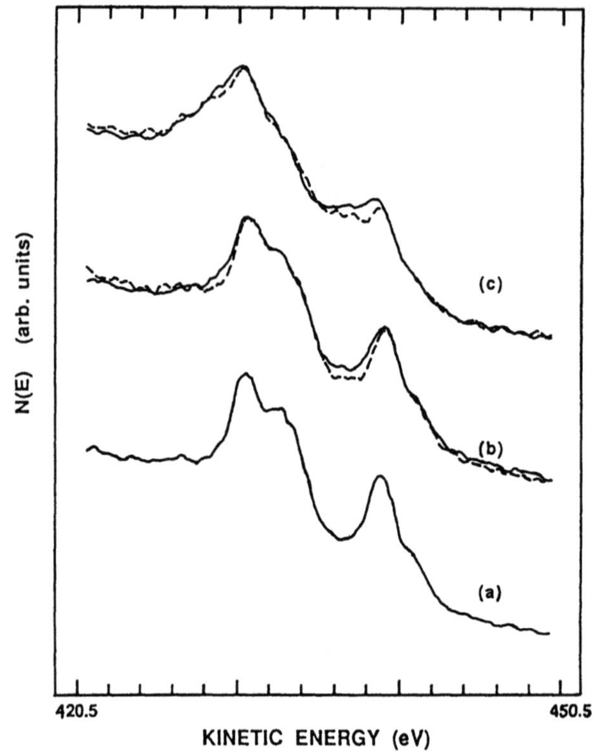

Figure 3. High-resolution Auger spectra taken after (a) annealing a Pt_3Sn sample at 200°C, (b) a 1000 L O_2 exposure at 25°C and (c) a 1000 L O_2 exposure at 400°C. The dashed curves in (b) and (c) are surface-sensitive AES spectra, and the solid curves in (a), (b) and (c) are bulk-sensitive AES spectra. For the 200°C annealed surface, there are negligible differences between the bulk-sensitive and surface-sensitive spectra so only the former is shown.

surface after a 1000 L O_2 exposure at 25°C, and curves C were taken after a 1000 L O_2 exposure at 400°C. The dashed curves are more surface sensitive than the solid curves. The more fully oxidized the tin is, the more altered the spectra in b and c are from that in a. The dashed curves in b and c are altered less than the solid curves indicating that the tin lying beneath the surface is more fully oxidized. It is interesting to note that even though the total amounts of oxygen adsorbed during a 1000 L O_2 exposure at 25 and 400°C are similar /7/, the tin in the near-surface region is more fully oxidized at 400°C. This observation and ESCA and ISS data led to the conclusion that two forms of adsorbed oxygen are present: oxygen adsorbed at the surface and a subsurface oxygen. The latter is assigned as oxidic which is responsible for large changes in the shapes of the Sn AES features.

3.3 ARELS

The same experimental configuration used to perform ARAES can also be used to perform ARELS. Like ARAES, the surface sensitivity of ARELS spectra is strongly dependent upon the incident and collection angles used, but in ELS the primary electron beam energy also has a large influence on surface sensitivity /8/.

ELS spectra are quite complex, and interpretation of the spectra is usually quite difficult. The features are due to excitation of surface or bulk plasmon oscillations or electronic excitation from filled to unfilled energy levels. It can be difficult to distinguish between the different types of excitations. Surface and bulk plasmons can often be identified by observing their behavior as surface sensitivity varies. Surface plasmons become more prominent as the surface sensitivity increases, and bulk plasmons become more prominent as the surface sensitivity decreases. In previous ELS studies the surface sensitivity has typically been increased by reducing the primary beam energy. However, this can lead to problems with interpretation because variations in cross sections with primary beam energy can be considerable. Thus, it is difficult to know if the peak size variation is due to a change in surface sensitivity or cross section. By using ARELS this problem can be avoided since the cross sections remain constant when a constant primary beam energy is used. The variation in surface and bulk plasmon losses and multiple plasmon loss peaks from a polycrystalline tin surface with both incidence and collection angles has been examined by HOFLUND and CORALLO /4/. This study shows that the surface sensitivity of ARELS can be increased so that bulk plasmon features are eliminated. The data also suggest that ARELS may be useful in studying surface roughness and sputter damage in the outermost few atomic layers.

Figure 4 illustrates a typical variation of ELS spectra which is obtained by changing the collection angle. The spectra were collected from a cleaned polycrystalline tin surface following a 3500 L O_2 dose

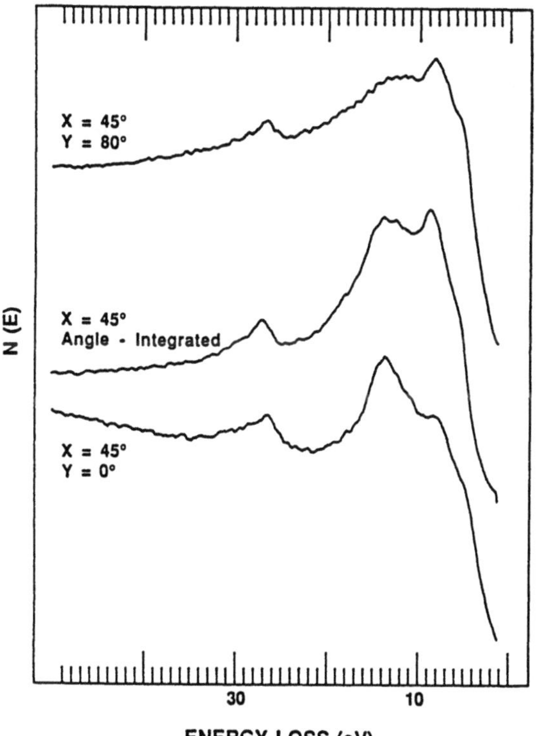

Figure 4. ARELS spectra obtained from polycrystalline tin following a 3500 L oxygen dose. The incident and collected beam angles are labelled.

using a primary beam energy of 200 eV. The incident beam struck the sample at an angle of 45°, and collection angles of (a) 0°, (b) all angles, and (c) 80° with respect to the sample normal were used. The variation of the spectral features as a function of surface sensitivity is useful both in assigning the spectral features and in determining the composition as a function of depth. A detailed discussion of these spectra is given elsewhere /5/.

4. SUMMARY

Using ARAES greatly increases the surface sensitivity of AES. Both the incidence angle of the primary electron beam and the collection angle influence the surface sensitivity. It is possible to determine a compositional depth profile of the near-surface region using ARAES. Combining high-energy-resolution AES and ARAES (HRARAES) allows the determination of depth-sensitive chemical-state information. ARELS can also be used to obtain depth-sensitive compositional information and is particularly helpful in assigning the ELS spectral features to plasmon or electronic transitions. It also appears that these angle-resolved techniques will be useful in studying surface roughness and damage phenomena.

5. REFERENCES

1. G.B. Hoflund, Austin L. Grogan, Jr., D.A. Asbury: In J. Catal. 109, 226 (1988)
2. G.B. Hoflund, D.A. Asbury, C.F. Corallo, G.R. Corallo: In J. Vac. Sci. Technol. A6, 70 (1988)
3. R.E. Gilbert, G.B. Hoflund, D.A. Asbury, M.R. Davidson: In J. Vac. Sci. Technol. A6, 2280 (1988)
4. G.B. Hoflund, G.R. Corallo: In Surf. Inter. Anal. 11, 00 (1988)
5. G.B. Hoflund, G.R. Corallo: to be published
6. H.H. Madden: In J. Vac. Sci. Technol. 18, 677 (1981)
7. D.A. Asbury, G.B. Hoflund: In Surf. Sci. 199, 552 (1988)
8. D.F. Cox, G.B. Hoflund: In Surf. Sci. 151, 202 (1985)

Part V

Complementary Techniques

BIS Investigation of the Co/Si(111) Interface Formation

J.Y. Veuillen[1], *M. De Crescenzi*[2], *T.T.A. Nguyen*[1], *R. Cinti*[1], *and J. Derrien*[3]

[1]LEPES/CNRS, Associé à l'Université J. Fourier, BP 166, F-38042 Grenoble Cedex, France

[2]Dipartimento di Fisica, Università di Roma II, I-00173 Rome, Italy

[3]CRMC2/CNRS, Luminy, F-13288 Marseille, France

1. INTRODUCTION

The Co/Si(111) interface formation at room temperature has been extensively studied during the last few years using various electron spectroscopies such as ultra-violet Photoemission (UPS) [1], X-Ray Photoemission (XPS) [1,2], Auger Electron Spectroscopy (AES) [3], Electron Energy Loss Fine Structure [4], as well as transport measurements [5] and electron microscopy [6]. The common result of all these investigations is that for very thin Co deposits (thickness ≤ 5 Å) an intermixing between Co and Si occurs even at room temperature, the physico-chemical properties of the intermixed phase being very similar to that of the cobalt disilicide $CoSi_2$ [7]. For thicker Co deposits the surface composition of the film becomes Co-rich and essentially an almost pure Co film grows at Co coverages larger than $\sim$ 10 Å. Co/Si(111) may be regarded as a well characterized interface and this is the reason why it has been chosen in this present paper as a standard to check the applicability of Bremsstrahlung Isochromat Spectroscopy (BIS) to interface studies.

BIS has proven to give detailed informations on the empty electronic states above the Fermi level (E_F). The principle of the experiment [8] is to excite the surface of the sample with a high energy ($\sim$ 1500 eV) electron beam and to detect the number of X-ray emitted by the sample at a given energy, as the primary electron energy is varied. For simple metals and transition metals [8], BIS spectra are satisfactorily interpreted in terms of the one-particle Density of States (DOS), without major complications arising from many-body effects or electron energy losses as long as the investigated region is limited to about 10 eV above E_F. Up to now, only little attention has been paid to silicides in BIS literature [10,11,12].

BIS technique, probing empty states is often considered as the counterpart to Valence Band (VB) XPS probing occupied states. In fact, BIS and XPS cross sections are more or less similar [9], and the surface sensitivity should be nearly identical for both techniques since the energy of the electrons and X-ray involved in the two processes respectively are the same. It follows that in interface studies BIS should give the same kind of information on empty VB states than XPS on occupied VB states. However, the rather large probing depth of the incident electrons ($\sim$ 50 Å) and the low count rate in BIS (a few c/s) may appear as major drawbacks for interface studies. We demonstrate in this paper that thanks to some improvements, BIS techniques may be a powerful probe for surface and interface phenomena.

The paper is organized as follows. In a first part, we describe the experimental set up. In the second and fourth part, we compare our spectra of core and Auger lines with previously published data from other groups in order to monitor interface formation. We will discuss in part 3 and 5 the results we obtained in VB XPS and BIS respectively.

2. EXPERIMENTAL SET UP

Experiments were performed in a Vacuum Scientific Workshop (VSW) ultra-high vacuum multitechnique system which consists of an analysis and a preparation chamber connected to a fast entry chamber. The analysis chamber is equipped with an unmonochromatized MgKα X-ray tube and a 150 mm hemispherical analyzer for AES and XPS experiments. For BIS studies we used an electron gun with a LaB_6 cathode (total target current $\sim$ 200 μA à 1500 eV) together with a large X-ray monochromator to detect 1486.6 eV X-ray photons emitted by the sample.

Si(111) was cleaned by cycles of Ar sputtering and annealing to obtain the 7x7 reconstruction. Co films were then sequentially deposited on the surface by evaporating Co from a metal rod heated by electron bombardment. The deposition rate was estimated to be about 0.7 Å/min by XPS core lines intensity variations. The cleanliness of the surface was checked by XPS at each stage of the experiment. XPS, Auger and BIS spectra were taken immediately after each deposit. Cobalt disilicide was formed by annealing a 70 Å thick Co layer at 650° C for 10 min. LEED diagram showed that an epitaxial $CoSi_2$ film was grown on Si(111).

3. CORE LEVEL SPECTROSCOPY

The evolution of the $Co2p_{3/2}$ XPS line with increasing Co coverage is shown in Fig. 1. Our results are essentially identical to those of PIRRI et al. [1] : first there is little change in the binding energy of the $Co2p_{3/2}$ line when the Co coverage increases, and secondly, its binding energy for small deposits (3 and 6 Å) is almost identical to that of bulk $CoSi_2$, which reflects the interfacial silicide formation.

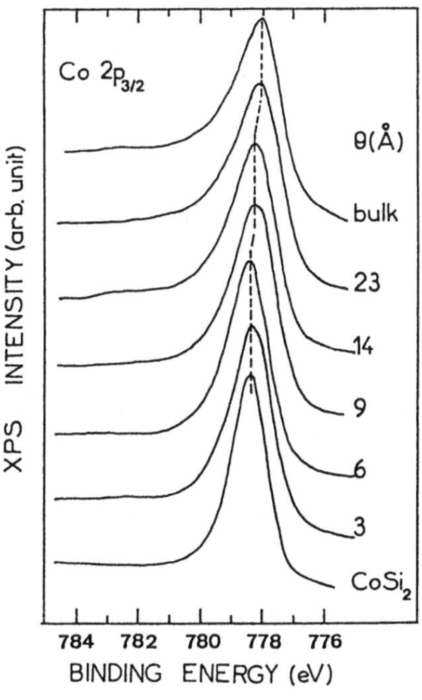

Fig. 1 : Evolution of the $Co2p_{3/2}$ core line upon Co deposition on Si(111)(7x7) for different Co coverage θ.

One notices that the $Co2p_{3/2}$ line is much narrower and more symmetric for $CoSi_2$ than for Co. We attribute this to a change in electron-hole excitation spectrum [13] due to a decrease in the DOS at E_F upon silicide formation.

4. VALENCE BAND XPS

XPS spectra for increasing Co coverage are shown in Fig. 2. The spectrum for the 3 Å thick deposit has a broad maximum at about 1.8 eV binding energy (BE), which shifts toward the Fermi level E_F for larger coverages. For pure Co the maximum of the VB is found 0.5 eV below E_F. These results are in good agreement with UPS data of PIRRI et al. [1].

In Fig. 3 we compare the VB spectrum of $CoSi_2$ with that of a thin (3 Å) Co film deposited on Si. The spectra show strong resemblance, obviously due to the thin (~ 10 Å) interfacial silicide formation. (The Co 3 Å/Si(111) spectrum cannot be accounted for by superimposing spectra of Si and

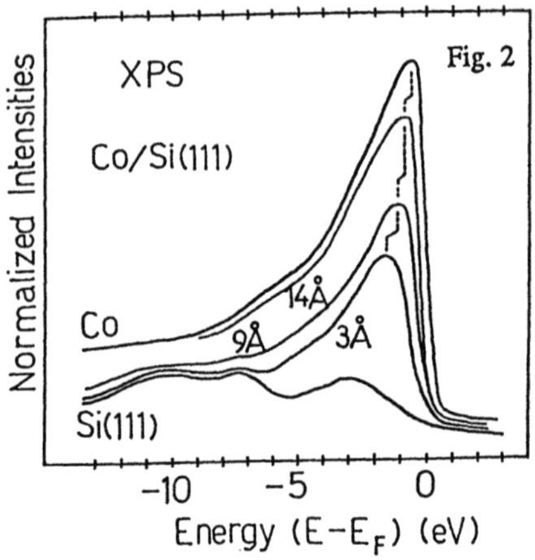

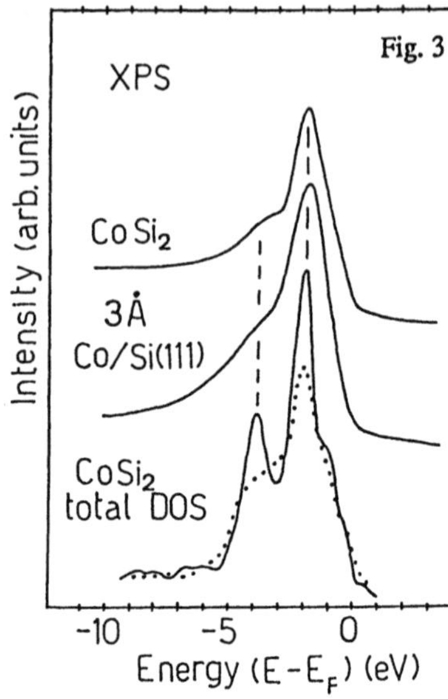

Fig. 2 : Evolution of the XPS valence band spectra of the Co/Si(111) interface with increasing Co coverage.

Fig. 3 : Top curve : $CoSi_2$ XPS valence band. Middle curve : XPS valence band for a 3 Å thick Co film on Si(111). Bottom curve : theoretical total occupied DOS [14] without broadening (full line) and after broadening by convolution with a Lorentzian line of HWHM 1 eV (dots).

pure Co, since this should give a peak closer to the Fermi level). The lower curve of Fig. 3 shows the theoretical $CoSi_2$ total DOS [14] (full line) and broadened by the estimated instrumental resolution (dots). The agreement in peaks energy position with experimental data is excellent. We may thus identify the structures in the XPS spectra from the calculated DOS : the peak at ~ 1.8 eV BE is assigned to non-bonding Co states and the shoulder at 3.7 eV BE to hybridized Co3d-Si3s3p bonding states. One notices that the intensity of the bonding structure is weaker in the experimental spectra than in the theoretical one, which is presumably due to cross-section and/or hybridization effects. The shift from 1.7 eV to 0.5 eV BE of the main peak with increasing Co deposition reflects the (eventually progressive) Co enrichment of the surface [1,2]. What we wish to underline here is that in spite of the large probing depth of VB XPS (~ 50 Å), experimental spectra clearly helps identifying the early stages of interfacial reaction, and we should observe the same behavior with BIS technique.

5. AUGER SPECTROSCOPY

The evolution of the CoL_3VV Auger line is shown in Fig. 4. Once again we notice a strict analogy in shape and position between the Auger lines for small deposits and $CoSi_2$. It should be noticed here that AES is in fact more sensitive to the interfacial reaction than XPS for the case of this particular interface since the shift in Kinetic Energy (KE) of the CoL_3VV line between $CoSi_2$ and Co is much larger (~ 1.1 eV) than the corresponding shift for the $Co2p_{3/2}$ line (~ 0.4 eV). Moreover, these shifts are in opposite directions (i.e. the shift toward lower BE of the $Co2p_{2/3}$ core line with increasing Co coverage should correspond to a decrease of the KE of the CoL_3VV Auger line if no other parameters could change). This is readily explained by the shift toward lower BE of the maximum of the Co VB when the Co coverage increases - which increases the KE of the CoL_3 VV line - without the need of changing significantly the effective interaction energy U_{eff} between two 3d holes [15].

In a previous paper [3], DERRIEN et al. have fitted the CoL_3VV Auger lineshape variation during the interface formation using the CINI-SAWATZKY model [16]. They used their experimental UPS valence band and XPS $Co2p_{3/2}$ line position as input parameters. The results are shown in Fig. 5. The same U_{eff} value was used for the various Co coverages and for $CoSi_2$. The agreement with experiment

Fig. 4

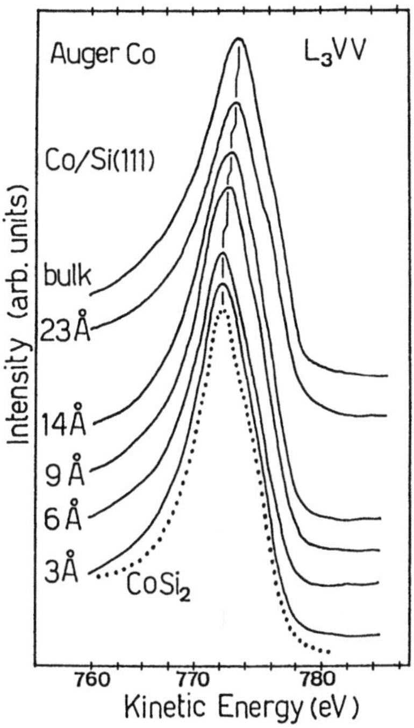

Fig. 5

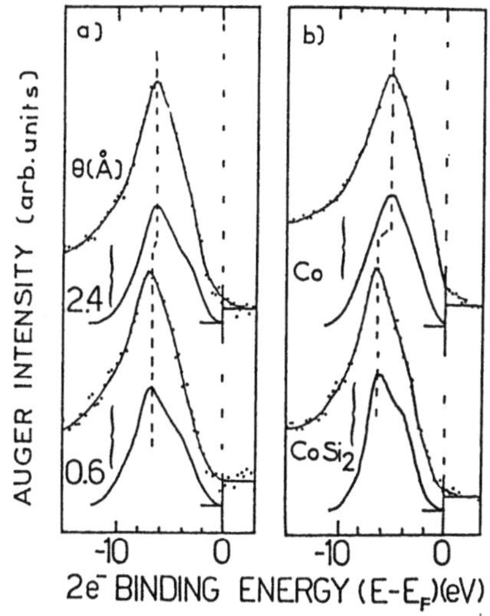

Fig. 4 : Full line : evolution of the CoL_3VV Auger line at Co/Si(111) interface with increasing Co coverages. Dots : CoL_3VV Auger line recorded on a $CoSi_2$ sample.
Fig. 5 : Comparison between experimental data (dots) of the CoL_3VV Auger line and the calculation within the CINI-SAWATZKY model (full line). The two holes spectra energy scale is referred to the Fermi level. Panel (a) is for the Co/Si interface (Co coverage θ) and panel (b) for bulk Co and $CoSi_2$. From Ref. [3].

is rather good, which suggests that neither the hole-hole 3d repulsion nor the relaxation energies are affected by interface or silicide formation relative to pure Co.

6. BREMSSTRAHLUNG ISOCHROMAT SPECTROSCOPY

The results of the BIS interface study are summarized in Fig. 6 and Fig. 7. Clean Si(111) spectrum in Fig. 6 is quite similar to those of Ref. [8] and [17] and shows a slowly rising front above E_F followed by a broad maximum at about 2-3 eV. This shape is compatible with DOS calculations [18]. In the early stages of deposition (3 and 6 Å) one notices an increase in overall signal as well as a change in the spectrum shape close to E_F due to an enhancement of emission in the 0-1 eV range. this shows that BIS spectra are indeed sensitive to very thin Co deposits (as expected from XPS results and in spite of the large electron mean free path at about 1500 eV), although the evolution of the spectra is much less dramatic than in XPS. The broad maximum at 2-3 eV remains unaffected in this coverage range. For further deposition (9-14 Å) the spectra become square-shaped and unambiguously metallic-like as a result of increasing emission close to the Fermi level (0-2 eV range). For larger deposits (≥ 23 Å) a sharp maximum is seen at about 1 eV and the characteristic spectrum of pure Co [8] is recovered for the 70 Å thick film. In this coverage range the spectra demonstrate that an essentially pure Co film is formed in the surface region since no known silicide BIS spectra show such a sharp and intense peak near the Fermi level [17].

We now return to the 3 Å Co/Si spectrum and compare it to that of bulk $CoSi_2$ in Fig. 7. There is a good correspondence between the two spectra. One observes in particular a shoulder at ~ 0.3 eV above E_F which originates from the high energy tail of the empty Co non-bonding states around E_F. A broad feature at ~ 3 eV which originates from the empty antibonding states between Si3s3p and Co3d

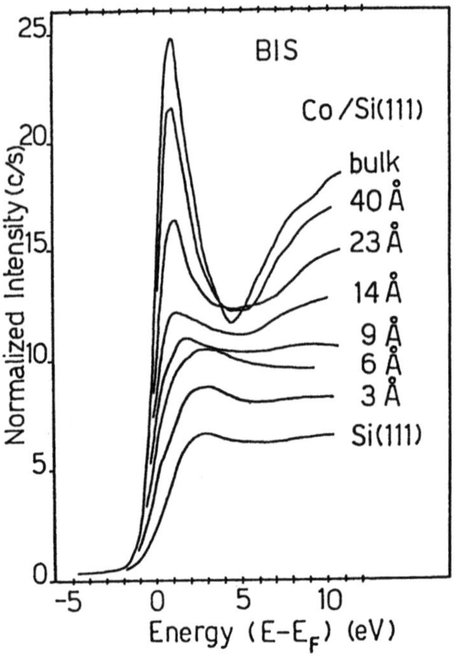

Fig. 6 : Evolution of the BIS spectra for increasing Co coverages

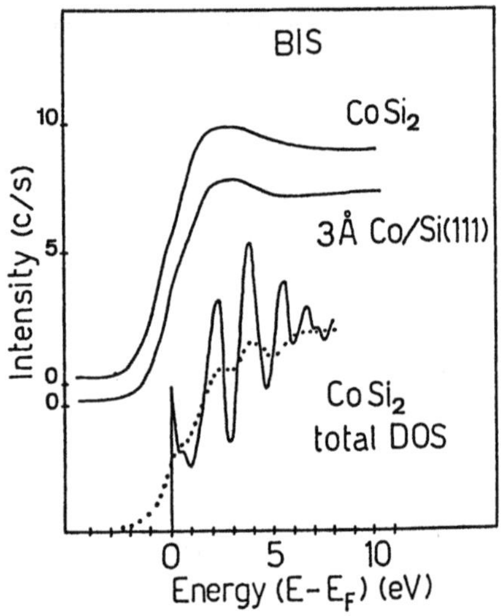

Fig. 7 : Top curve : $CoSi_2$ BIS spectra. Middle curve : BIS spectra for 3 Å Co deposited on Si(111)7x7. Bottom curve : Theoretical empty DOS [14] without broadening (full line) and after broadening by convolution with a Lorentzian line of HWHM 1 eV (dots).

electrons with a larger Si character. The experimental curves agree nicely with theoretical results [14]. The presence of the 0.3 eV structure in the 3 Å Co/Si spectrum is important to establish the BIS efficiency in interface studies since it appears in the present case as the major difference between the spectrum of bulk Si and that of bulk (or interfacial) $CoSi_2$ (above 2 eV the spectra of Si and $CoSi_2$ are in fact very similar). However, although sensitive to interface formation the BIS study alone may not be a definite proof of the interfacial silicide formation, but in combination with XPS it shows that the total DOS (empty as well as occupied states) of the interfacial layer is identical to that of $CoSi_2$.

7. CONCLUSION

Our main objective in the present work was to test on a model system the feasibility of interface studies using high-energy BIS. The various stages of interface formation were monitored by core-level XPS and Auger spectroscopy, and are entirely compatible with previously published data. As a conclusion, we have found that the complementary techniques of VB XPS and high energy BIS were able to give a rather accurate insight in the total DOS evolution during interface formation.

LITERATURE.

1. C. Pirri, J.C. Peruchetti, G. Gewinner, J. Derrien, Phys. Rev. B 29, 3391 (1984).
2. F. Boscherini, J.J. Joyce, M.W. Ruckman, J.H. Weaver, Phys. Rev. B 35, 4216 (1987).
3. J. Derrien, M. De Crescenzi, E. Chaînet, C. d'Anterroches, C. Pirri, G. Gewinner, J.C. Peruchetti, Phys. Rev. B 36, 6681 (1987).
4. E. Chaînet, M. De Crescenzi, J. Derrien, T.T.A. Nguyen, R.C. Cinti, Surface Science 168, 801 (1986).
5. J.Y. Veuillen, J. Derrien, P.A. Badoz, E. Rosencher, C. d'Anterroches, Appl. Phys. Lett. 51, 1448 (1987).
6. C. d'Anterroches, Surface Science 168, 751 (1986).
7. J. Derrien, Surface Science 168, 171 (1986).

8. W. Speier, J.C. Fuggle, R. Zeller, B. Ackermann, K. Szot, F.U. Hillbrecht, M. Campagna, Phys. Rev. B 30, 6921 (1984) ; H.J.W.M. Hoekstra, W. Speier, R. Zeller, J.C. Fuggle, Phys. Rev. B 34, 5177 (1986).
9. W. Speier, J.C. Fuggle, P. Durham, R. Zeller, R.J. Blake, P. Sterne, J. Phys. C 21, 2621 (1988).
10. S.J. Oh, J.W. Allen, J.W. Lawrence, Phys. Rev. B 35, 2267 (1987).
11. D.D. Sarma, W.Speier, L. Kumar, C. Carbone, A. Spinsanti, O. Bisi, A. Iandelli, G.L. Olcese, A. Palenzola, Z. Phys. B 71, 69 (1988).
12. D.D. Sarma, F.U. Hillbrecht, M. Campagna, C. Carbone, J. Nogami, I. Lindau, T.W. Barbee, L. Braicovitch, I. Abbati, B. De Michelis, Z. Phys. B 59, 159 (1985); O. Bisi, O. Jepsen, O.K. Andersen, Europhys. Lett. 1, 149 (1986).
13. G.K. Wertheim and P.H. Citrin, In Photoemission in Solids 1, ed. by M. Cardona and L. Ley, Topics in. Appl. Phys., Vol. 26 (Springer, Berlin, Heidelberg 1978).
14. J. Tersoff, D.R. Hamann, Phys. Rev. B 28, 1168 (1983).
15. R. Weissmann, K. Müller, Surface Science Reports 1, 251 (1981) ; G.A. Sawatzky, A. Lenselink, Phys. Rev. B 21, 1790 (1980); E. Antonides, E.C. Janse, G.A. Sawatzky, Phys. Rev. B 15, 1669 (1977).
16. M. Cini, Solid State Comm. 24, 681 (1977); G.A. Sawatzky, Phys. Rev. Lett. 39, 504 (1977).
17. M. Campagna (Private Communication).
18. D.A. Papaconstantopoulos, Handbook of the Band Structure of Elemental Solids (Plenum, New York, 1986).

Photoemission and Scanning Tunneling Microscopy

H. Neddermeyer and St. Tosch

Institut für Experimentalphysik der Ruhr-Universität Bochum,
Postfach 102148, D-4630 Bochum, Fed. Rep. of Germany

1. Introduction

A combination of results obtained by direct or inverse photoemission spectroscopy (PES and IPES) with those of scanning tunneling microscopy (STM) is very attractive, since it ideally allows the assignment of electronic states (as deduced from PES and IPES data) to specific locations of a sample surface. It is clear that PES and IPES alone does not simply provide information on the electronic structure in real space, although the intimate relation between electronic states and structure is often used to derive structural details of the sample surface. Another interesting aspect of comparing PES and IPES data with STM results is that in case of ordered surfaces PES and IPES are favorably described in the reciprocal lattice, whereas the atomic and local nature of STM is apparent.

The use of photoemission techniques for an evaluation of the surface and bulk electronic structure is already known for several decades and a large number of publications exists on concepts and systems, both theoretically and experimentally. On the other hand, STM as a spectroscopical tool is just a few years old and only a small number of surfaces have been studied, so far. Therefore, the discussion of STM for a determination of the local electronic structure will be the main issue of the present work. Since a general theory for the spectroscopical application of STM does not yet exist we are constrained to a more phenomenological description of the employed methods. In addition, we restrict ourselves to examples we have obtained previously on clean surfaces of Si and noble metal/Si interfaces by using both STM and PES. Results from a bulk noble metal single-crystal surface (Au(111)) will also be discussed.

2. Photoemission and Inverse Photoemission

In case of ordered surfaces the conservation of the wavevector component parallel to the surface during the entire photoemission process has proven to be most useful for data evaluation /1,2/. For bulk single-crystals or epitaxial films with a thickness of several monolayers the three-dimensional energy band structure may be derived from the energy distributions of photoemitted electrons (EDC's) if an angle-resolving electron energy analyser is utilized for the measurements. For two-dimensional systems, e.g., for surface states or layered structures the two-dimensional dispersion relations are obtained. For PES studies the use of monochromatized synchrotron radiation is preferable against excitation with laboratory light sources, since it allows a simple analysis of normal emission data. Moreover, polarization and cross section effects provide additional information on the experimental system. For polycrystalline or disordered materials and by using an angle-integrating photoelectron spectrometer the EDC's reflect properties of the density of states (more exactly, of the joint density of optical states). PES delivers results for the occupied part of the band

Springer Series in Surface Sciences, Vol. 18 **Auger Spectroscopy and Electronic Structure**
Editors: G. Cubiotti · G. Mondio · K. Wandelt

structure and the unoccupied states above the vacuum level E_V. The important energy region between Fermi level E_F and E_V is accessible by IPES /3/.

3. Scanning Tunneling Microscopy and Surface Electronic Structure

In the atomic picture of STM developed by TERSOFF and HAMANN for metal surfaces /4/ and in the limit of negligible sample bias voltage U the tip follows lines of constant density of states at E_F upon scanning of the surface under the condition of a constant tunneling current I. With increasing U additional states below or above E_F (depending on the polarity of U) contribute to the tunneling current, which means that the density of states of both the sample and the tunneling tip (in addition to other factors) determines the tunneling current. The relation between tunneling current and density of states factors constitutes the basis for a determination of the local electronic structure of the sample surface.

The first spectroscopical application of STM has been reported by BINNIG et al. /5/, who determined the slope of the I/U characteristic by means of lock-in technique and regulating the average tunneling current to a constant value. The latter condition had the consequence that the sample-tip distance was not constant during sweeping of U, which produced resonance-like structures in the dI/dU characteristics. These structures (which are known as field-emission resonances) can be followed to very high orders /6/ but are apparently not directly related to the local surface electronic structure of the surface and do not exist in PES or IPES results. Another difficulty is that the lateral resolution drastically changes during sweeping of U. To obtain information on the local electronic states the sample-tip distance should therefore be kept fixed during variation of the sample bias, which has first been realized by HAMERS et al. /7/. By stabilization of the sample-tip distance for a small time interval and measuring the I/U characteristics subsequently at every lateral tip position HAMERS et al. succeeded in obtaining information on the atomically resolved local electronic structure of the surface.

The derivation of density of states values from the I/U characteristics remains a major problem. In order to understand the difficulties let us consider the most simple approximation of the local tunneling current I according to the theory developed by TERSOFF and HAMANN. In the limit where the tip is replaced by a point probe, negligible sample bias and low temperatures the tunneling current may be expressed by

$$I \propto \sum_{\nu} |\Psi(r_0)|^2 \delta(E_\nu - E_F),$$

where we have to sum up the contributions of the states Ψ_ν of the sample surface at the position r_0 of the point probe. The right-hand side of the equation corresponds to the charge density of states at E_F measured at the location r_0. Inelastic effects are not taken into account in this description of the tunneling process.

For nonnegligible sample bias voltage, tunneling occurs between states of the tip and the sample surface, which extend over a range of energies. This means that one has to integrate the various contributions dI in the relevant energy range to obtain the total tunneling current I. Experimentally, one measures a kind of averaged charge density of states. In an accurate description of the tunneling current, the decay length of the various states into the potential barrier between sample and tip has to be considered, as has been shown in an explicit calculation of the voltage-dependent tunneling current for graphite /8/.

In a more classical description of the tunneling process the tunneling current is related to a product of the density of states of the sample and the tip, which is weighed by the transmission probability of the tunneling electrons. This approach has been found useful for an explanation of the energy distribution of field-emitted electrons /9/. It has been shown in ref. 9 that the transmission probability, which varies exponentially with U, essentially determines the shape of the field emission energy distribution curves. Since for large sample bias voltages (U larger than a few V) the tunneling barrier between sample and tip corresponds to the field emission regime, the energy-dependent transmission probability has certainly to be taken into account in STM. This means that dI/dU doe not only contain density of states effects but also contributions due to the energy-dependent transmission probability. However, the latter effect should smoothly vary with U and does not produce fine structure details in the dI/dU characteristics. Therefore, structures in dI/dU should indeed reflect density of states properties of the sample. If the latter varies locally the effect of tip density of states may be ruled out, since it should contribute equally well to every position of the sample surface. Moreover, since the contributing free electron-like sp states of the tip, which is mostly prepared from a W wire, do not show considerable variations within a few eV near E_F, structures in I/U characteristics by tip effects are not expected for sample bias voltages up to a few V.

According to the planar electrode model of STM /10/ the tunneling current may be written as

$$I \propto U \exp(-\text{const}\,\Phi^{1/2}\, d),$$

where Φ is the height of the rectangular potential barrier and d its width. Differentiating this equation leads to dI/dU=I/U and consequently to (dI/I)/(dU/U)=d(lnI)/d(lnU)=const, independent of the width of the potential barrier or (within the approximation of the planar electrode model) of the sample-tip distance. Since density of states maxima should be visible in dI/dU by onsets the division by I/U should lead to a quantity, which is independent of the sample-tip distance. In other words, density of states effects may be derived from the measurements <u>independent of the actual sample-tip distance</u> and the strongly with d varying transmission factor.

Another problem is, how under the condition of a fixed sample-tip distance the changes of the potential barrier introduced by variations of the sample bias voltage, i.e., the effects of the <u>energy-dependent transmission probability</u>, influence the I/U and the dI/dU characteristics. It has been proposed by STROSCIO et al. /11/ that even most of the voltage-dependence of the tunneling current may be removed by computation of the ratio of differential to total conductivity, (dI/dU)/(I/U). However, this procedure does not seem to be justified theoretically, since in case of an exponential dependency of the tunneling current on the sample bias voltage (which, to a good approximation, is found for larger values of U) the quantity (dI/dU)/I (or d(lnI)/dU) should be constant, instead. It should be noted that in the literature mostly the quantity d(lnI)/d(lnU) is compared with density of states curves. This procedure has certainly the advantage that density of states effects obtained with low (high) tunneling current are enlarged (scaled down) thereby removing steep slopes of the conductivity curves. We believe, however, that major density of states features are already visible in the dI/dU characteristics. In some cases it is even advisable to analyze the I/U dependencies more closely since they also contain direct information on the density of states.

It has to be noted that the spectroscopical information of STM results is already most simply contained in the topographical images themselves.

For example, topographically imaging of Si(111)7x7 with positive and negative sample bias voltage corresponding to tunneling into empty and from filled states of Si, respectively, has revealed distinct differences in the height of the Si adatoms /7,12,13/, which are explained by a different spatial distribution of the metallic-like surface states on the adatoms. Another typical example has been reported by STROSCIO et al. for clean Si(111)2x1 /11/, where the atomic-like protrusions in the images did not correspond to the position of atomic nuclei but rather reflected the spatial distribution of occupied and unoccupied surface states.

The STM images measured by regulating the tunneling current to a constant value are usually called constant current topographies (CCT's). If the I/U characteristics are obtained at a fixed sample-tip distance (which is determined by the sample bias voltage U_0 used in the stabilization period) the local tunneling current I may be displayed in an image, which we call current image (CI). The stabilization voltage U_0 (which determines the corrugation followed by the tip on scanning of the surface) and U are parameters of the CI. For comparison with PES and IPES data the local I/U and dI/dU (or the d(lnI)/d(lnU)) characteristics may be used.

4. Results

In the present work photoemission and STM results from three typical systems will be described and discussed. We begin with Si(111)7x7, where STM has proven to be particularly useful, since the dangling bonds of the Si atoms in the topmost atomic layer are sufficiently separated and easily resolved both in the topographical and spectroscopical mode of STM. As next example we have chosen the dense-packed Au(111) surface, which shows an sp-like surface state, whose dispersion relation crosses the Fermi level and where the d electron states are fully occupied and located not too deep below E_F. Finally, we consider the system Ag/Si(111)7x7 for room temperature Ag deposition, which is interesting both with regard to an analysis of the growth mode of the Ag films and the formation of regular Ag clusters during the initial stage of deposition.

4.1 Si(111)7x7

The STM observation of the 7x7 reconstruction of Si(111) by BINNIG et al. was one of the milestones in the history of STM /14/ and established the role of Si(111) as a model system for the development of this new technique. The possibility of observing local electronic states was also exemplified by using this surface /7/. In Fig. 1 we show the typical CCT pattern of Si(111)7x7 obtained by using a positive sample bias voltage U (tunneling into empty states of the sample), where the 12 adatoms appear at the same height in the entire unit cell. In Fig. 2 we reproduce I/U (bottom) and dI/dU (top) characteristics /15/, which have been recorded at specific locations of the unit cell (see right bottom part of the Fig.) by employing the technique as reported by HAMERS et al. /7/. It has to be emphasized that the actual sample-tip distance is determined by the condition that at the stabilization voltage U_0 the tunneling current is 4 nA at every lateral tip position, which means that the differences in the characteristics are indeed caused by local electronic effects. The analysis of these curves is not straightforward, since their exact shape is dependent on the stabilization conditions /13/. However, some general properties are apparent, namely the occurrence of a nonzero slope of I/U at E_F on the adatoms, which is indicative of their "metallic" character and a distinct maximum at -1 V in dI/dU on positions 5 and 6 (corresponding to the dangling bonds of the second monolayer in terms of the dimer adatom stacking fault (DAS) model of Si(111)7x7 /16/). The "metallic" character of the ad-

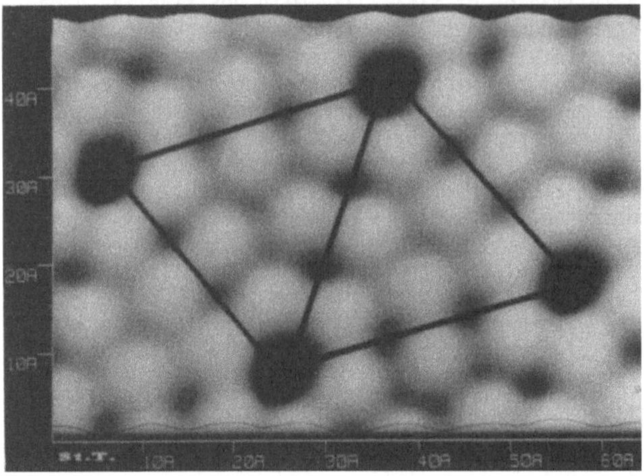

Fig. 1 CCT from Si(111)7x7 obtained at positive sample bias. The halves of the 7x7 unit cell are indicated by the black lines

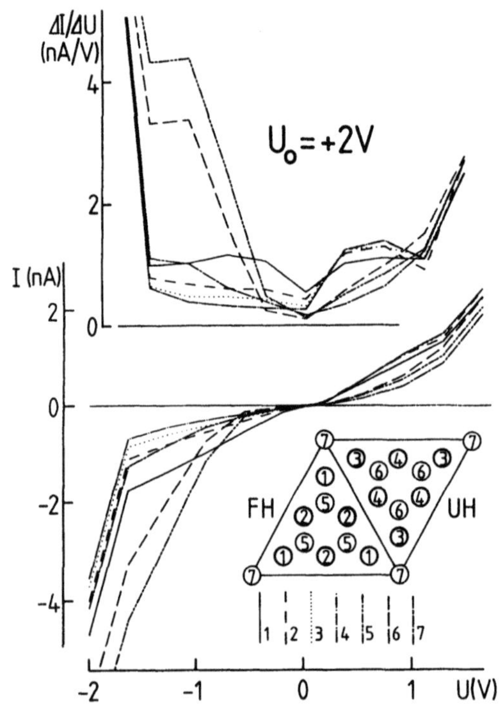

Fig. 2 Local I/U (below) and dI/dU (above) characteristics from Si(111)7x7 obtained at a stabilization voltage of 2 V /15/

atoms may be deduced from the possibility of obtaining a tunneling current at low sample bias voltages (typically 0.1 V) via states, which must be located in the band gap region of Si.

Photoemission results of Si(111)7x7 are displayed in Figs. 3 and 4. The EDC's have been measured by using an angle-resolving spectrometer, which means that they cannot be directly compared with the density of states.

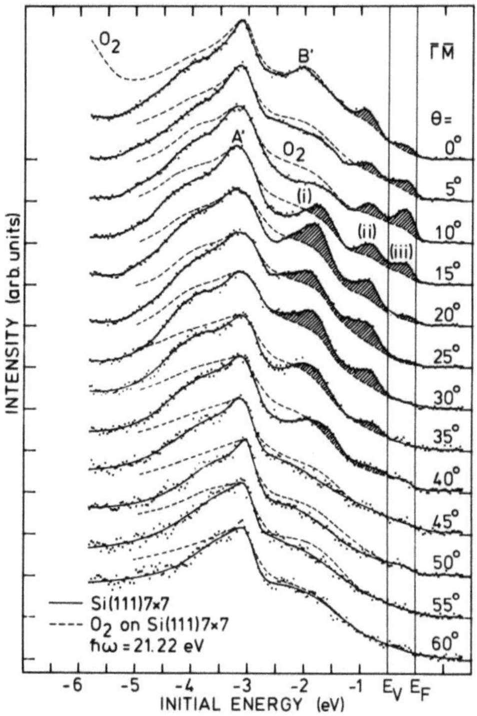

Fig. 3 Photoemission from clean and O_2 covered Si(111)7x7. The exposure with activated O_2 was approximately $2 \cdot 10^{-10}$ mbar·s

However, since the results do not show dramatic effects in the position of peaks (see Fig. 3), the energy position of density of states features may be obtained in an approximate way by averaging over the polar (and more exactly also the azimuthal) angle. Adsorption of activated O_2 was used to identify occupied surface states (see dashed areas in Figs. 3 and 4). Three surface states denoted (i), (ii) and (iii), respectively, are recognized. The structure (iii) is located at E_F and ascribed to a metallic-like surface state. This surface state extends into the unoccupied part of the band structure as has been shown with IPES /17/.

It should be noted that PES studies of the metallic-like surface state does show at most a weak dispersion /18/ corresponding to a particularly heavy mass. This observation finds its natural explanation in the STM localization of the metallic-like surface state /7/ on the adatom sites of the DAS structure. Since the distance between the adatoms is quite large their interaction is correspondingly small and the metallic-like surface state only weakly dispersing with the parallel component of the wavevector ($k_{||}$). Another interesting consequence may be deduced from the fact that the adatoms of the 7x7 reconstruction are locally arranged in a 2x2 periodicity. This implies that the metallic-like surface need not show characteristic behavior in the 7x7 two-dimensional Brillouin zone but rather in some other periodicity due to an arrangement of atoms in twice the distance of the ideal lattice parameter. The PES results indeed show (Fig. 4) that the intensity of the metallic-like surface state is concentrated in rings, which are centered at the $\bar{\Gamma}$-points of the extended zone scheme of the 1x1 lattice. The diameter of these rings exactly corresponds to a twofold periodicity of the ideal Si lattice and confirms the local arrangement of the adatoms as seen in the STM images.

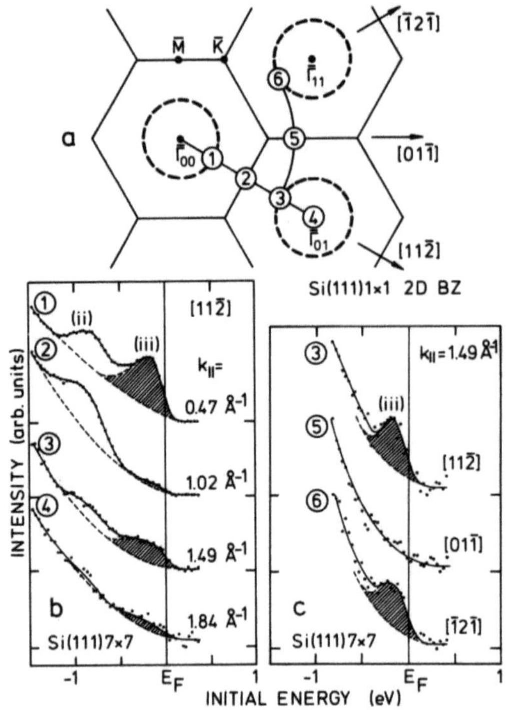

Fig. 4 Photoemission from clean Si(111)7x7 at various positions in the two-dimensional Brillouin zone (extended zone scheme). The dashed lines correspond to the assumed contribution from bulk states

The energy position of surface state (ii) (Fig. 3) agrees with the maximum in the dI/dU curves on the dangling bonds of the second atomic layer (see Fig. 2). The same surface state has therefore been identified in both STM and PES results. As to surface state (i) at -2 eV (Fig. 3), the use of STM for larger sample bias voltage is connected with some difficulties. CCT's normally loose its atomic resolution for larger sample bias voltages, since the sample-tip distance increases and a correspondingly large surface area contributes to the tunneling current. On the other hand, if the tunneling current is allowed to increase simultaneously thus retaining a small sample-tip distance, the voltage drop in the sample may become too large or thermal heating of the tip may lead to unwanted tip changes and instabilities. We therefore conclude that in case of semiconductors atomically resolved STM is most suitable for sample bias voltages smaller than 2 V, but can in this energy range ($-2\ \mathrm{eV} \leq E \leq 2\ \mathrm{eV}$) indeed be used for a determination of local density of states features.

4.2 Au(111)

The electronic structure of the (111) faces of the noble metals is characterized by an sp-like surface state in the vicinity of the Fermi level /19/ and by occupied d states, which energetically overlap with nearly-free electron states. By means of angle-resolved PES the bulk and surface electronic structure can be determined with high accuracy /20,21/ and it is an interesting question, whether the electronic states and their spatial properties may be explored by STM. KAISER and JACLEVIC have reported on spectroscopical STM results from Au(111), where the dI/dU characteristic showed a distinct peak at the energy location of the occupied sp-like surface state /22/. The measurements did not extend to the d states, which lie in the energy range between 2 eV and 6 eV below E_F.

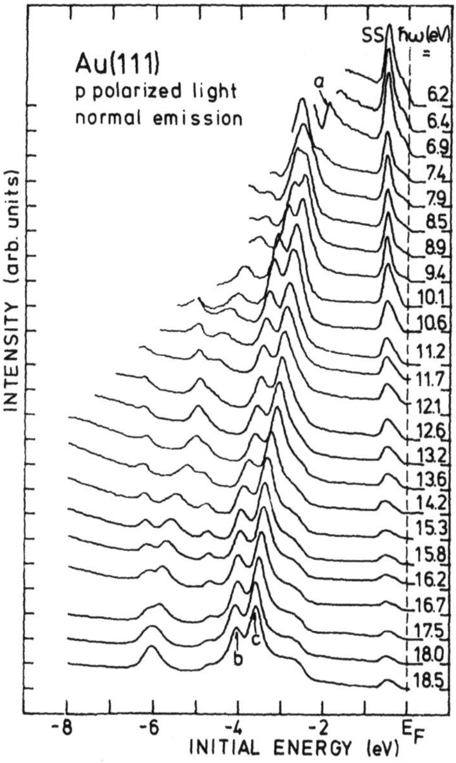

Fig. 5 Angle-resolved photoemission from Au(111) obtained by excitation with monochromatized synchrotron radiation /23/

In Fig. 5 photoemission results from Au(111) as obtained by using synchrotron radiation of BESSY (Berlin) are displayed for photon energies between 6.2 eV and 18.5 eV /23/. The dispersing peaks a and b,c are due to direct transition from the bulk sp and d bands, respectively, and peak SS immediately below E_F is explained by emission from the above mentioned sp-like surface state.

STM topographical images from Au(111) showed flat parts as well as regions with a high density of monoatomic steps /24/. Spectroscopical information on Au(111) has been collected in form of local I/U characteristics in the range of $-10\ V \leq U \leq 10\ V$ at a fixed sample tip distance by using various values for the stabilization voltage. This was found necessary since at small sample tip distance (as defined by the usual small sample bias in the mV range for metal surfaces) the tunneling current for larger values of U increases considerably and probably leads to an unacceptable temperature increase of the tip by resistive heating. Using the local I values we have plotted CI's and found that they did not show atomic-like structures. Therefore, a large number of local characteristics have been averaged for improvement of the signal-to-noise ratio. One example of an I/U characteristic is shown in Fig. 6, which covers the energy range between E_F and -10 eV /24/. The position of the d bands and that of the occupied surface state is indicated in the Fig. The I/U curve decreases linearly for a sample bias between 0 V and approximately -2 V. For larger (absolute) values of U the magnitude of the tunneling current increases dramatically, which is caused by the energy-dependent transmission probability of the tunneling barrier. Neither an onset in the tunneling current at the position of surface state SS nor one due to d electron contributions are visible. Since a stabilization voltage of -10 V has been employed

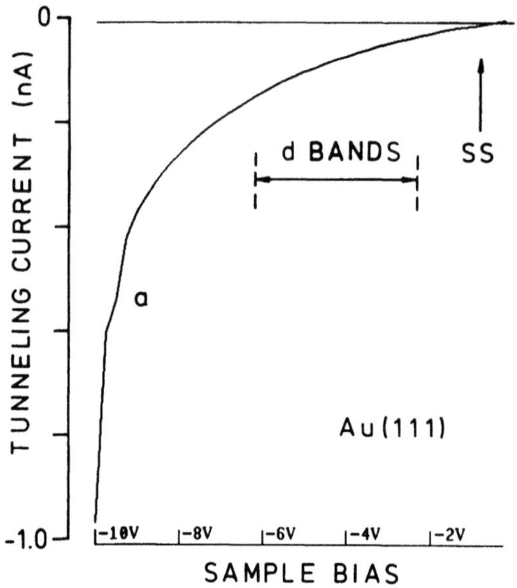

Fig. 6 I/U from Au(111) obtained at fixed sample-tip distance and using a stabilization voltage of -10 V. Structure a is an artefact due to the time dependency of the current amplifier

absence of a d electron contribution is not surprising because of the large sample-tip distance, which only leads to tunneling into states extending well into the tunneling barrier. Absence of a surface state contribution may be explained by an unsufficient selectivity of the tunneling process to states with small parallel component of the wavevector. We note that the dispersion relation of the SS reaches E_F already at $k_{||} = 2.8\ nm^{-1}$. If the electrons are tunneling from occupied states with larger values of $k_{||}$ (corresponding to bulk nearly-free electron states) then a feature due to SS is averaged out and does not appear in the I/U curve. We note that also a dI/dU curve obtained by numerically differentiating I/U did not show any indication of a surface state contribution.

The measurements have shown that the I/U characteristics of Au(111) do not contain distinct density of states effects except for their general behavior, namely the nonzero slope of I/U in the vicinity of U = 0 V indicating ohmic resistivity of the tunneling barrier for small sample bias voltage. As a matter of fact, absence of states near E_F cannot lead to such a dependency and was already found for certain positions on Si(111)7x7. Metallic or nonmetallic character (on a local scale) can therefore well be evaluated from the I/U's and differences or a transition between both states may be seen in the STM results as is demonstrated by the following example.

4.3 Ag on Si(111)7x7

The initial stage of metal atom condensation on a clean semiconductor substrate belongs to the kind of problems, which may advantageously be studied by means of PES and STM. In the present work we show the effects on Si(111)7x7 upon low temperature deposition of Ag in the submonolayer range. Low means in this context that the substrate temperature is not yet sufficient for the $\sqrt{3}x\sqrt{3}$ phase transition, which is observed for temperatures above a few 100 ^{o}C and an Ag coverage of at least 1/3 monolayer. The general nature of the growth process is already known from low energy electron diffraction and Auger electron spectroscopy /25/. Up to a coverage of a few monolayers of Ag the metal deposit grows nearly in a layer-by-

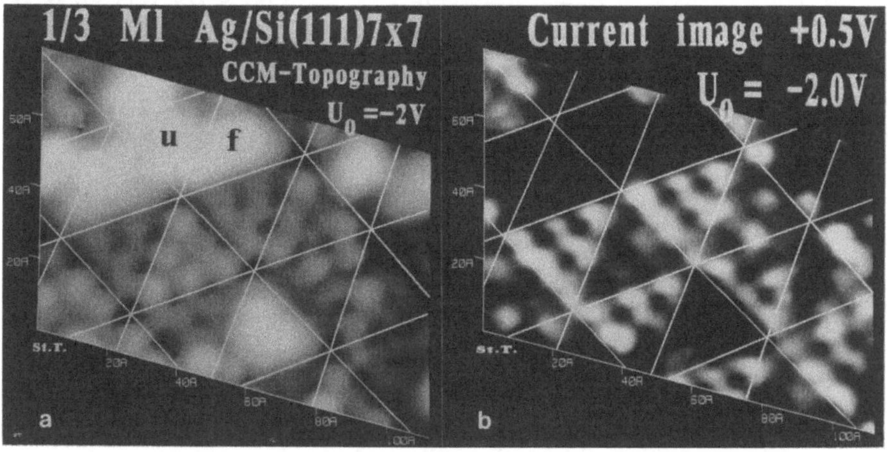

Fig. 7 CCT (a) and CI (b) acquired simultaneously on 1/3 monolayer Ag on Si(111)7x7 /26/

layer type growth mode as may be deduced from the nearly exponential dependency of the Si and Ag Auger intensities. With respect to the initial stage one is mainly interested in the adsorption sites of the Ag atoms, in eventual changes of the substrate atomic structure and in the local electronic effects upon metal atom adsorption. The question whether the Ag atoms may be identified in the STM images also plays a major role in such studies.

In Fig. 7 we reproduce from a recent work /26/ a CCT (a) and a CI (b) acquired simultaneously under the condition of a fixed sample-tip distance on Si(111)7x7, where 1/3 monolayer of Ag has been deposited at a temperature of 130 ^{o}C. The lateral resolution of the images was worse than in case of Fig. 1. However, other measurements have shown and it may also be recognized in Fig. 7 (a) that the Ag deposit mostly forms triangular islands on top of the faulted halves (denoted with f in Fig. 7 (b)) of the 7x7 unit cells. For a sample bias voltage of 0.5 V corresponding to tunneling into unoccupied states of the sample, the positions of the Ag islands appear as dark areas in the CI (Fig. 7 (b)). The same behavior is found for a CI measured at -0.5 V (not shown here). Compared with the uncovered 7x7 unit cells the position of the Ag islands is therefore connected with a reduced density of states at least 0.5 eV below and above E_F.

The reduction of the density of states in the energy region near E_F may more directly be inferred from the local I/U (Fig. 8 (a)) and dI/dU (Fig. 8 (b)) characteristics obtained in the same measurement and are compared with photoemission results from a room temperature deposited Ag film on Si(111)7x7 (Fig. 8 (c)). The tunneling current on the Ag islands (full line in Fig. 8 (a)) is particularly small for $-0.5\ V \leq U \leq 1\ V$ in contrast to the local current on the Si adatoms of the uncovered part of the surface (dashed and dashed-dotted lines). Correspondingly, dI/dU from the Ag islands is small in the same range of voltage, whereas the uncovered Si atoms show nonzero dI/dU in the entire range of sample bias. Note that the Si adatom positions show a little difference in dI/dU when compared with results from clean Si(111)7x7. For -1 V Ag/Si shows a maximum (Fig. 8 (b)), which is practically absent for clean Si(111)7x7 (Fig. 4). We ascribe these differences to the different lateral resolution in both measurements and to the different polarity of the stabilization voltage.

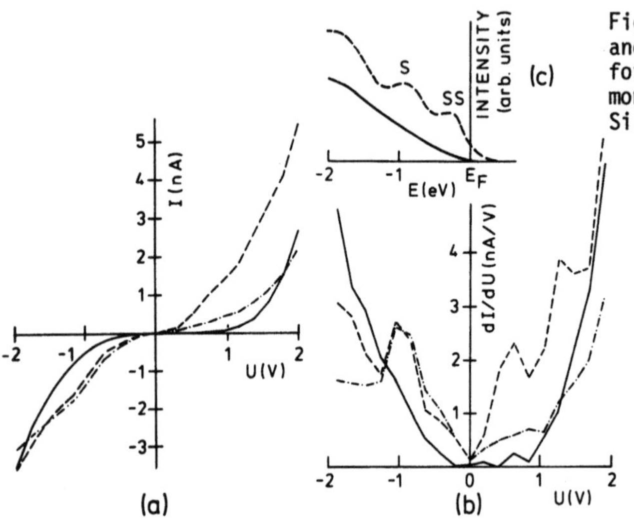

Fig. 8 I/U (a), dI/dU (b) and photoemission results for Ag (approximately 1 monolyer) covered Si(111)7x7 /26/

A reduction of the density of states near E_F by Ag adsorption is also visible in the PES results (Fig. 8 (c)). We have to mention that the EDC's from clean and Ag covered Si(111)7x7 (dashed and full lines, respectively, in Fig. 8 (c)) have been approximately normalized to their maximum in the valence band region. This procedure is somewhat arbitrary due to different photoionization cross sections of the Ag d states and the sp states of clean Si. However, the disappearance of the metallic-like surface state (labeled ss in Fig. 8 (c)) and absence of a metallic-like edge is apparent. The disappearance of the Si surface states upon Ag adsorption, which is deduced both from the STM and the PES results, may be explained by an appreciable Ag/Si interaction leading to a saturation of the dangling bonds of the Si adatoms in the 7x7 structure. That such an interaction exists may also be concluded from regularly shaped Ag clusters, which are frequently observed during the initial stage of Ag adsorption /26/.

5. Discussion

Our comparison of PES and STM spectroscopical results has shown experimentally that a close connection between the dI/dU curves and the EDC's exists. In case of Si(111)7x7 one surface state appears as a peak both in the PES and STM results. Another example, the metallic-like edge of EDC's, has to be indirectly inferred from the nonzero slope of I/U characteristics in the limit of negligible sample bias. The sp-like surface state of Au(111) did not produce structure in I/U or dI/dU.

In a recent theoretical approach for the voltage dependence of the tunneling current NOGUERA has shown that the tunneling current should normally not contain contributions from surface states /27/. This may be understood by the nature of the wave functions of these states, which decrease exponentially normal to the surface and therefore cannot carry flux in this direction. NOGUERA has proposed that a current parallel to the surface may eventually be responsible for the experimentally observed tunneling current, for example, in band gaps of semiconductors. Actually, in case of Si(111)7x7 the surface states (ii) and (iii) of Fig. 3 are not located in the center of the two-dimensional Brillouin zone and are therefore characterized by a "propagating" wavevector parallel to the surface. On the other hand, the effective mass of both states (and in particular that of the metallic-like state) is large and should not be associated with a strong current. Absence of structure in I/U ot dI/dU in the region of the sp-like surface state for Au(111) is not surprising in terms of the model given by NOGUERA, since this state is located in the center of the two-dimensional Brillouin zone and cannot carry a considerable current parallel to the surface.

That contributions in the local tunneling current (either from bulk or surface states) due to states with nonzero $k_{||}$ are indeed present may be deduced from the measurements of Au(111). For example, for a sample bias near 0 V, where a tunneling current may readily be obtained no states exist for this surface with $k_{||} = 0$, because of the presence of the bulk band gap extending down to about -1 eV below E_F and the low energy position of the surface state.

The preceding paragraphs have shown that a derivation of density of states features from STM measurements may be more problematic than has been suggested previously. One is tempted to compare the situation with the development of field emission spectroscopy, where the determination of density of states features is also not a straightforward procedure. However, an important difference and advantage of STM against field emission is that STM delivers information on local electronic properties and that indeed appreciable differences in the local I/U or dI/dU curves are present, which clearly have to be associated with electronic effects. Another very simple and powerful application of STM for spectroscopical use is the analysis of CCT's acquired at different polarities of sample bias and relating the observed shape of the pattern to the spatial distribution of bonding (i.e., occupied) and antibonding (i.e., unoccupied) states of the sample with particularly high sensitivity to the energy region near E_F, which may then be related to density of states information derived from PES and IPES measurements.

Acknowledgment

This work has been supported by the Deutsche Forschungsgemeinschaft and by the German Federal Minister for Research and Technology (BMFT) under the contract number 05 234 NMP. We thank P. Rupieper and W. Radlik for photoemission measurements on Si(111)7x7 and Au(111), respectively. We are grateful to Wacker-Chemitronic (Burghausen), who provided us with the Si wafers.

References

1. Photoemission in Solids I, ed. by M. Cardona and L. Ley, Topics Appl. Phys., Vol. 26 (Springer, Berlin, Heidelberg 1978); Photoemission in Solids II, ed. by L. Ley and M. Cardona, Topics Appl. Phys., Vol. 27 (Springer, Berlin, Heidelberg 1979).
2. R. Courths and S. Hüfner, Phys. Reports 112, 53 (1984).
3. V. Dose, Surface Sci. Repts. 5, 337 (1985).
4. J. Tersoff and D.R. Hamann, Phys. Rev. B 31, 805 (1985).
5. G. Binnig, K.H. Frank, H. Fuchs, N. Garcia, B. Reihl, H. Rohrer, F. Salvan and A.R. Williams, Phys. Rev. Lett. 55, 991 (1985).
6. J.H. Coombs and J.K. Gimzewski, to be published in Proc. 3rd Int. Conf. STM, Oxford, 1988.
7. R.J. Hamers, R.M. Tromp and J.E. Demuth, Phys. Rev. Lett. 56, 1972 (1986)
8. A. Selloni, P. Carnevalli, E. Tosatti and C.D. Chen, Phys. Rev. B 31, 487 (1985).
9. J.W. Gadzuk and E.W. Plummer, Rev. Mod. Phys. 45, 487 (1973).

10. N. Garcia, IBM J. Res. Develop. 30, 533 (1986).
11. J.A. Stroscio, R.M. Feenstra and A.P. Fein, Phys. Rev. Lett. 57, 2579 (1986).
12. R.M. Tromp, R.J. Hamers and J.E. Demuth, Phys. Rev. B 34, 1388 (1986).
13. Th. Berghaus, A. Brodde, H. Neddermeyer and St. Tosch, Surface Sci. 193, 235 (1988).
14. G. Binnig, H. Rohrer, Ch. Gerber and E. Weibel, Phys. Rev. Lett. 50, 120 (1983).
15. Th. Berghaus, A. Brodde, H. Neddermeyer and St. Tosch, J. Vac. Sci. Technol. A 6, 483 (1988).
16. K. Takayanagi, Y. Tanishiro, M. Takahashi and S. Takahashi, J. Vac. Sci. Technol. A 3, 1502 (1985).
17. Th. Fauster and F.J. Himpsel, J. Vac. Sci. Technol. A 1, 1111 (1983).
18. H. Neddermeyer, U. Miße and P. Rupieper, Surface Sci. 117, 405 (1982).
19. P. Heimann, H. Neddermeyer and H.F. Roloff, J. Phys. C: Solid State Phys. 10, L17 (1977).
20. R. Courths and S. Hüfner, Phys. Reports 112, 53 (1984).
21. R. Courths, H.-G. Zimmer, A. Goldmann and H. Saalfeld, Phys. Rev. B 34, 3577 (1986).
22. W.J. Kaiser and R.C. Jaklevic, IBM J. Res. Develop. 30, 411 (1986).
23. H. Neddermeyer, in Electronic Band Structure and Its Applications, Lecture Notes in Physics Vol. 283, ed. by M. Yussouff (Springer, Berlin, Heidelberg, 1987) p. 164.; see also A. Baalmann, M. Neumann, H. Neddermeyer, W. Radlik and W. Braun, Ann. Israeli Phys. Soc. 6, 351 (1983).
24. A. Brodde, St. Tosch and H. Neddermeyer, to be published in Proc. 3rd Int. Conf. STM (Oxford, 1988).
25. M. Hanbücken, H. Neddermeyer and P. Rupieper, Thin Solid Films 90, 37 (1982).
26. St. Tosch and H. Neddermeyer, Phys. Rev. Lett. 61, 349 (1988).
27. C. Noguera, to be published in 3rd Int. Conf. STM (Oxford, 1988).

Index of Contributors

MIX
Papier aus verantwortungsvollen Quellen
Paper from responsible sources
FSC® C105338
FSC
www.fsc.org

If you have any concerns about our products,
you can contact us on
ProductSafety@springernature.com

In case Publisher is established outside the EU,
the EU authorized representative is:
Springer Nature Customer Service Center GmbH
Europaplatz 3, 69115 Heidelberg, Germany

Printed by Libri Plureos GmbH
in Hamburg, Germany